Workers, Owners and Politics in Coal Mining

Workers, Owners and Politics in Coal Mining

An International Comparison of Industrial Relations

Edited by
Gerald D. Feldman and Klaus Tenfelde

Berg

New York / Oxford / Munich

Distributed exclusively in the US and Canada by
St Martin's Press, New York

First published in 1990 by
Berg Publishers Limited
Editorial offices:
165 Taber Avenue, Providence, RI 02906, USA
150 Cowley Road, Oxford OX4 1JJ, UK
Westermühlstraße 26, 8000 München 5, FRG

© Gerald D. Feldman and Klaus Tenfelde 1990

British Library Cataloguing in Publication Data

Workers, owners and politics in coal mining: an
International comparison of industrial relations.
1. Coal industries. Industrial relations
I. Feldman, Gerald D.   II. Tenfelde, Klaus
331'.0422334

ISBN 0-85496-603-X

Library of Congress Cataloging-in-Publication Data

Workers, owners, and politics in coal mining : an international
comparison of industrial relations / edited by Gerald D. Feldman and
Klaus Tenfelde.
    p. cm.
Includes bibliographical references.
ISBN 0-85496-603-X
1. Coal miners.   2. Industrial relations.   I. Feldman, Gerald D.
II. Tenfelde, Klaus.
HD6976.M615W67   1989
331.89'0422334—dc20
                                                                89-35882
                                                                    CIP

Printed in Great Britain by
Billing & Sons Ltd, Worcester

Contents

Figures and Tables

Figures and Tables

1
On the History of Industrial Relations in Mining

Klaus Tenfelde*

I It was the availability of raw materials within national borders that substantially accelerated the process of the industrialization of the Western nations in the nineteenth century, if indeed it was not one of the most important prerequisites for industrialization at all. In this process, bituminous coal attained a prominent importance because the national economies needed massive quantities of a readily available energy source free from problems of location and weather conditions. They needed it for the production of a large quantity of good quality iron and steel, for the expansion of railroad transportation into industrial regions, for the production and the operation of machinery, and finally for general public and private heating needs. Everywhere where bituminous coal was plentiful, industrialization excelled quite rapidly. This is especially true for the second phase of industrialization in the latter half of the nineteenth century as bituminous coal mining was on the verge of exceeding the century-old ore mining in importance. In principle, the technology of coal mining was known long before thanks to the example of ore mining, even though there existed important differences due to the differing locations of deposits and methods of extraction 'at the pit face'. Moreover, the rapid expansion of coal mining would not have been conceivable without the invention, improvement and widespread usage of the steam engine as the most decisive piece of equipment – until the end of the nineteenth century – for transforming heat energy into kinetic energy.[1]

---

* Translated by Dona Geyer.
  1. A summary and comparative economic history of coal mining in the important industrial countries does not exist; however, for an earlier attempt see Ferdinand Friedensburg, *Die Bergwirtschaft der Erde. Bodenschätze, Bergbau und Mineralienversorgung der einzelnen Länder*, Stuttgart 1938. For individual comparative studies, see the bibliography, and the important study of economic history in Western Europe, with repeated reference to coal mining, by David S. Landes, *The Unbound Prometheus*, Cambridge 1969.

As their coal industries developed into 'leading sectors' of the national economies, the Western industrial nations found themselves confronted at the same time with a social and political problem of unprecedented dimensions. Because of the degree of investment necessary in advance of the point of production, coal mining has an especially great need for capital and is really profitable only under the planning, transportation and production conditions of a large-scale enterprise. Because the locations of coal and ore fields often bordered on one another, and also and above all due to the great need of coke by the smelting works, which were eager to save on the transportation costs either of ore or coke, the mining and the metal industries became so closely associated with one another that they began to converge into a vertical concentration of the so-called heavy industries. While the respective conditions for such concentrations varied in degree, several giant companies emerged in the coal and steel industry towards the end of the century. These large-scale enterprises and operations attracted several thousand, usually underqualified, workers. Agglomerations of populations shaped by heavy industry and entire urban landscapes appeared; the occupational attraction of these new centres was almost always the employment offered in mining and the metal industries.[2] In those areas where mining was able to attain only a solely local or regional importance due to the deposit locations, there evolved the typical example of a mono-industrial mining community. The process of forming industrial classes accelerated in these mining and foundry cities and in the population centres of heavy industry; indeed, heavy industry became the prototype *par excellence* for class formation. The industrial communities very quickly expanded beyond the original settlement borders; new ones also sprang up near the major plants; infrastructural development was slow in coming; the working population was not only highly mobile and originated for the most part in rural regions, but was poorly qualified and had only a slight chance for advancement. The middle class was represented to a lesser degree in the population centres of heavy industry, especially in the phase of rapid growth. The numerical presence of a propertied upper class was not at all

2. On this, see especially E. A. Wrigley, *Industrial Growth and Population Change. A Regional Study of the Coalfield Areas of North-West Europe in the later Nineteenth Century*, Cambridge 1962, and Michael Haines, *Fertility and Occupation: Populations Patterns in Industrialization*, New York 1979 (a comparison between the mining regions of England, Prussia and the USA).

pronounced, since the concentration of capital had by the end of the century often crowded out even the self-managing entrepreneurs and allowed for a new elite of managers to take their place, an elite which strove to control the markets on assignment from the owners of capital.

Briefly stated, the problem of the industrial nations with mining in this phase was twofold.[3] For one, it was often acknowledged with concern that mining, as a basic resource industry, had advanced into a role that was nearly that of an essential functional agency of the national economy. By the second half of the century, too many secondary areas of the economy appeared to be dependent on it; the control over this important branch of the economy had been placed in too few hands during the process of capital concentration under the prevailing liberal theory of economics. A problem of major concern in Germany was that, especially in cases of international conflict, the manoeuvrability of the state's leadership appeared to be restricted by the concentration and degree of power belonging to heavy industry. The other side of the problem was the social issue. The regions of heavy industry were not threatening to become the seething pots of revolution, yet in view of the enhanced political strength of the working class, they did threaten to become of strategic importance in the flow of goods. A mining strike always posed the danger of becoming a mass or even a general strike. If the coal supply stopped, then the rest of the economy was forced to come to a standstill eventually since there was no adequate alternative source of energy anywhere in sight at the time, and the energy needs of the national economies in this phase of industrialization were especially great.

The struggle of governments to obtain a trouble-free energy supply was not over once coal mining had lost its essential importance as the only such source. The decline of mining began to become obvious worldwide as early as the 1920s – even if mining entrepreneurs were not willing to acknowledge this at first – and reached its lowest point following the Second World War as petroleum pressed to the fore and the supply of coal became relatively less important. Since the 1950s, coal production has declined throughout the 'old' industrial nations,[4] and this decrease has in the meantime also reached the 'new' coal nations, such as

3. For a more accurate discussion of this, see the contribution by G. D. Feldman below, chapter 8.
4. Cf. the table on coal production in appendix I.

Australia and South Africa. The political sensitivity continued, especially where the closing of mines increased the threat of unemployment. Contributing to this was the fact that an extensive network of auxiliary and related industries had emerged in the course of the technological development of mining, making the dependency of the coal and steel regions on mining even more apparent. Despite all this, the idea of relative autarky for the nation in its energy supply remained very active: in the case of a worldwide energy crisis or any other foreign complication – such as the oil crisis which started in the 1970s – the national energy supply should be able to rely on at least a 'remainder' of native coal. Moreover, it appeared obvious in some places that mining would not be in a position on its own to handle the necessary, painful market adjustments through capacity reduction – at least not without paying a politically unjustifiable price in the form of serious social repercussions.

Since the decades preceding the First World War, national governments, with the USA being perhaps the least typical, have repeatedly turned their economic and socio-political attention to the problems of mining, as opposed to or, in any case, more so than to those of other branches of the economy. At different phases of development, organizations were formed with which the national governments co-operated, namely the early regional entrepreneur associations, the national workers' unions beginning in the last two decades of the nineteenth century, and, as a reaction to these, the national employer associations of mining entrepreneurs. Even though 'industrial relations' at the level of the mining operations had always possessed great importance for conflict settlement, this was overshadowed more and more by industrial relations at the higher level of measures enacted by associations, and general domestic and economy policy.

II   Research on industrial relations covers those individual and collective institutions and associations that influence and shape labour relations in their entirety and their interaction with one another. In the German sociology of labour and industry, this field is also often labelled 'industrial labour relations' (*industrielle Arbeitsbeziehungen*); however, in the Anglo-American field of 'industrial relations' and also in the French '*relations industrielles*' the term is understood to be much more comprehensive both in terms of research and the institutions involved. Industrial relations are studied as a field of industrial sociology in all industrial nations; the Anglo-American field has gone so far as to concentrate research in independent university 'departments for industrial relations'. These special departments also furnish higher educational and practical training for those who will be employed in responsible positions in the involved associations and bureaucracies, enabling them to contribute to shaping the labour relations of individual industries. In Germany such training is less common, a fact that is indicative not only of a different research tradition, but also of the loss of certain other research traditions.[5]

Among the most important structural characteristics of industrial relations[6] are: firstly, the framework conditions and leeway in decision-making set by the state for the actions of both the government and the autonomous associations of the opposing parties in the production processes; in the centre of the study of industrial relations are, secondly, the type of conflict settlement (such as through wage contracts, compulsory arbitration, forms of confrontation, etc.); and thirdly, the institutional framework of this conflict settlement (associations, chambers, trade courts, etc.). The fundamental prerequisite is the assumption and acknowledgment that a conflict of interest proceeds from the very composition of the production process in private capitalism; and indeed conflicts also occur in nationalized industries, and even in socialized industries, though in

---

5. It is characteristic that such a widely read German encyclopedia as *Meyers Enzyklopädisches Lexikon* lists industrial relations under the English term: vol. 12 (Mannheim 1974), p. 569. On the loss of the research traditions of political economy in the German sociology of industry, see e.g. Friedrich Fürstenberg, *Industrielle Arbeitsbeziehungen: Untersuchungen zu Interessenlagen und Interessenvertretungen in der modernen Arbeitswelt*, Vienna 1975, pp. 7f. The expansion of the subjects of research in English and French-speaking fields is especially evident in the relevant journals of the discipline, above all *Relations Industrielles – Industrial Relations* (dating from 1946, published by the Département des Relations Industrielles, University of Québec), as well as the *British Journal of Industrial Relations* (dating from the 1960s, by the London School of Economics), and *Industrial Relations: A Journal of Economy and Society* (dating from 1961, published by the Department for Industrial Relations at the University of California at Berkeley). A loose-leaf collection is: Roger Blainpain (ed.), *International Encyclopedia for Labour Law and Industrial Relations*, Deventer 1977ff. Cf. also George S. Bain, *A Bibliography of British Industrial Relations*, Cambridge 1979, as well as Albert A. Blum (ed.), *International Handbook of Industrial Relations*, London 1981. A further suggestion is: Klaus Tenfelde, 'Sozialgeschichte und vergleichende Geschichte der Arbeiter', in Tenfelde (ed.), *Arbeiter und Arbeiterbewegung im Vergleich. Berichte zur internationalen historischen Forschung*, Munich 1986, pp. 13–62. As an attempt to include the history of industrial relations into a comparative history of the modern welfare state, see Gerhard A. Ritter, *Der Sozialstaat: Entstehung und Entwicklung im internationalen Vergleich*, Munich 1989.

6. Cf. Fürstenberg, *Industrielle Arbeitsbeziehungen*, pp. 8f.

a different shape. Of course, the company structure of modern industries always includes a relationship of ruling and submission. Further it is expected that these conflicts cannot be neutralized, but can be fought out without the use of violence and can be settled by arriving at an acceptable compromise, usually for a limited period of time, by way of negotiation under the supervision of one of the instruments of conflict settlement. As a rule, one can differentiate between institutional stabilization and the settlement of conflicting interests at the company and at the inter-company levels. Further differences can be emphasized to different degrees in various societies. In addition to established traditions and accepted rules of procedure, it is especially here that the state's regulatory competence makes an impact. Not only does the state contribute to the shaping and efficiency of industrial relations in a narrow sense by legislative regulation of the negotiating partners and the results of their negotiation, but it also contributes in a broader sense. In modern welfare states, this broader sense encompasses generally the entire social policy in so far as it is a policy of social security and distribution of wealth, the legal framework regulating labour relations through commerce legislation, and also those areas, such as the entire public service sector, in which the state and its organizations are not only active in setting the framework conditions, but also as negotiating partners and interested parties.

It thus becomes clear that industrial relations do not occur in a sort of power vacuum within government that is filled by opponents negotiating a conflict autonomously. On the contrary, the constellations of political power in modern presidential or party democracies are reflected rather well in the impact they have in shaping the framework conditions, in taking a stand and often even by intervening during actual conflicts in the state of affairs and development of industrial relations. The liberal-capitalist model of resolving conflicts solely through marketplace forces without governmental interference has failed everywhere in historical terms. Since the end of the nineteenth century, the governments of all the industrial nations have been in a continual, decades-long process – interrupted and yet also accelerated by the world wars – of retrieving the regulatory competence that they relinquished, albeit in differing degrees, during the rule of liberal economic thinking. Nowhere has this happened as a rectilinear process. Phases of a relatively undisturbed conflict management by well-equipped organizations and institutions changed into phases of radicalization usually accompanied by shifts within the balances of political power.

Conditioned by convention and its institutional entrenchment, various forms and models of conflict settlement have come about. In England, for example, the legal tradition of common law has been used to justify the form of 'legally unrestrained industrial democracy' which was practised until recently, while simultaneously the unions were acknowledged from very early on as representative authorities.[7] The direction taken by German development via the union policy position of 'economic democracy' (*Wirtschaftsdemokratie*)[8] led more towards a legally guaranteed codetermination at the company and intercompany level, a development that required the strength of a unified trade union in its actions and negotiations. An explanation for this development can be found in the Continental, Roman legal tradition and its tendency towards positivist definitions of legal relations, but also in a later organizational and institutional establishment of the trade union movement and stronger traditions of a corporative state, which also helped in founding the idea of a social partnership in Austria under the influence of Christian social thought.[9] It appears that since the Second World War, industrial relations in Western industrial nations have converged in some way or another in both the term and the model of 'industrial corporatism' featuring workers' participation, the degree and forms of which have remained very controversial.[10] Thus it is often taken into account that basic system

7. Cf. Ulrich von Alemann, 'Auf dem Weg zum industriellen Korporatismus? Entwicklungslinien der Arbeitsbeziehungen in der Bundesrepublik Deutschland und in Großbritannien', *Gewerkschaftliche Monatshefte* 30, 1979, pp. 552–65, on p. 554.

8. In 1928 *Wirtschaftsdemokratie – ihr Wesen, Weg und Ziel* by Fritz Naphtali first appeared as an article; see the new 4th edn, ed. Rudolf F. Kuda, Frankfurt 1977.

9. On this point, see Anton Pelinka, *Modellfall Österreich? Möglichkeiten und Grenzen der Sozialpartnerschaft*, Vienna 1981, pp. 1–4 as well as various contributions in Gerald Stourzh and Margarete Grandner (eds), *Historische Wurzeln der Sozialpartnerschaft*, Vienna 1986.

10. In addition to the essay by Alemann cited in note 7, see for example Ulrich Nocken, 'Corporatism and Pluralism in Modern German History', in Dirk Stegmann *et al.* (eds), *Industrielle Gesellschaft und politisches System*, Bonn 1978, pp. 37–56; Werner Abelshauser, 'The First Post-Liberal Nation: Stages in the Development of Modern Corporatism in Germany', *European History Quarterly* 14, 1984, pp. 285–318; and especially Hem C. Jain and Anthony Giles, 'Workers' Participation in Western Europe. Implications for North America', *Relations Industrielles* 40, 1985, pp. 747–74, an essay in which the application of West European forms of

changes in the form of nationalization, socialization, or even co-operative concepts do not alter in principle the need for regulation of industrial relations, and that a public solution to the question of control and ownership of property does not necessarily resolve all the tensions that are created by production processes based on the division of labour.

III   The aim of this book is to depict and comparatively to evaluate the different directions that the more important Western industrial states have taken to date in shaping the industrial relations of an especially significant and, for many reasons, quite easily compar-able branch of industry, namely coal mining. The selection of the countries to be studied is justified naturally by the role that mining played and plays in the industrial history of these nations, but also by the distinctiveness of the solution found (as in the case of Austria), and last but not least, out of practical considerations. Not all countries can be treated. However, since the conclusion of this project it has become clear to the editors that industrial relations and their own individual variations in countries under British influence such as South Africa and Canada, on the one hand, and coal mining as an example of Japan's 'authoritarian corporatism', on the other hand, would have deserved consideration.

The period of study has been very deliberately set at the turn of the century. An abundance of studies exists on the often highly complicated and self-sacrificing protests and actions of the mine workers' communities and the processes leading to the formation of the unions. These concentrate predominantly on the nineteenth century. Even from a comparative perspective, this phase of coal mining has attracted methodological as well as empirical studies.[11] Nevertheless, the emphasis of these studies has almost always been

---

co-determination (especially following the examples of the Netherlands, the Scandi-navian countries, Austria and the Federal Republic of Germany) to industrial relations in the USA is primarily discussed. For an earlier history of co-determination in Germany, see Hans Jürgen Teuteberg, *Geschichte der industriellen Mitbestimmung in Deutschland: Ursprung und Entwicklung ihrer Vorläufer im Denken und in der Wirklichkeit des 19. Jahrhunderts*, Tübingen 1961; also Teuteberg's 'Ursprung und Entwicklung der Mitbestimmung in Deutschland', in Hans Pohl (ed.), *Mitbe-stimmung – Ursprünge und Entwicklung*, Wiesbaden 1981.
   11. Cf. M. I. A. Bulmer, 'Sociological Models of the Mining Community', *Sociological Review* 23, 1975, pp. 61–91; Klaus Tenfelde, 'Comparative Research in the History of Mining Workers: Some Problems and Perspectives', in Gustav Schmidt (ed.), *Bergbau in Großbritannien und im Ruhrgebiet: Studien zur vergleichenden Geschichte des Bergbaus 1850–1930*, Bochum 1985, pp. 18–35.

placed – certainly rightly or at least understandably so – on investigating labour history, the history of the workplace, as well as the
history of the process of the formation of the unions, and the more
specialized field of union history. The latter has been all too often
made the dominant subject of study 'in its own right' in all the
countries presented here. However, the history of entrepreneurs
has remained underdeveloped, let alone any general history of
conflict settlement in industry and industrial societies. As far as the
history of labour, life at the workplace and workers' culture are
concerned, completely new fields of study have been opened up
and interpreted often without regard to the presence of more
important political power arenas in which institutional frameworks
are set, standards of living are negotiated, and positions of power
are defined and balanced. Jonathan Zeitlin has correctly pointed out
recently that in such studies, unions sometimes appear 'as direct
expressions of workplace conflict; sometimes as remote institutions
of little relevance to workers' everyday concerns; and sometimes as
bureaucratic constraints on rank-and-file militancy'.[12]

Labour history, '*Arbeitergeschichte*' and '*histoire du travail*' should
not be limited to the histories of the union or workers' movements.
That would lead to insights that are much too narrow and
especially to an under-evaluation of those traditions, institutions
and influences that existed outside the workers' movement, yet
helped shape these quite considerably. Labour and working-class
history did not and does not take place in a sphere free of politics.
Nor does it occur in some sort of political sphere other than that
which is articulated in a society and which shapes this society. This
book makes use of the concept of industrial relations, although it is
very seldom described as such in the literature on social history in
the Federal Republic of Germany, or in such literature in the other
industrial nations covered here.[13]

The associations and institutions crucial for influencing industrial
relations in coal mining within the countries addressed here had taken
on their basic shape by the turn of the century. By this time, too,
national governments everywhere had more or less acknowledged

12. Jonathan Zeitlin, 'From Labour History to the History of Industrial Relations', *Economic History Review* 40, 1987, pp. 159–84.
13. In certain respects, England and the United States are more advanced,
compare e.g. the collected volumes edited by Chris Wrigley that are mentioned in
the bibliography, as well as the literature mentioned above in note 5. Yet Zeitlin's
criticism (see note 12) is directed precisely at the Anglo-American literature.

a need to regulate the state of industrial relations. Therefore, all the contributions in this volume commence with a brief review of the degree of organization that had been achieved by the politics of industrial interest at the turn of the century. All these studies also include the formation and change of employer interests – albeit, in the editors' view, to a perhaps still insufficient degree. However, this circumstance also certainly reflects the state of research on the history of entrepreneurs in mining, which has received much less attention than that paid to the history of workers.

All the authors have extended their studies as far into the present as is possible and justifiable in light of the availability of sources. In this way, it should be possible to give the reader a greater understanding of current conflicts and their dimensions – the major British miners' strike of 1985–6 comes to mind – and to furnish the discussion in industrial sociology by means of historical studies of branches of industry with some focus. The outcome has been more of an implicit than explicit comparison of the value of each of the different forms and institutions of industrial relations that have evolved; special attention has been paid to the West German case of the 'Montan' co-determination, as this was achieved under extraordinary political conditions and has withstood the most serious crisis of recent mining history – namely the crisis of decline since the end of the 1950s. The first step towards such a comparative evaluation is offered in the essay of one of the editors following the country reports (chapter 8).

IV   This project was able to be completed after two years of good co-operation in which national borders were no barrier. The authors assembled here enjoy, without exception, an outstanding reputation in mining, industrial, and working-class history of their countries and the editors are pleased to have gained their participation for this project. Without the decisive financial support of the project by the Hans-Böckler-Foundation in Düsseldorf, all this would not have been possible. By financing this project, the foundation, which is closely associated with the unions, has done more than just provide the funding: it has enabled the documentation of a comprehensive, diverse, scientific approach towards interpreting the history of the unions which also includes the history of entrepreneurs. It has thus made a not insignificant contribution to a modern 'culture' of industrial relations. The intent of this volume is the same; it is therefore appearing along with other publications on the occasion of the centennial anniversary of the German industrial union for mining and energy.

With the aid of the Wissenschaftskolleg in Berlin, a conference of the authors was held there at the beginning of July 1988 in order to co-ordinate the contributions. The editors would like to thank sincerely the director of the Wissenschaftskolleg, Professor Dr Wolf Lepenies, and its secretary, Herr Joachim Nettelbeck, as well as Dr Jonathan Zeitlin from Birkbeck College of the University of London, who showed the editors and co-authors during this conference new ways in which the history of industrial relations can be meaningfully studied.

2
Employers, Trade Unions and the State, 1889–1987: The Origins and Decline of Tripartism in the British Coal Industry[1]

Roy Church

The pre-eminence of the British coal industry before the First World War was based on an almost continuous expansion of output and employment; at the peak in 1913 a workforce of more than a million workers (10 per cent of the male labour force) produced 287 million tonnes of coal. As the overwhelmingly predominant source of light, heat and power, and as a major export, in many respects coal was the pivot upon which the economy rested. But between 1913 and 1946 output fell by some 27 per cent, while employment dropped from 1.1 million to 697,000 in the same period; in 1938 British coal exports comprised only 38 per cent of the international coal trade compared with 55 per cent in 1913, and had fallen to an insignificant level by 1946 when the industry passed from private ownership. Setting aside for a moment the two war periods, the structure and organization of the industry, except in degree, remained broadly unaltered until nationalization. Neither the relations between owners, managers and workers, nor the institutions through which industrial relations were conducted experienced fundamental change.

Before 1893 industrial relations in Britain were affected directly neither by government nor by the law, although more than in any other industry indirect influence was present as a result of legislation affecting safety, the age and sex of workers, the hours of work of women and young children and the payment of wages. Wages and terms and conditions of employment were the outcome of voluntarist bilateral bargaining between employers and miners

---

1. I am grateful to Dr David Smith, of the University of East Anglia, Norwich, for helpful comments on an earlier draft.

organized to varying degrees within regions and districts, where local influences and diverse bargaining procedures produced contrasting working practices and wage levels.[2] Fundamental to an understanding of the history of British coal mining is an extreme regional diversity based on differing geological characteristics. These in turn determined the types of coal produced, influenced the methods and organization of production and the market orientation of the various coal mining regions, and affected the degree to which they participated in expansion or decline. The consequence was that industrial relations also exhibited regional differences, in part reflecting variations in prices and wages; throughout the period intermittent attempts to encompass these within national agreements occasioned considerable political turbulence.

Clegg described the role of the state before 1914 as one of abstention from industrial relations, other than for legislation affecting only the framework within which they were conducted.[3] This disregards the unprecedented intervention by government in the major 1893 dispute between the newly formed Miners' Federation of Great Britain (MFGB) and the coal-owners' associations, which heralded an era of growing, if erratic and intermittent, direct and indirect government intervention, which at times involved the military and the police.[4] From this period emerged a tripartite relationship which culminated in, and continued after, the nationalization of coal mines in 1946. It is not surprising, therefore, that a dominant theme in the history of industrial relations during this period is the changing relative importance of market influences and market discipline compared with the role of the state. This raises the question whether the notion of 'corporatism' promotes or hinders an understanding of the development of tripartite relations between employers, trade unions and government.

Among the various definitions of the corporatist paradigm, that offered by Panitch seems to be the most complete: 'a political structure within advanced capitalism, which integrates organised socio-economic producer groups through a system of representation and cooperative mutual interaction at the leadership level of

2. Roy Church assisted by Alan Hall and John Kanefsky, *The History of the British Coal Industry*, vol. 3: *1830–1913. Victorian Pre-eminence*, Oxford 1986, chapter 8.
3. H. A. Clegg, *The Changing System of Industrial Relations in Great Britain*, London 1979, pp. 290–305.
4. See, for example, D. Smith, 'Tonypandy 1910: Definitions of Community', *Past and Present* 87, 1980, pp. 158–84.

industrialization and social control at the mass level.'⁵ Initially a
concept formulated to describe industry–state relations at a macro
level and in a later period, it has since been applied by some
historians to the period before the First World War. Taking as its
context a growing conviction, signalled by the majority report of
the Royal Commission on Labour 1891–4, that collective bargaining held the key to industrial peace, a number of agreements
between employers and trade unions between 1893 and 1914 have
been interpreted as 'a compromise between the full collectivism
which the world crisis [of capitalism] seemed to some to demand,
and the *laissez faire* fragmentation of pure Gladstonian democracy'.⁶

In a perceptive review of corporatist literature, Zeitlin drew
attention both to common strands and principal divergencies. That
there existed a tendency for the state to co-opt and subordinate
interest groups in order to further government objectives is a
widely held view. None the less, theorists disagree over the importance to be attached to the respective power of employers and
unions in their ability to influence government policy; the nature of
state goals and priorities – political or economic – is also a matter
for disagreement. Zeitlin has pointed to the lack of a unified or
coherent body of theory called 'corporatism', and underlined the
diverse economic, political and ideological influences which over
time can alter the balance of power between the major interest
groups and the state. He noted a similar variation in the priorities
which the state attaches to competing objectives, which may differ
in the short and longer term and may be perceived differently by
different branches of government.⁷ His critique, therefore, questions the validity of corporatism as a useful concept, yet at the same
time points to the need for empirical research into the history of
those tripartite relations which the 'theory' of corporatism has so
far failed to explain. It also prompts the question whether in
characterizing government relations with employers and unions, a
form of industrial politics which we shall label 'tripartism' offers a

5. L. Panitch, 'The Development of Corporatism in Liberal Democracies', *Comparative Political Studies* 10, 1977, pp. 61–90.
6. R. Currie, *Industrial Politics*, London 1979, p. 48. See also R. Price, *Masters,
Unions and Men*, Cambridge 1980, for an analysis of the building industry within
this framework.
7. Jonathan Zeitlin, 'Shop Floor Bargaining and the State: a Contradictory
Relationship', in Steven Tolliday and Jonathan Zeitlin (eds), *Shop Floor Bargaining
and the State*, Cambridge 1985, pp. 16–37.

more satisfactory shorthand. Inevitably, therefore, it will be necessary to explore the nature of tripartism and its relevance to understanding the history of industrial relations in coal mining.

Organization, Structures and Influences under Private Enterprise

Throughout the pre-nationalization era coal mining continued to
be an extremely labour-intensive process, hand-cut coal accounting
for 90 per cent of production in 1913 and greater than 50 per cent
until the late 1930s. The conveyance of coal was even less
mechanized.[8] Hewers, who cut the coal, were equivalent in status
to skilled workers in other industries, though that skill owed more
to experience of geological and mining conditions than to technical
competence, and depended much upon physical strength in the use
of hand tools or manually operated machines. In some coalfields
hewers were sub-contractors employing two or three assistants, the
hewers themselves receiving their pay from a larger employer.
Underground, miners worked in independent groups in small
'rooms' or 'stalls' where effective supervision was difficult. This
was also true even under longwall conditions before the 1930s,
when most coal-faces were of limited length – numbered in tens,
rather than, as later, hundreds of metres – and the workers manning
the face relatively few in number. Work organization underground
thus offered a considerable degree of relative autonomy, especially
for the hewers, who enjoyed the highest material rewards and
status in the industry and comprised the major force in mining
trade unionism. One effect of mechanization was to extend longwall working which, partly through the adoption of multiple shifts
to exploit investment fully by maintaining co-ordinated moving
progression across the coal-face, increased the scope for specialization within longwall teams and both necessitated and facilitated
closer supervision of workers.[9] Consequences of this progressive
division of labour, according to some interpretations, included not
only a differentiation of skills and extensive deskilling, but an
increase in rank and file worker activity in reaction to a skilled elite

8. Barry Supple, *The History of the British Coal Industry*, vol. 4: *1913–1946. The
Political Economy of Decline*, Oxford 1987, table 9.6.
9. Church, *History*, pp. 274–6; Supple, *History*, pp. 433–8.

which dominated official trade unionism, a growth of industrial militancy, and radical socialist politics.[10]

The history of coal mining before the Second World War offers no support for this thesis. The actual extent of mechanization in this period was limited, division of labour was not new in some coalfields even before mechanization, and mechanization actually demanded skills which were different but no less specialized than those of the hewer.[11] A comparison between the two most strike-prone coalfields reveals only 3 per cent of coal cut by machinery in South Wales by 1921, whereas the figure for Scotland was 36 per cent. In the east Midlands and Lancashire the proportion of coal cut by machinery was above the national average of 14 per cent, yet these were not the regions recording high levels of miners' militancy.[12] Even in the 1930s the essential mixture of autonomy and discipline, equality and hierarchy, individuality and group cohesion, skill and physical application characterized mineworking, the precise combinations prevailing varying between regions.[13] Within the workforce, therefore, except for the consequent introduction of multiple shifts, mechanization and its effects on work organization were not central to changes in industrial relations.

The variable pace and extent of mechanization and of longwall working between coalfields underlined the fragmentation of industrial organization and of the industry's productive and commercial structure. The number of productive mines in the 1880s exceeded 3,000, falling to barely 1,900 before the Second World War. The trend in mine size was upwards, reaching an average of 408 workers by 1913, and 427 in 1938. The number of mines employing more than 2,000 workers increased until the mid-1920s, but the proportion of workers employed in such pits, which was 17 per cent in 1913, after rising to nearly one-third in 1924 returned to roughly the 1913 proportion by 1938. To the extent, therefore, that labour relations were affected by the size of a colliery's workforce, there was no marked change in scale over the long term, the

10. The literature is reviewed by Alistair Reid, 'The Division of Labour and Politics in Britain, 1880–1920', in W. J. Mommsen and H. G. Husung (eds), *The Development of Trade Unionism in Great Britain and Germany*, London 1985, pp. 150–65.

11. Supple, *History*, pp. 432–8.

12. Roy Church, 'Edwardian Labour Unrest and Coalfield Militancy, 1890–1914', *Historical Journal* 30, 1987, table 2, p. 851.

13. Supple, *History*, pp. 431–2.

medium-sized pit employing between 500 and 1,999 workers continuing to be characteristic.

Numerous writers have drawn attention to connections between size of establishment and industrial conflict.[14] Whereas size of pit is relevant in considering the organization of production and the implications for industrial relations, the size of firm, its ownership and management structure, is important in relation to broader matters of policy and implementation. That the size of colliery undertakings was growing throughout the period there is no doubt, but the level of concentration measured by the proportion of total coal production mined by the largest enterprises did not alter substantially. The same is true of trends in corporate structure. Before 1913 the largest fifty companies employed about one-third of all colliery workers. Another third were employed by 211 firms each with between 1,000 and 1,999 men on their payrolls. During the interwar period large firms became relatively more important, but the level of concentration changed little. Thus, whereas 94 firms employed slightly less than half the workforce in 1913, the 1938 figure was 66 undertakings employing slightly more than one-half.

Firms employing between 1,000 and 2,999, numbering 150 in 1938, accounted for one-third of the workforce. Large firms were of greatest relative importance in South Wales (where the trend continued between the wars) and in the north-east (where the trend towards larger firms was reversed). In Yorkshire, the east Midlands and Scotland a proportionately smaller number of firms extended their control over an increasing share of the regional industry. Overall, however, the long-term trend towards industrial concentration of mines and of firms was gradual and suffered some reversal, notably in the late 1920s. It is not surprising, therefore, that the nineteenth-century pattern of family enterprises, partnerships and private companies should have continued to be characteristic of the industry throughout the period of private enterprise. Large-scale public joint stock companies with dispersed shareholding and subject to 'impersonal' buying and selling of quoted shares was exceptional before 1914, limited mainly to a handful of coal-producing iron and steel companies and coal-exporting enterprises,

---

14. For example, see S. Wellisz, 'Strikes in Coal Mining', *British Journal of Sociology* 4, 1956, pp. 346–66; B. J. McCormick, 'Studies in the Yorkshire Coalfield, 1947–1963', *Journal of Economic Studies* 4, 1969, pp. 171–98.

principally in South Wales and Scotland, though increasingly in the
east Midlands and Yorkshire thereafter. Even after 1914 the trend
in some other industries towards a bureaucratized corporate structure, in which professional managers took major decisions relating
to business rather than to merely production matters, was to be
found in no more than a handful of the largest coal-producing
companies. One implication, therefore, of the highly competitive
structure of the industry, its self-imposed reliance on resources
within the industry for capital and finance, and the dominance of
owner-management or something similar, was the perpetuation of
long established administrative structures, policies and practices,
manifest in individualistic, and for the most part unrestrained
competition.

This is not to deny that corporate relationships existed in the
1920s and 1930s, ranging from extensive amalgamated colliery
groupings to more diffuse alliances consisting of a network of
interlocking shareholdings and directorships, but the complexity of
the matrix which resulted was not conducive to widespread and
effective co-operation and collusion. The history of association
among coal-owners had been regional and continuous in the major
coal-producing areas since the 1880s; however, the competitive
structure of the industry, even at the regional level, ensured that
successful association occurred only when colliery owners were
presented with effective organization among colliery workers. The
range and complexity of interests which rendered inter-firm co-
operation problematic, even when investment or participant links
existed, also undermined coal-owners' attempts to achieve sustained policy co-ordination at either a national or regional level.
Supple has drawn attention to the apparent paradox presented by
juxtaposing the individualism which continued to characterize the
industry throughout the interwar years with the public and political
reputation of the owners as one of determined and successfully
maintained solidarity.[15]

Who were the business leaders in the industry before 1914? Most
were descended from colliery-owning families or families in which
mining engineering or colliery management consultancy was the
fathers' (and in several cases grandfathers') profession. Metallurgi-
cal enterprise was another significant source of recruitment to the
industry's business elite. Many qualified as mining engineers after a

---

15. Church, *History*, pp. 386–92, 434–48; Supple, *History*, pp. 361–77.

period of apprenticeship; few received university education, and the inbred nature of recruitment from within families ensured that even in some of the largest companies key decisions lay in the hands of owner-managers lacking formal training. Excluding aristocratic colliery enterprise, which was a small part of the industry even in the mid-nineteenth century, coal was the basis of few large fortunes by comparison with other forms of business activity. Few owners or managers played an active role in politics, though a public role in the local community was not unusual. Increasingly the line between professional and proprietorial interests and influences were blurred, as within colliery-owning families professional training was substituted for learning by doing.

D. A. Thomas (later Lord Rhondda), for example, chairman and managing director of the nation's largest colliery combine, the Cambrian Collieries in South Wales, was the son of a wealthy coal-owner, read mathematics at Cambridge, spent time in the sales office of his father's company and in a stockbroker's office before gaining experience underground.[16] The architect of the Cambrian Combine, the result of a series of amalgamations and mergers, he served briefly as an MP but quit to return to business, which he described as: 'a modern equivalent of war. Business attracts the man who loves conquest, who loves to pit himself against vast odds, who could not live without strain of effort.'[17] Thomas, one of the industry's few millionaires, earned notoriety for his relentless opposition to organized labour and for the scale of the lockout at the Cambrian Collieries in 1910. It was also in the South Wales coalfield that the militancy of miners was substantially greater than elsewhere. In 1916 *The Times* leader accused the South Wales coal-owners of pursuing a 'grasping policy', and criticized their mismanagement, obstinacy and greed, acknowledging, however, that in no other coalfield did there exist such strife and distrust of the owners.[18]

From this cockpit in 1919 Evan Williams, a formidable colliery owner in the Thomas mould, was elected president of the Mining Association of Great Britain (MAGB). In this capacity for twenty-five years he continued to dominate the politics of coal, in a period when the public role and impact of the MAGB as a national body

16. Church, *History*, pp. 449–68.
17. Quoted in Church, *History*, p. 464.
18. Quoted in Supple, *History*, p. 75.

was transformed.[19] Because of this development, the battleground metaphor employed by Thomas seems relevant not merely to an understanding of industrial relations in South Wales but to the public reputation of colliery owners elsewhere. As their national representatives, Sir Evan Williams, or Sir Adam Nimmo and Wallace Thorneycroft of Scotland, or Lord Gainford and Reginald Guthrie of Durham, or W. Benton Jones of Yorkshire, adhered strictly and uncompromisingly to crude *laissez-faire* ideology, whether in relation to costs or industrial reorganization. Whatever the merits of their economic analysis of the industry, the unyielding persistence of the MAGB in reiterating this doctrine between the wars – regardless of the social consequences of the policy measures it implied – revealed a failure to appreciate the psychological benefits that improved industrial relations might bring and the possibility that 'a distasteful public presence' might ultimately weaken both their political and economic position.[20]

Before 1919 the public presence of the MAGB was negligible, for since its formation in 1854 its principal concern had been with legislation affecting the industry. Matters relating to prices, wages and conditions of employment were dealt with within the district coal-owners' associations and by individual coal-owners. These associations took permanent root during the late nineteenth century in response to the successful organization of miners' trade unions, which were similarly established on a local basis. Within these several districts (twenty-three constituent coal-owners' associations existed in 1912) differences in geological conditions, market conditions, working methods, payment systems and employment practices cut across the common interests of coal-owners in restraining the miners' unions; consequently, coal-owners' divergent commercial interests presented them with serious difficulties in securing sufficient support to implement policy even at the regional level. The formation of the Miners' Federation of Great Britain (MFGB) in 1889 offered some incentive to colliery owners to promote greater unity, but not until the 1893 strike, which affected all the English coalfields, did the associations come together to formulate a national strategy. Henceforward, the MAGB existed as a loose federation of owners and managers for the purpose of

19. Barry Supple, '"No Bloody Revolutions but for Obstinate Reactions"? British Coalowners in their Context, 1919–20', in D. C. Coleman and Peter Mathias (eds), *Enterprise and History*, Cambridge 1984, pp. 227–34.
20. Supple, *History*, pp. 417–24.

dealing with the MFGB. None the less, co-ordination and collaboration in the formation of policy and its implementation continued to elude the MAGB even after wartime shortages, postwar dislocation and an intensification of industrial politics forced upon it a national role.

Coal-owners continued to protect their local autonomy after 1919, even though their representatives on the executive committee of the MAGB agreed to adopt national principles in setting wages and fixing hours after 1921. This concession to national agreements indicated a recognition that the centralized role which the pressure of wartime shortages and postwar dislocation forced on the MAGB also implied a greater dependence on the national body to defend its employer-members against unions and government in order to maintain constituent associations' freedom of action, most importantly in actual wage determination.[21]

Local freedom of action in determining wages and conditions of employment was seen to be vital in an industry in which wages accounted for roughly three-quarters of the average production cost per tonne of coal, and in which the variety and immense complexity of payment systems produced enduring local differences. Piece-rates were paid to face-workers, since their production or advancement could be measured and was susceptible to incentive payment. Others tended to receive day rates. 'Standard wages' consisted of rates paid at a given date at every pit in a particular district for a specified task, but these also embodied variations depending on changes in working conditions, principally due to variations in geological factors. When new mines were sunk or seams or faces opened up, or when new conditions arose, price lists of standard or, later, basis rates would have to be drawn up. Further complexity arose from the practice of making additional allowances or deductions which varied from pit to pit.[22]

Where trade unions existed, price lists were the outcome of bilateral agreement achieved through the operation of district boards of conciliation and arbitration. They originated in those coalfields where miners had established relatively strong unions – in South Yorkshire beginning in 1866, Durham in 1872, and West Yorkshire and Northumberland in 1873 – in each case the initiative to form district joint committees having come from the unions

21. Church, *History*, p. 652; Supple, *History*, pp. 417–24.
22. Church, *History*, pp. 556–7; Supple, *History*, pp. 36–9.

who persuaded coal-owners to co-operate. Even where sliding scales were introduced, the connection between the movements of coal prices and wages was hardly closer than where no formal scale existed. In practice, the particular scale adopted merely reflected the balance of industrial forces at the time of the agreement; rarely did an agreed scale survive longer than two years. Even so, the union leaders in the export coalfields of South Wales and the north-east, which were most affected by price volatility, favoured sliding scales long after the inland coalfields had abandoned them for conciliation boards. Although the supersession of the former by the latter has hitherto been regarded as a failure of sliding scales, a recent detailed analysis of the machinery of sliding scales and boards concluded that the only major distinguishing feature between them was the incorporation in board agreements of maximum and minimum limits; that continuity is more evident than historians have supposed.[23]

Sliding scales reflected regional differences, as did the locally determined portion of the total pay received by miners, and it is within this context of immense regional and intra-regional diversity that the history of industrial relations, even after the formation of the MFGB, must be seen. The 'centralized' district unions which made up the membership of the federation differed in the degree of executive control exercised within each district. At one extreme was the Durham Miners' Association, possessing a highly centralized organization in which power lay with the executive council made up of pit lodges each entitled to a single vote. At the other extreme was the South Wales Miners' Federation, which came into permanent existence only in 1898 and comprised twenty fairly autonomous district unions. When the SWMF affiliated to the MFGB in 1899, only Northumberland and Durham – the other major exporting coalfields – remained outside. When those coalfields joined in 1907, the MFGB embraced thirty separately constituted unions, which meant that the basic structure of mining trade unionism was the familiar pattern of pit lodge, district and county union held together by a federated national assembly. From 1889, common wage scales were adopted by the MFGB, but the centrifugal forces generated within the regions continued to affect indus-

23. J. H. Porter, 'Wage Bargaining under Conciliation Agreements, 1860–1914', *Economic History Review* 23, 1970, pp. 460–75; John G. Treble, 'Sliding Scales and Conciliation Boards: Risk-Sharing in the Late 19th Century British Coal Industry', *Oxford Economic Papers* 39, 1987, pp. 679–98.

trial relations. The MFGB was formed in a boom period when trade union membership rose dramatically to reach 42 per cent of the workforce in 1892. After a temporary drop in the mid-1890s, the percentage rose to an average of 66 per cent between 1899 and 1909, and to 72 per cent in 1910–13, when the MFGB membership exceeded 0.7 million.[24] Throughout the prewar period the dominant ideology within which organized mining labour worked was firmly labourist, moderate and reformist. The political traditions which endured in mining districts took root long before the emergence of socialism in the 1880s. The leadership was politically moderate and Liberal in complexion; strikes were regarded as the major tactically effective weapon, to be avoided if possible, but they were employed in what the leaders regarded as appropriate circumstances to secure realistic objectives. It is significant that in South Wales, where trade unionism was late to establish a permanent hold, the labourist tradition was relatively weak and a new radical socialist, and later syndicalist, ideology took root. Elsewhere, democratic religious nonconformity strengthened the link with peaceful struggle and with Liberalism.[25]

Consistent with this distinction, often found in the literature, between a politically moderate and liberal leadership and, by implication, a more volatile and militant rank and file is a perception of relations between leaders and rank and file in which tensions arose from conciliation and arbitration agreements which were promoted and supported by full-time officials often in opposition to the views and interests of members.[26] Griffin's detailed study of conciliation board agreements in Leicestershire, however, led him to conclude that they did serve the wage bargaining interests of members; moreover he rejects the view that the institutionalization of wage bargaining enhanced the 'respectability' and nurtured political moderation among trade union leaders. Passivity, moderation and even obstructive attitudes towards militancy he found to be more characteristic of rank and file rather than of union leaders.[27]

24. Church, *History*, pp. 708–11.
25. Ibid., pp. 711–12.
26. K. Burgess, *The Origins of British Industrial Relations*, London 1975, pp. 189–91, 217.
27. C. P. Griffin, 'Conciliation in the Coalmining Industry before 1914: The Experience of the Leicester Coalfield', *Midland History* XII, 1987, p. 77; see also Treble, 'Sliding Scales and Conciliation Boards', pp. 679–98.

Yet the years between 1893 and 1912 brought a series of major crises in industrial relations which for the first time occasioned government intervention and legislation directly affecting the conduct and outcome of industrial relations. Government intervention in the industry before that time had affected directly employment, safety and the payment of wages. Women were prohibited from working underground in 1842, although landed coal-owners, well represented and vocal in Parliament, had resisted all but the prohibition of young children below the age of ten underground. In the absence of strong support from within the mining districts, not until 1887 was a legal minimum age of twelve adopted. Unintentionally, framers of the safety legislation between 1855 and 18⁷2 actually afforded legal reinforcement of colliery rules hitherto included in workers' contracts with employers. Acts passed in 1855, 1860 and 1872 to regulate mines in accordance with a general safety code also incorporated special rules designed to meet local conditions. These were drawn up by coal-owners on a local basis subject to the approval of a mines inspector, and it was within these special rules that coal-owners inserted clauses concerning workmen's hours of work, wages, quantity and quality of work which had little relation, if any, to safety considerations. Of particular concern to miners was the imposition of fines for various infringements, and the power retained by colliery owners to confiscate tubs below weight or containing dirt. A piece of countervailing legislation was the Checkweighman's Act of 1860, which gave legal recognition to the position of a full-time person who, although working at a colliery, was paid by employees to protect workers from fraud perpetrated by employers. However, it was left to miners to persuade individual colliery owners to agree to an appointment, and not until 1887 did it become possible for miners to appoint a checkweighman regardless of an employer's wishes. In the intervening period the appointment or dismissal of a checkweighman was a frequent issue between colliery owners and workers. Truck payments, however, a common method of preventing workmen's mobility in the early nineteenth century and the subject of repeated legislation, occasioned fewer disputes and by the end of the century finally disappeared.[28]

Some colliery owners, notably in the north-east and Scotland, continued to rely on colliery housing to retain workers, though

28. Church, *History*, pp. 188–98, 263–73.

probably no more than one-third of all colliery workers lived in company houses during the mid-nineteenth century, and perhaps only 12 per cent by 1913. The threat of eviction to counter industrial action by workers was a not infrequent tactic employed by colliery owners, in the north-east and Scotland especially, before the 1870s. Thereafter, the diminishing financial returns from housing investment, the growth of coal mining in urban districts and the rise of trade unions strong enough to resist coercion in housing constrained the effectiveness of housing provision as one form of regulating labour. None the less, the opening of new mines in rural areas in the early twentieth century was accompanied by investment in colliery housing, together with attempts to create colliery communities which would attract and retain 'respectable' workmen.[29] This policy of constructive paternalism, long established in practice, was an attempt to substitute company loyalty and harmony for industrial conflict. A leading textbook on coal mining, published in 1896, observed that a manager's duty included the implementation of decisions unpopular with the men, which was why he should 'cultivate relationships with them . . . by taking a personal interest in their reading rooms and institutes, their athletic clubs, their musical band, or in some of the various institutions which usually exist in colliery villages.' Until the widespread growth of mining unions from the 1880s, company welfare policies were almost invariably coupled with the proscription of trade unions, and in a few instances with the formation of company unions.[30] Even though vestiges continued to survive, by the 1890s the development of a formal structure of collective bargaining had eroded paternalist influence on industrial relations.

The pattern of industrial relations before 1893 reveals similarities across the regions. Employers' policies were mainly determined by the extent to which miners accepted the frequent wage adjustments deemed necessary by employers to compensate for price changes. Especially in regions where sliding scales were not in operation, strikes and lockouts, mainly of brief duration, punctuated the trade cycle. By the 1890s coal-owners were less inclined to recruit non-union labour as a method of replacing strikers, preferring to contribute financial support to owners of the collieries chosen by

29. M. J. Daunton, 'Miners' Houses: South Wales and the Great Northern Coalfield, 1880–1914', *International Review of Social History* XXV, 1980, pp. 143–75; Church, *History*, pp. 278–81, 559–602.
30. Church, *History*, pp. 274–99.

the union as targets to test the coal-owners' response. Mutual co-operation, supplemented by the dismissal of trade union activists, were common characteristics of disputes in this period.[31] One effect of this strengthening in organization, both among trade-unionists and coal-owners, was the spread of boards of conciliation and arbitration, perhaps the outstanding change to have taken place in employer–worker relations in the industry between 1889 and 1914.[32] Contributing to this growth in orderly collective bargaining by agreement between miners through the MFGB and the federated employers was the mutual acceptance of the joint negotiation of wages over the entire federated area. This is illustrated by the advance of collective agreements negotiated by the MFGB, which by 1910 applied to one-half of all coal-miners, whereas the comparable figure for the entire British workforce was one-fifth. The most common method of settling disputes before 1914 was through conciliation and pit negotiations. The proportion of disputes settled by conciliation machinery involving official arbitration or by the intervention of third parties was less than 5 per cent between 1888 and 1913.[33]

At the same time, even by the 1870s coal had the reputation of being the most strike-prone of all industries, a record which can be measured with some precision from the 1890s. Undoubtedly, without the operation of boards, the number and proportion of disputes settled by strikes and lockouts would have been much greater; but fundamental to relations between employers and workers was the apparent impossibility of finding a way to reconcile movements in wage rates with cyclical coal price variations. A further intractable problem was encountered in devising a system of piece-rate earnings which could accommodate, and be adapted to, the short-run changes that occurred in working conditions, mainly the result of geological factors. The recorded objectives of strikes and the issues which precipitated direct industrial action were overwhelmingly related to piece-rates and working arrangements (which commonly amounted to wage adjustments in return for changes in working practice). Most strikes were of short duration and limited in geographical extent. However, after 1889

31. Ibid., pp. 672–4.
32. John Benson, *British Coalminers in the Nineteenth Century: A Social History*, London 1980, p. 200.
33. Church, *History*, p. 734.

the growth of increasingly powerful federated organizations and the conduct of collective bargaining across the coalfields had the effect of enlarging the scale of disputes. Although minor issues continued to be resolved mainly at local level within boards of conciliation by joint committees, the potential for major regional confrontations arising from changes in prices and working conditions is illustrated by the lengthy stoppages that occurred in Durham in 1892, Scotland in 1894 and South Wales in 1898. Furthermore, in 1893 the industry experienced a stoppage affecting all English coalfields, excluding only collieries in the north-east; 1912 brought a national strike.[34]

The origins of the 1893 dispute lay in the extreme fluctuation in prices and wages in 1888–91 which, having carried miners' earnings sharply upwards, brought a price fall and a demand from the federated owners for a 25 per cent cut in wage rates. The MFGB countered with a demand for an eight-hour day, which it was argued would restrict output and thereby restore coal prices and remove the pressure to reduce wages. In the course of preparing their campaign, the MFGB formulated not only the objective of securing an eight-hour day, but also of establishing a minimum 'living' wage. There is a suspicion that the coal-owners intended to destroy the MFGB, for not only was the size of the cut called for very large but owners rejected out of hand the offer from the mayors of six major cities to conciliate. After the lockout (which was accompanied by blacklegging, rioting, violence and the intervention of the military) had been effective for fifteen weeks, the Prime Minister took the unprecedented step of appointing the Foreign Secretary, Lord Rosebery, to call a conference including coal-owners and leaders of the MFGB. The Rosebery Conference agreed that pre-strike wage levels should be restored for four months, pending arbitration. The outcome was a 10 per cent wage reduction to apply until the end of 1895, and the establishment of a National Conciliation Board. After 1895, variations (upwards only) from the new rate of wages up to a maximum of 45 per cent on the 1888 standard could be achieved only through the board; a minimum level to base-rate wages (not in itself a 'minimum wage') was also accepted by coal-owners. The minimum period during which an award could apply was seven months. The settlement reduced the intensity of wage fluctuations. However, the board

28   *Roy Church*

functioned in such a way that the decisions it took implied that a variant of the sliding scale was the guiding factor in reaching them, and therefore favoured employers during the prosperous years of the industry down to 1914. In a wider context, the formula employed by government to settle the 1893 dispute led directly to the first Conciliation (Trades Disputes) Act passed in 1896.[35]

While the events of 1893–4 cannot be described simply as either a victory for the owners or for the miners, the degree of inter-regional solidarity achieved among miners across the English coalfields encouraged the MFGB to strengthen its campaign for an eight-hour day and a proper minimum wage, both of which had been denied in 1892 and 1893. Though opposed by owners in most regions, the Eight Hours Act was passed in 1908, leaving the minimum wage and nationalization as the remaining major objective of MFGB policy. When miners' representatives had argued that working in abnormal places implied the relevance of a minimum wage, employers had always resisted the notion of such a concept.[36] Furthermore, owners regarded the minimum wage campaign as an attack on the piece rate system. None the less, coal-owners in the inland coalfields expressed less hostility than did the exporting coal-owners of South Wales and Northumberland, whose trade was most vulnerable to price fluctuations. The federated owners sought safeguards against a diminution of miners' work effort, though by 1912 they accepted the principle of a minimum wage. By this time both the miners of South Wales and of Northumberland and Durham had affiliated to the MFGB; consequently a united national union of mineworkers confronted a divided array of owners, and this the national strike, called in support of specific proposals for a minimum wage in 1912, was intended to exploit.

For the second time, government intervened, though the counter-proposals put forward were rejected by both sides. This was followed by the introduction by government of a bill which embodied the principle of a minimum wage, though it (unlike the MFGB proposal) omitted figures. This, too, was rejected by the union. In the fourth month of the strike, the support from miners in South Wales (where the effects of the strike were most extreme) and in the Midlands (a high productivity, high wage area where the

35. Ibid., pp. 736–9.
36. M. W. Kirby, *The British Coalmining Industry, 1870–1946*, London 1977, p. 19.

minimum wage was not a real issue) was withdrawn. When the required majority needed to continue the campaign was not forthcoming, the strike was called off. The government's terms were accepted, which left the actual wage levels to be decided by joint boards in each region, a formula which allowed employers to avoid introducing uniform minimums. In effect, with few exceptions, average wages were barely affected, the minimums granted in some regions falling well below miners' demands.

Although the first national miners' strike of 1912 made little difference in material terms, this does not warrant the conclusion that it was also unimportant. The campaign demonstrated the ability of the MFGB to organize and sustain a lengthy national stoppage and to threaten the economy to such a degree that government intervention, as in 1893, was deemed necessary to avoid further damage, and it prompted legislation on wages.[37] Phelps Brown suggested that in securing the Eight Hours Act in 1908 and the Minimum Wage Act in 1912 the MFGB had secured national settlements which would have been impossible without having manipulated government by intimidation. The 1912 episode he regarded as a particularly important precedent, and the first of numerous occasions when governments were manoeuvred into making concessions to demands which employers otherwise found completely unacceptable,[38] a process which intensified under wartime conditions after 1914.

Wars and Depression: From Private Enterprise to Dual Control

Success in war depended fundamentally on a high level of mobilization to fight the enemy, and maximum coal production to supply virtually all the Allies' fuel demands in addition to domestic needs. These two priorities came into conflict in 1914/15, when miners enlisted in large numbers, resulting in a net loss of miners employed in the industry of 16 per cent. This prompted government measures to restrict the recruitment of miners and in 1916 to exempt underground and certain surface workers from conscription under the

37. Church, *History*, pp. 739–46.
38. E. H. Phelps-Brown, *The Growth of Industrial Relations, 1906–14*, London 1959, pp. 325–8.

Military Services Act. None the less, by January 1917, 26 per cent of the industry's original labour force of 1.1 million had entered the forces.[39] Until 1918 the recruitment of mining labour, heavily diluted by the replacement of those who had enlisted by juveniles and inexperienced non-mining workers, took precedence over military manpower, but as the military situation deteriorated during the German spring offensive, arrangements were made to conscript 75,000 miners. Later in 1918 an estimated shortfall of 25 million tonnes (10 per cent of annual output) tipped the balance of priorities back towards mining labour supply in order to avoid the predicted alternative of industrial rationing and the risk of half a million unemployed. The War Cabinet agreed to release 50,000 miners from the armed forces.[40]

Controlling the labour supply was easier for government to achieve than was securing a commensurate increase in output. This was one of the problems which in 1915 prompted the establishment by government of a joint Coal Mining Organization Committee comprising equal numbers of representatives from the MFGB and the MAGB. Apart from advising government on pricing control, export and distribution, it saw the key elements in production strategy as the reduction of absenteeism and the improvement of industrial relations. Yet virtually no progress was made; in the last resort levels of absenteeism were controlled by the miners themselves, while the MFGB pursued a policy of non-cooperation with government, subordinating war priorities to sectional interest. Anxious to ensure continuous production and therefore avoiding the provocation of miners, government acquiesced in the MFGB's refusal to accept that the Munitions of War Act should apply to miners as to all other munitions workers. Miners remained exempt, therefore, from the statutory regulation of labour and the outlawing of strikes and lockouts; their leaders obstructed changes which might have increased the supply of juvenile and female labour, resisted alteration to the Eight Hours Act, and rejected the application of compulsory arbitration to the industry.[41]

The counterpoint to this strategy of scarcity reinforcement which secured for the miners a uniquely privileged position was a sustained and equally successful campaign to maximize improve-

39. Supple, *History*, pp. 43–8.
40. Ibid., pp. 93–8.
41. Ibid., pp. 50–7, 66, 98–9.

ments in miners' wages and material conditions. Demands presented to the owners were not only substantial in scale but
contained some novel features. As in some other industries the
union sought a national settlement, but the basis was to be earnings
rather than wages, and the standard adopted was cost of living
rather than coal prices.[42] During the war, miners received substantial wage increases, government intervening on each occasion to
compel owners to concede most, if not all, demands – though some
part of the final award was left for negotiation at the district level.
The final concession occurred in 1918 at the height of the crisis – for
forces in the field and in coal supplies – when the War Cabinet
declined to wage war with the miners. The concessions were made
in the face of strike threats and the prospect of disastrous implications for the war effort. So critical was the continued supply of
the Admiralty with coal from South Wales that in 1916, when
owners refused to consider miners' demands for wage increases,
government assumed financial control of the industry under the
Defence of the Realm Act. A few days after, assuring the SWMF
that control did not involve legal discipline or wage reductions, it
conceded the miners' claims, which owners would have to pay.[43]
In effect, the government acknowledged that, at least in South
Wales (though coal control was extended to all coalfields three
months later), private enterprise was unworkable in the national
interest. A major piece of socialization by comparison with previous experience, public control stemmed from the need to secure
labour's co-operation in production, in part by reassuring the
miners that owners would not benefit from coal shortage.[44]

That private enterprise was unworkable was a view strongly
endorsed by the MFGB (which voted unanimously for nationalization in 1918) and especially strongly by the miners of South Wales,
where socialist and syndicalist advocacy was intense. When the war
ended, however, the MFGB placed its weight behind a campaign
for substantial material improvements rather than for structural
reform. The basis of miners' social need advanced by the MFGB
was their contribution to the economy and their historically high

42. Ibid., p. 63; H. A. Clegg, *A History of British Trade Unions Since 1889*, vol. 2,
Oxford 1987, pp. 84, 128–30.
43. Supple, *History*, pp. 64–7, 70, 74–7, 99–102.
44. Barry Supple, 'Ideology or Pragmatism? The Nationalization of Coal', in
Neil McKendrick and R. B. Outhwaite (eds), *Business Life and Public Policy*,
Cambridge 1986, p. 230.

living standards compared with other occupational groups, factors used to justify their demand from government of a 30 per cent increase in basic earnings, reduced working hours and preferential treatment for miners in the process of demobilization. Faced with a strike threat, Lloyd George's Cabinet established a royal commission of inquiry under the chairmanship of a leading judge, Sir John Sankey, which was committed to produce an interim report on wages and hours within a 24-day deadline.[45] The Sankey Commission included representatives of miners, owners and 'independents', but under forceful chairmanship the inquiry soon assumed the character of what Beatrice Webb described as a 'state trial' of the coal-owners which culminated in the question 'why not nationalize the industry?'[46]

In the face of interrogation along unexpected lines by a combination of forceful trade union leaders and politically sympathetic intellectuals, the coal-owners' evidence was inept and unconvincing. Even their potential supporters, including the *Colliery Guardian* and *The Times*, expressed no confidence in the coal-owners' defence of the prevailing system, which was widely regarded as unsatisfactory, even though nationalization was thought to be too extreme.[47] Resentful of their humiliation by the Sankey Commission, fearful of the threat which it presented to private ownership and angered by the government's acceptance of the report's recommendations for increased wages, reduced working hours and profit limitation, the MAGB responded by reforming its representative structures and strengthening the national executive. Under the formidable leadership of Evan Williams and with some success, the MAGB began to employ systematic propaganda to persuade politicians and the public of their case against any form of dual control or socialization. Intensive lobbying of Tory MPs and the diminished intensity of the miners' campaign following material concessions on wages, hours and profit restrictions provided the context in which the Cabinet rejected nationalization.[48] Even Lloyd George's proposals for restructuring by the formation of large regional corporations were discarded after being criticized by the miners' president, who expressed his members' preference for the

45. Supple, 'No Bloody Revolutions', pp. 216–18.
46. *The Diary of Beatrice Webb*, ed. Norman and Jeanne Mackenzie, London 1982, p. 337.
47. Supple, 'No Bloody Revolutions', p. 220.
48. Ibid., pp. 220–3.

old system of private enterprise over trusts which he regarded as against the public interest.[49] The shortlived campaign by the MFGB to obtain 'Mines for the Nation' failed. The Trades Union Congress (TUC) declined to take direct action in support of nationalization. Meanwhile, the government proceeded to prepare for decontrol of the industry by conceding another large wages settlement under the dual control regime.[50] As the international market for coal slumped, the government retreated from control of – and from responsibility for – the industry forthwith. Faced with the prospect of a £5 million deficit per month on the Coal Charges Account, no alternative to withdrawal seemed possible.[51]

Into the lap of owners and miners was thrust the issue of national versus district negotiations and agreements, a recurrent theme in the industry for ever after. Termination of the existing wages agreement coincided with decontrol, whereupon the coal-owners simultaneously proposed swingeing reductions in wages of up to 45 per cent in the least productive areas, for example South Wales where almost 90 per cent of production was at a loss. Abandoned by government, which ruled out subsidies and favoured district standards, the miners' rejection of district settlement on a temporary basis led to the collapse of the 'triple alliance' with railwaymen and transport workers, who decided against sympathetic strike action. The miners went on strike alone – and after three months a lockout precipitated a swift defeat. The National Wages Agreement which followed formed the foundation of the industry's wage structure until the Second World War. 'Basis' rates were to be determined locally (including a rate for piece-work in particular pits or seams and district-wide rates for entire categories of day workers). Percentage additions were to be determined for a district as a whole, though they were ultimately dependent on profitability. The agreement did, however, provide for a minimum.[52]

After the brief recovery of demand in 1923, resulting from the French occupation of the Ruhr, in 1924 the MFGB again tempted fate by making new demands – even as the international market collapsed once more and coal-owners went on the offensive. No accommodation proved possible between owners and miners, who made it clear that they would accept neither long hours nor wage

49. Ibid., p. 234.
50. Supple, *History*, pp. 148–50.
51. Ibid., pp. 153–7.
52. Ibid., pp. 157–63, 216–17.

cuts and demanded national bargaining. Government intervened
yet again by offering a subsidy for nine months, which would
ensure current wage rates would be sustained while a royal commission inquired into wages, labour relations and every other
aspect of the industry.

Like its predecessor in 1919, the 1926 Samuel Commission,
which consisted essentially of non-experts, lacked even near unanimity in its recommendations, though the method of approach
and the evidence it received resulted in an informative analysis of
the industry's problems; it could not, however, conceal the dead-
lock between employers and workers which had prompted government to initiate the inquiry. The commission acknowledged the
reality that, in the context of market demand then and in the
forseeable future, the industry was overmanned, that most of the
coal produced incurred a loss and that structural change was necessary to achieve more economic production. In the short term,
longer hours were rejected but 10 per cent average wage reductions
were offered as the only method of lowering costs and enhancing
competitiveness. Samuel ruled out both subsidies and nationalization. The presentation of these recommendations fudged the issue,
enabling both owners and mineworkers to justify inaction until
steps were taken by the opposing interest. Neither coal-owners nor
the MFGB produced proposals for restructuring. Meanwhile, the
miners were joined by the whole of the trade union movement in a
general strike in support of the miners' demands. The government's superiority in strategy and strength, however, contrasted
with the TUC's failure to plan and the unwillingness of trade union
leaders and leaders of the Labour Party to continue the campaign
when the Prime Minister transformed the industrial dispute into a
political challenge over constitutional government.[53]

After nine days, the Trades Union Congress accepted an unofficial offer from Lord Samuel (who hinted at subsidy) to negotiate,
while a complete refusal on the part of the miners' leaders to
compromise left the MFGB to continue the strike alone. This time
government remained inflexible in the face of the miners' intransigence. Almost seven months later, the coal-miners returned to
work, the MFGB instructing the districts to seek local settlements
on the principles contained in the National Wages Agreements of

53. K. Middlemas (ed.), *Whitehall Diary by Thomas Jones*, vol. 2, *1926–30*,
London 1969, p. 51.

1921 and 1924. The settlement represented a setback to miners' aspirations to achieve wage settlements negotiated on a national basis, and brought about wage reductions. Thereafter, miners' real incomes dropped drastically as employment fell; between 1928 and 1936 no fewer than 24 per cent of coal-miners were recorded as wholly or partly unemployed. By 1938 miners were among the lowest paid of all major groups of industrial workers.[54]

These years have been seen as a period when the perception of the miners by coal-owners, government and the public altered in ways which were to constrain the confrontational character of industrial relations, at least until the Second World War. This change owed something to the effects of the swingeing defeats of 1921 and 1926 on the miners' self-perception; they were followed by a period in which a 'bitterness of acquiescence'[55] inhibited all industrial action other than sporadic attempts to recover some lost ground. After 1920 the strategic power of the miners vanished and the apprehension with which that power had been regarded by government and owners was replaced by a recognition of vulnerability which for nearly twenty years the government found embarrassing, and which even after 1926 the coal-owners refrained from exploiting to maximum short-term advantage. Supple has argued that such circumstances are explicable neither in terms of collective bargaining machinery nor of the principles underlying settlements, for these remained largely unchanged, the National Wages Agreement of 1921 continuing to form the foundation of the industry's wages structure until the Second World War.[56] The explanation for changing attitudes includes the effect of the widespread publicity given to the shocking working and living conditions revealed by the commissioners' investigation of industrial unrest in the South Wales coalfield in 1919 – which coincided with the coal-owners' complacent and insensitive performance before the Sankey Commission. The vivid accounts of appalling social conditions down the pits and in the mining villages were echoed in subsequent commentaries on the life in the coalfields during the slump, when high levels of unemployment further depressed miners' material standards and morale.

The virtual cessation of collective bargaining between the mid-1920s and mid-1930s was in contrast to the initiatives undertaken to

54. Supple, *History*, pp. 225–54, 443–9, 569.
55. Ibid., p. 487.
56. The following paragraphs are based on Supple, *History*, pp. 461–87.

improve industrial relations in their wider definition, a recognition by government that in part the scandal of coalfield conditions reflected a failure of private enterprise and – especially in South Wales – of the coal-owners. Meanwhile, investment in the new large pits in south Yorkshire, the east Midlands and Kent had been accompanied by investment in housing and amenity provision, much of it through the Industrial Housing Association. Formed by colliery companies in these regions, the association was able to borrow at favourable rates from the government-funded Public Works Loan Board, and obtained housing subsidies under the 1923 housing legislation.

For its part government responded to the widespread feelings of sympathy for miners among the public by introducing measures to promote social welfare, and these were incorporated in the Coal Mining Industry Act of 1920. This created a miners' welfare fund financed by levies on coal royalties, the proceeds from which were spent on measures to improve the social and educational facilities for miners and their families, both at the pithead and in the pit villages. Most of the expenditure went on recreational and health facilities, the latter, in the form of convalescent homes, hospitals, ambulances and nursing staff, absorbing the bulk of the funds in the late 1920s. The lack of provision of pithead baths, to which the Samuel Commission drew attention in 1926, pointed to another symptom of failure for which government and coal-owners shared responsibility. For even though the need for baths had been recognized in legislation passed in 1914, the voluntary principle combined with limits on costs effectively rendered this measure redundant. The establishment of what came to be known as the 'bath fund' in 1926 was financed from a 5 per cent levy on royalties. In 1937 the acknowledgement of the inadequacy of this source led to an agreement by Cabinet to make greater financial provision directly in order to accelerate the bathing programme.

Though limited in extent, the miners' welfare fund contributed to an amelioration of miners' social conditions, and the miners' welfare committees, through joint participation at national and local and district level, demonstrated the possibility of successful collaboration between miners and coal-owners. Joint control of a welfare scheme, however, contrasted with a lack of co-operation among coal-owners in tackling the fundamental economic problems highlighted by the Samuel Commission: the need for structural transformation to increase efficiency in production and render marketing and distribution more effective. Before the 1926 General strike, successive governments had sought either to negotiate a *modus vivendi* mutually acceptable to employers and miners, or to disengage from the industry entirely in order to avoid unforeseen political consequences of interventionist policies.[57] Especially after 1926, the industry's uniquely appalling labour relations produced widespread public sympathy for the miners, the result of continuing exasperation with the employers' unwillingness to contemplate measures other than wage cuts to improve industrial performance.

Against this background, the Board of Trade moved closer to the view, advocated for some years by the well-informed and increasingly influential Mines Department, that if rationalization was to occur government would have to intervene.[58] The ineffectiveness of the 1927 Mining Industry Act, a permissive measure which sanctioned the principle of the coercion of minorities in amalgamations, was matched by the failure of similarly voluntary schemes for market control to regulate prices and output. When in 1930 the minority Labour government's commitment to reduce underground workers' hours without cutting pay seemed possible only if the majority of coal-owners received protection, intervention occurred in the form of the Coal Mines Act. Membership of a coal sales cartel, operated and subsequently strengthened by a Conservative-dominated National government, became compulsory. A Coal Mines Reorganization Commission (CMRC) was given statutory powers to enforce amalgamations, which justifies the description of the Act as 'the most controversial piece of industrial legislation' between the wars, though it was not until the Coal Act of 1938 that the reorganization commission's powers for compulsory amalgamation were truly effective.[59] Under the same Act, nationalization of mining royalties became the responsibility of the CMRC, which became the Coal Commission. In 1942 the Mines Department was replaced by a Ministry of Fuel and Power, but in Sir Ernest Gowers, former permanent under-secretary at the Mines Department and subsequently chairman of CMRC, the department's accumulated experience and expertise continued to be an important strategic influence in the plans for industrial reorganization

57. Ibid., pp. 297–8, 604–5; M. W. Kirby, 'The Politics of State Coercion in Inter-war Britain: The Mines Department of the Board of Trade, 1920–42', *Historical Journal* 22, 1979, pp. 373–82.
58. Kirby, 'The Politics of State Coercion', pp. 382–5.
59. Ibid., pp. 382–92.

– and specifically the role to be played by the state – after the war.[60]
Meanwhile, complete nationalization for the purpose of waging
war, urged once more by the MFGB, was rejected by a wartime
coalition government unwilling to risk losing the coal-owners'
co-operation, regarded as vital to the war effort. In Churchill's
words: 'everything for the war, whether controversial or not, and
nothing controversial that is not *bona fide* needed for the war.'[61]
Miners' co-operation was sought by a system of dual control of
industry, in which the new Ministry of Fuel and Power could
instruct mine managers to adopt certain policies even though
ownership was retained by coal-owners. As managers depended on
the latter for employment and prospects, in practice the minister's
effective power was limited in an arrangement both unsatisfactory
and ambiguous.[62]

At the onset of war the shortage of mining labour was immediately evident and measures were taken to increase the number of
miners to 720,000. Some enlisted miners were recalled from the
armed forces; discharge procedures were tightened up by the Mines
Medical Service established in 1942; compulsory recruitment was
introduced into the mines, applying to 20 per cent of those con-
scripted for armed service from 1943 (the 'Bevin boys'). However,
the target figure was never achieved; furthermore labour productivity measured by output per man shift (OMS) fell. As was the
case during the First World War, government treated the miners'
opinions and claims with sensitivity, relying on exhortations appealing to patriotism to raise performance. Such appeals implied
the need for miners to increase effort and reduce absenteeism.
Meanwhile, the Essential Work Order obliged miners to stay
working at the same pit, for which they received partial compensation in the form of some guaranteed earnings. The implementation
of this restriction became the responsibility of the state, which
thereby assumed the disciplinary function. This was the consequence of depriving coal-owners and managers of the power of
dismissal, a step which inevitably undermined managerial authority to some degree. Towards the end of the war a mines inspector
remarked on the virtual control of the mines by miners themselves,
which produced a 'multi-willed control exerted by thousands of

60. Ibid., pp. 392–6.
61. Quoted in M. W. Kirby, *The British Coal Industry, 1870–1946: A Political and
Economic History*, London 1977, p. 88.
62. Supple, *History*, p. 241.

individual miners, each of whom has the power to temper his efforts to suit his own conscience'.[63] Rarely did the state activate the legislation implemented to discipline miners, for absenteeism, for example, which approached 20 per cent among face-workers by 1945; neither did the newly formed pit production committees succeed in affecting productivity.[64]

As in the First World War, the MFGB used its strength to resist increases in working hours. With the support of the Secretary of State for Mines, the union also secured a closed shop agreement with the MAGB. Sectional advantage seemed to override national interests, and elicited critical comment from a team of American engineers and economists who toured British mines in 1944. Their report, which expressed dismay at the appalling state of industrial relations, the poor morale of the workforce and the absence among owners, managers and miners of a sense of urgency in relation to the war effort, was regarded as potentially too damaging to publish. Again, as in the First World War, union pressure for higher wages, though resisted by coal-owners, was forced on them by government intervention. The Greene (1942) and Porter (1944) awards together resulted in a threefold rise in real wages at nearly twice the rate of improvement shown by national average wage rates. The Greene award included not only large wage increases for miners but recommended minimum weekly earnings and an output bonus. By establishing a national negotiating and conciliation scheme, the Greene award was also significant in the mode of proceeding as well as in its substantive recommendations.

The decision to recommend acceptance of the miners' claims was clearly taken on political grounds, fearful of the consequences for the conduct of war if they were rejected. The limited evidence presented to the tribunal was an inadequate basis for a well-considered adjudication, while the claim itself was unjustified on strictly economic criteria. Furthermore, the particular formula adopted by the National Reference Tribunal under Lord Porter's chairmanship demonstrated such an insensitivity to the effect of an increase in the minimum wage on differentials, so valued by the face-workers, that both the MAGB and the MFGB in the Joint Negotiating Committee agreed that a further rise in piece-rates to be determined at district level was necessary. Government declined to guarantee

63. Ibid., p. 564.
64. Ibid., pp. 556–65.

to finance any such additional payments – and faced striking miners in South Wales and Yorkshire as a result, threatening gas supplies at a time when preparations were in progress for the Normandy invasion. The government capitulated; the minister took over wage negotiations and conceded an extensive compromise, restoring much of the differential between piece-rate and day wage earnings and introducing some measure of further reward related to output (rather than to profitability as hitherto).[65]

These developments were significant, not only because coal prices had to be increased to finance the increased expenditure from the Coal Charges Account, but also because the new national wages agreement ran to December 1947, in effect a medium-term government commitment. Government had been manoeuvred into taking responsibility not only for prices but for wages too. Reinforcing this impetus towards policy centralization was the new constitution adopted by the MFGB in 1944, marking the creation from 1 January 1945 of the National Union of Mineworkers (NUM).[66]

Placed within a corporatist conceptual framework, the history of tripartite relations in the coal industry is ill-fitting, yet the established interpretation of industry–union–government relations in this period has been described as a form of the corporatist model of industrial relations, particularly during the First World War.[67] Hinton's study of engineering workers was the basis for wider generalizations concerning the effects of changes in industrial organization involving dilution of labour, and leading to revolutionary aspirations for 'workers' control' in British industry. This was a period, according to Hinton, when the state, including a closer alliance of big business and the civil service, increased its powers of compulsion and, by appearing to concede organized labour's demands, successfully incorporated trade union leaders in the interests of a business-dominated state. One result was a reaction by workers against their elected leaders in the form of rank and file organization in opposition to official trade-unionism. Other his-

65. Ibid., pp. 574–7.
66. Ibid., pp. 413, 577.
67. Alistair Reid, 'Dilution, Trade Unionism and the State in Britain during the First World War', in Steven Tolliday and Jonathan Zeitlin (eds), *Shop Floor Bargaining and the State*, Cambridge 1985, pp. 46–9.

torians broadly accept the growing friction between trade union leadership and rank and file, but cast the state in the role of a neutral institution, seeking to avoid internal disorder by incorporating the leaders both of employers and the trade-unionists – while Reid's research into shipbuilding suggests that industrial relations was more the outcome of an essentially pragmatic process by which government 'muddled through'.[68]

Some similarities may be found between the shipbuilding and coal-mining industries.[69] In neither did extensive dilution occur, while the MFGB exploited the labour shortage prevailing throughout the war by defending traditional hours of work and weekly work patterns, by obstructing dilution through more flexible employment of women and boys, and by a lack of co-operation in the matter of recruitment or conscription of miners. Absenteeism continued undiminished. All this notwithstanding the 'incorporation' of leaders of the MAGB and the MFGB, who were equally represented on the Coal Mining Organization Committee established by government as an advisory board to the Coal Controller, and despite some increase in the state's powers of compulsion. The MFGB secured exclusion from the operation of the Munitions of War Act in 1915 which was intended to outlaw strikes and lockouts and imposed compulsory arbitration.

Throughout the war coal-mining continued to be the most strike-prone of all industries, an improvement in material conditions (including a reduction in hours) continuing to be the main strike issue – though there was no stoppage at national level and only one serious district strike, that occurring in South Wales in 1915. So advanced was the deterioration in industrial relations in the principality that, from December 1916, under the Defence of the Realm Act, the Board of Trade assumed control over all collieries in the South Wales coalfield, an acknowledgement by the government that the industry's workforce was, in effect, withholding co-operation from private enterprise and that only financial control by the state offered a method of reducing tensions and ending the perpetual state of unrest. It was from South Wales, where the rift between union officials and rank and file is clearly evident, that a demand originated in 1918 for 'joint control' and admin-

---

68. Ibid., p. 68.
69. The shipbuilding industry is treated in detail by Reid, ibid., pp. 47–70. Details for the coal industry are based on Supple, *History*, pp. 62–111.

istration of a nationalized coal industry by workers and the state. None the less, despite the unusual strength of socialist and syndicalist influences in South Wales, the Commission of Inquiry into Industrial Unrest conducted in 1917 did not envisage an imminent disturbance of revolutionary proportions, though a threat to social dislocation was not ruled out after the war ended.

In other coalfields official trade union representatives remained in command, though aspirations towards some alternative to private enterprise were widespread among the leaders of the MFGB, as they were among rank and file. This is hardly surprising since by 1918 the extent of state control over industry sales allocation and many of the financial functions of colliery enterprise pointed in the direction of nationalization, even if the actual operation of colliery enterprises remained in the hands of managers. Most important from the miners' standpoint was that labour policy, including decisions on wages and hours, effectively lay with government. That this had considerable implications for colliery owners can be seen from a minute of the War Cabinet in 1917, which recorded that 'The government could not embark on a conflict with the miners unless they were certain to bring it to a victorious issue',[70] an unlikely eventuality throughout the war and immediate postwar period. The balance of political and industrial interests, therefore, was favourable to organized labour, which benefited from the substitution by government of national security for private profit as the principal objective of industrial activity. National security, however, involved balancing priorities between maintaining coal production commensurate with the needs of war and the recruitment of physically strong males into the armed forces. Inevitably, therefore, the planning priorities of the Coal Control sometimes conflicted with those of the Ministry of Munitions, National Service, Labour, Shipping, the War Office and the Admiralty. The overriding objectives were clear, but the methods of achieving them depended on the successful reconciliation of competing departmental priorities.

After the war the balance of advantage remained temporarily with organized labour. The collapse of the market for coal in 1920 removed the miners' ability to repeat the substantial wages settlement conceded by the Coal Control earlier in the year and precipitated immediate decontrol. In doing so, government's political and

70. Supple, *History*, p. 101.

industrial policy was broadly congruent with that of the coal-
owners. It connived in their persistent advocacy of *laissez-faire*,
justifying heavy wage cuts and longer working hours, coupled
with a determined resistance to undertake industrial restructuring.
However, the Coal Mines Act of 1929/30 marked a movement
back towards control of the industry, albeit a modest step, and
throughout the interwar period government also intervened to
mitigate the social consequences for the miners of its own industrial
policies and to ameliorate living conditions on the coalfields. State
intervention in the coal industry was considerable, and historians,
placing it within a broader context of government initiatives – not
only in wage bargaining agreements but including the Whitley
Report, the National Industrial Conference, the Sankey Commission, courts of inquiry into the dock and road passenger transport industries and the creation of the Ministry of Labour – have
been tempted to interpret these developments within a quasi-
corporatist framework, though no consistent interpretation has
emerged.[71] Middlemas has interpreted government policy as stemming from a concern to avert a potential challenge from the
working class which might threaten the political order, a policy
which involved subordinating economic performance as a priority.[72]
   But the history of the coal industry provides no support for this
view. State intervention was the outcome of the industrial politics
of crises, in response to strike threats or lockouts over issues
concerning material conditions and rewards, or occasioned by
market failure, or prompted by the moral imperative to provide a
safety net for miners and their families suffering demoralization in
their communities and on the coalfields.[73] In a wider context, it is
clear that neither the employers nor trade unions in a variety of
industries were consistent in their responses to the compromises
and initiatives of government; their sectional interests, bargaining
strengths and weaknesses determined which initiatives to welcome
and which to resist. Perhaps nowhere was this more evident than in
the tensions and conflicts which surfaced among those unions

   71. See, for example, R. Charles, *The Development of Industrial Relations in Britain*,
London 1973; G. R. Garside, 'Management and Men: Aspects of British Industrial
Relations in the Inter-War Period', in Barry Supple (ed.), *Essays in British Business
History*, Oxford 1977, pp. 244–67; J. A. Cronin, 'Coping with Labour, 1918–1926',
in J. A. Cronin and Jonathan Schneer (eds), *Social Conflict and the Political Order in
Modern Britain*, London 1982, pp. 113–45.
   72. Keith Middlemas, *Politics in Industrial Society*, London 1979, pp. 16–21.
   73. Supple, 'Ideology or Pragmatism?', p. 240.

belonging to the triple alliance. Ambiguity, therefore, is characteristic of the relations between the state, employers and trade unions
before the Second World War. The year 1942 was a turning point,
when wartime demands and the return of national security as the
government's major priority also saw the balance of power shift so
firmly in favour of the miners that they could persuade government
to introduce dual control – just as in 1945 nationalization became
the condition for securing miners' co-operation and continuity of
production. Given such variable and chronologically unpredictable
interrelationships between the three parties in this period, 'tripartism' seems aptly to describe the continuity of their relatedness,
which is the only constant factor. Nationalization in 1946 marked a
new stage.

From Nationalization to National Strike

The 1930 Coal Mines Act, the acquisition of mineral rights in 1938
(vested in the Coal Commission in 1942) and the introduction of
dual control of the mines in 1942 provided the basis for extending
state intervention, which when war ended was favoured among
contemporaries holding widely differing political views. The conception of a nationalized coal-mining industry as a feasible political
objective had originated in 1892, when miners' leaders raised the
matter for discussion at the annual Trades Union Congress; in 1906
nationalization of the mines was part of the Labour Party's programme, and received endorsement by Lord Sankey in 1919.
Between 1893 and 1942 no fewer than eight nationalizing bills were
introduced into Parliament, but it was only with the exceptional
conditions of dual control, which offered experience of an alternative to private enterprise, and the imminent threat of a postwar
fuel crisis that nationalization was regarded as the only way to
achieve a rapid transformation of the industry in peacetime.[74]
   The effects on the coal industry of two world wars was to impose
a limited, but hitherto unprecedented, degree of state control over
production, finance and wages, and to bring a transfer of expenditure for the social benefit of miners and mining communities.
The coal industry in general and labour relations and the workforce
in particular were regarded by most people to be legitimate matters

74. E. E. Barry, *Nationalisation in British Politics*, London 1965, pp. 109–25.

for public concern and public policy. After decades of embittered industrial relations and at least thirty years of declining productivity, a consensus emerged that constructive planning was needed to achieve radical change, and that centralization, implying some form of state control, was likely to be the most effective method of transforming the industry's practices and performance. The miners' unions, the Labour Party, Sir Ernest Gowers, chairman of the Coal Mines Reorganization Committee and the authors of the Tory Reform Committee's publication, *A National Policy for Coal*, favoured the establishment of a central authority in some form similar to that recommended in the Reid Committee's devastatingly critical report on the industry in 1945. Leading colliery director representatives on the Reid Committee stopped short of endorsing state ownership, but did accept the need for a national policy involving compulsory amalgamations and the technical modernization of mines, and looked to government for an initiative.[75] Capitalism and competition had failed, producing a lack of congruence between private and public interest, which compelled coal-owners to acknowledge the desirability of industrial transformation yet at the same time the inability to achieve it through private enterprise. Specifically, it was acknowledged that the achievement of harmonious industrial relations and full co-operation of the workforce were necessary conditions for securing maximum production after the war, and in the long term for restructuring and modernization. Thus the logic of the industry's history pointed to the 'solution' of nationalization, which coincided with the principled and practical commitment to public ownership by the new postwar Labour government.[76]

In no sense was nationalization the outcome of extremist pressure for workers' control, the policy advocated fiercely by the South Wales Reform Committee. Neither the MFGB, nor the Trades Union Congress, nor the Labour Party offered support for this objective, favouring instead the formation of a public corporation, the National Coal Board (NCB), to replace private ownership. The NUM did not press for direct representation on the board during the negotiations which preceded its formation, and took no part in the nomination of members; the responsibility for this lay with the Minister of Fuel and Power, Emmanuel Shinwell, whose long

75. Supple, *History*, pp. 599–626.
76. Barry, *Nationalisation*, pp. 110–23; Supple, 'Ideology or Pragmatism?', p. 248.

career as a robust Labour politician made him a popular choice with
the miners.[77]
The composition of the first board established immediately
following the Coal Industry Nationalization Act of 1946 consisted
of the chairman, Lord Hyndley, formerly managing director of
Powell Duffryn, Britain's largest colliery company, who had been
commercial adviser to the Mines Department between the wars and
controller general in the Ministry of Fuel and Power during the
Second World War. The deputy chairman was Sir Arthur Street, an
outstanding senior civil servant; members were two mining engineers who were also managers and directors of colliery companies, an industrialist, an accountant, a scientist and two trade union
figures – one was Ebby Edwards, formerly general secretary of the
MFGB, the other was Walter Citrine, formerly general secretary of
the Trades Union Congress. The last two were assigned responsibility for labour relations, manpower and welfare. In addition, a
leading lawyer was appointed as the board's legal adviser based on
experience in dealing with some of the larger colliery companies as
clients. In later years one miners' leader became a divisional chairman and later chairman of the NCB, while several NUM district
officials took staff appointments with responsibility for industrial
relations.[78]
Overall objectives were formulated by government, and from
the outset placed important limits on the NCB's investment and
pricing policies. Long-run strategic policies initiated by the board
and specified in tripartite agreements between the NCB, the ministry and the mining unions were published in the form of successive Plans for Coal. Implementation at divisional (regional) and area
(district and pit) levels was the task of line managers operating
within a formal hierarchical committee structure. G. D. H. Cole
analysed the NCB in detail in 1948 and, with minor reservations,
approved its structure and organization;[79] his analysis reveals that it
conformed closely to a model centralized bureaucracy.[80] In contrast
to the quasi- and from 1948 non-functional character of the board,

77. B. J. McCormick, *Industrial Relations in the Coal Industry*, London 1979, p. 47;
William Ashworth with the assistance of Mark Pegg, *The History of the British Coal
Industry*, vol. 5: *1946–82, The Nationalised Industry*, Oxford 1986, pp. 122–8.
78. Ashworth, *History*, pp. 122–8.
79. G. D. H. Cole, *The National Coal Board*, Fabian Research Series 129, 1949,
pp. 5–6.
80. M. Barratt Brown, 'Coal as a Nationalised Industry', in David M. Kelly and
David J. C. Forsyth (eds), *Studies in the British Coal Industry*, London 1969, p. 100.

functional administrative control was the practice at divisional level, although the chairmen of the eight divisional boards possessed no special expertise. Area general managers, in the degree of authority they held, were equivalent to the production managers or company agents under private enterprise, and most of them, as then, were engineers. Colliery managers experienced a significant change in their role. Whereas under private enterprise they had enjoyed considerable independence in the conduct of recruitment and labour relations, and could rely on the support of company directors in the event of disputes with miners, henceforward they enjoyed only the right to appeal to the NCB area Labour Officer, and he, although subordinate to the area general manager, none the less commanded status in the hierarchy. As Labour Officers tended to be former mining trade-unionists this arrangement weakened the colliery managers' authority at the face. In other respects, however, the certificated colliery managers continued to exercise authority in matters affecting safety and day-to-day working, though control over the latter in particular was, as in the past, incomplete. From the miners' standpoint, however, personnel relations appeared to have been unchanged by nationalization. The managers were the same managers, and the overmen, deputies and other supervisory officials remained as before; according to Cole, the rank and file miner regarded the NCB as 'the old coalowner writ large', and the same ambiguity in attitude which miners displayed towards the certificated colliery manager – upholding his position in the face of external control by more senior officials, yet identifying him with authority – persisted. In effect, nationalization had brought about the emergence of 'State Capitalism'.[81]

Implementation of the Fleck Committee's report after 1953 brought increased measures of centralization and reinforcement of the managerial structure, reducing the power of the divisional boards – and of the trade unionists appointed to them – though in a period of acute labour scarcity this was not taken up as an issue by the unions.[82] In another way, however, the potential for workers' influence in the organization increased with the introduction of the 'ladder plan', the NCB's training scheme aimed at ensuring internal promotion of young employees to supervisory and managerial positions, together with the promotion of a broad range of craft

81. Cole, *National Coal Board*, pp. 15–17, 23–5; Supple, *History*, p. 649.
82. Barratt Brown, 'Coal as a Nationalised Industry', p. 100.

apprenticeship, day release and graduate training programmes.
Ahead of the rest of British industry, except possibly Imperial
Chemical Industries (ICI), this policy was a positive step towards
improving morale during the early years of nationalization.[83]
However, the comprehensive joint consultation machinery from
pit to national level set up under the Nationalization Act did not
fulfil the aspirations of its founders. Rarely did consultation committees meet before board decisions had been made, and even at
colliery level the fact that each committee was chaired by the
manager responsible for implementing decisions and serviced by a
colliery official underlined the purely advisory character of these
committees. As a result, until the dramatic downturn in the demand for coal in the late 1950s, when consultation over contraction
became a reality, the committees limited deliberations to such
matters as housing, safety and absenteeism.[84]

The miners had taken steps to rationalize their organization two
years before nationalization. The NUM, as the MFGB became in
1945, formed a more unitary structure than its predecessor, having
merged the constituent local, county associations into their area
unions.[85] The interwar years had seen rival unions spring up in
Nottinghamshire, South Wales and Scotland, where differences in
geology, technology, working practices, productivity and markets
generated differing priorities from those adopted by the national
executive. The splinter unions proved to be ephemeral; none the less
they signalled to the new national union that the unity built into the
constitution might not endure in reality. McCormick concluded
that, despite the renaming of the county unions ('areas') and the
creation of units to cover non-mining workers (a total of twenty-
one sub-entities altogether), the problem of fragmentation remained, mainly because financial resources were retained within
the areas – where until 1966 piece-rates continued to be negotiated
locally.[86] At the outset the NUM aspired to represent all workers in
the industry, but it failed to recruit significant numbers from the
supervisory and managerial grades, while the colliery winding-

83. Ibid., pp. 100–2.
84. Ibid., pp. 105–8.
85. I. W. Durcan, W. E. J. McCarthy and G. P. Redman, *Strikes in Post-War
Britain*, London 1983, pp. 264–7.
86. McCormick, *Industrial Relations*, p. 61.

men's craft interests caused friction within the larger union. The National Association of Colliery Overmen, Deputies and Shot-firers (NACODS) was the second largest union, but accounted for no more than about 5 per cent of all union membership. Within the NUM, continuing regional differences in South Wales, Scotland and Kent were combined with factionalism, for in these coalfields communists dominated policy-making, and help to account for the considerable number of unofficial strikes (not sanctioned by the national executive) which characterized the industry after 1945.[87]

Although unofficial, this lack of worker co-operation at the pit level was not completely inconsistent with the decision of the NUM to distance itself from direct participation in the manage-ment of the NCB and instead to press hard for immediate improve-ments in wages and conditions. Collective bargaining through agreed negotiating procedures was the preferred route by which to obtain greater material rewards – though wartime experience had also taught the union how to add political pressure in enforcing their demands. Having secured a guarantee of virtual closed shop rights, including the automatic deduction of union members' sub-scriptions at source, doubtless it was also assumed that an organ-ization in which ex-union officials held positions with responsibility for industrial relations would prove to be more sympathetic to miners' claims than had the coal–owners between the wars.[88]

Widespread unofficial strikes in Scotland and threats of industrial action in South Wales during the 1947 fuel crisis were signs of the extent to which sectional interests largely discounted the signifi-cance of the introduction of social ownership. However, the NUM, too, had rejected the use of national conciliation machinery in pursuing substantial demands in 1946. Instead the government was presented with the 'miners' charter' containing large wage claims, a five-day week without pay reductions, an extra week's annual holiday and six additional paid holidays. Anxious to ensure conti-nuity of production and hopeful of reversing the long established decline in productivity, the minister expressed a sympathetic view of these demands. In effect this undermined the bargaining position of the NCB, which was pressured by government into conceding the miners' principal demands even before the physical assets of the 1,500 collieries had passed into the board's ownership, bringing

87. Durcan *et al.*, *Strikes*, pp. 264–7.
88. Supple, *History*, pp. 670–9.

with them the statutory regulation of the industry. This development signalled the probability that, as in war, under nationalization 'purely economic considerations would continue to be
relegated in the interest of political expediency'.[89]
The postwar machinery for resolving disputes continued to
operate at district and national levels. The Joint National Negotiating Committee consisted of sixteen representatives from the union
side and the same from the NCB. The National Reference Tribunal
comprised four assessors and three part-time members; it acted as a
permanent arbitration body to which, until 1961, all disputes not
resolved by the Negotiating Committee within an agreed time
limit were necessarily referred. District-level machinery paralleled
that at national level, with the important exception that compulsory binding arbitration occurred in the event of a failure to agree.
A pit-level conciliation scheme was introduced in 1947, its principal
characteristic in the event of a dispute remaining unresolved within
fourteen days, being a rapid progression to an umpire for a final
and binding decision – procedures which, with slight modifications, continued in operation until the 1970s. Durcan *et al.*
conclude that because the system of national, district and pit
negotiations were kept separate, disputes could not be extended
beyond their place of origin. Moreover, a compulsory binding
arbitration system at pit and district level virtually precluded stoppages unless they were unconstitutional and unofficial. One consequence of this was the pressure placed on union officials to direct a
dispute into the machinery available.[90] The result was that between
1947 and 1971/2, when an overtime ban and subsequent strike
occurred, not a single official national dispute is recorded.[91] None
the less, following a brief fall in unofficial strike activity in the late
1940s, the period to 1957 saw a rise in the frequency of strikes
which, although they were unofficial, brief in duration and limited
in extent, confirmed the coalfields by comparison with other
industries as 'the traditional battleground'.[92]
   Thus, although nationalization institutionalized potential conflicts in industrial relations, the fundamental difficulties remained
unresolved. Piece-rates continued to be the prime cause of most
disputes. Even after the introduction of a national day wage struc-

89. Ibid., pp. 685–9.
90. Durcan *et al.*, *Strikes*, pp. 260–3.
91. Ashworth, *History*, p. 595.
92. Durcan *et al.*, *Strikes*, p. 240.

ture in 1955, when the NCB and the NUM agreed new national minimums and national additions to wages, the final remuneration of 40 per cent of the workforce continued to depend on piece-rates. Minorities dissatisfied with their representation in the new pay structure also contributed to a relatively high strike record in the 1950s, as did the closure of certain 'uneconomic' pits. As a test of the worker co-operation which the architects of nationalization had hoped for, the industry's record of stoppages in the 1950s indicated false expectations. The same was true of the record of absenteeism which, after falling immediately following Vesting Day, remained at only slightly lower levels than the exceptionally high wartime rates, and increased steadily from the late 1950s.[93] Positive consequences of nationalization, however, included a three- to fourfold decline in accident rates by the 1980s, improvements in pithead facilities, guaranteed pensions and more extensive compensation for industrial diseases.[94]

Absenteeism and difficulties in recruitment (the NUM was uncooperative over the introduction of foreign miners) after the immediate postwar period also contributed to stagnant labour productivity in the 1950s. A turning point occurred in 1957 when home consumption fell from its postwar peak of 221 million tonnes, heralding a rapid contraction in the demand for coal both at home and overseas, and a decline in the mining workforce, which since nationalization had remained fairly steady at between 772,000 and 795,000.[95] In part, contraction was the consequence of competition from oil and gas. In the late 1940s coal accounted for over 90 per cent of UK inland energy consumption, and was still above 80 per cent in 1957, but nearly one-third of the increase in energy consumption in the interim had been supplied by oil and gas. By 1970 coal supplied less than 50 per cent, and by 1972 the figure was 37 per cent, stabilizing into the 1980s.[96]

Market forces, however, form only a partial explanation for the industry's contraction, for the NCB's response to these developments was constrained by the requirements laid down in the Act at its creation. That required the industry to produce all the coal of the

93. Ashworth, *History*, pp. 167–9, 210–11.
94. Ben Fine, Kathy O'Donnell and Martha Prevezer, 'Coal after Nationalization', in B. Fine and L. Harris, *The Peculiarities of the British Economy*, London 1985, p. 168.
95. Durcan *et al.*, *Strikes*, p. 248.
96. Ashworth, *History*, pp. 39–42.

quality which the public needed, and to break even, taking one year
with another (though without any obligation to be profitable).
Furthermore, the Act limited the scope for differential pricing or
for borrowing capital, other than from government, to finance
investment.[97] When the market for coal began to contract, the
board found itself, therefore, lacking in investment capital and
limited in its ability to satisfy the heightened expectations of the
NUM, to which over 80 per cent of miners belonged.[98] Under
these circumstances, it was clear even during the first ten years of
nationalization that the board had succeeded neither in achieving a
substantial increase in labour input nor in stemming rising production costs, and was dependent on continuing government subsidies to permit some of the investment necessary to secure higher
production. Constrained by the allocation of financial resources
from government, the NCB's capacity to expand and improve
competitiveness was limited, forcing more production from existing high-cost mines. This left the industry lagging behind in
achieving the targets set in the 1950 Plan for Coal, vulnerable to
competitive fuels and by 1960 to a revitalized European coal industry.

Thereafter, these external market pressures, added to those produced by the financial framework of the industry and the slowing
down in productivity, led to the rationalization programme of the
1960s.[99] Later in that decade, fewer than half the eighteen mining
areas were profitable – and fewer than one-third after taking
interest charges into account. The strategies adopted by the board
to reverse these trends were a major programme of investment in
mechanization, an acceleration of pit closures and a relocation of
production (though at a lower level) in the more profitable areas,
policies endorsed by the unions whose co-operation was crucial to
their success. Through pit consultative arrangements and discussions at high levels within agreed joint national procedures, the
unions negotiated generous financial assistance for miners transferred as part of this process.[100]

97. E. F. Schumaker, 'Some Aspects of Coal Board Policy, 1947–67', in Kelly
and Forsyth (eds), *Studies in the British Coal Industry*, pp. 3–29, especially notes to
table 15; Ashworth, *History*, pp. 48–9.
98. McCormick, *Industrial Relations*, p. 63.
99. Schumaker, 'Some Aspects of Coal Board Policy', pp. 3–8; Ashworth,
History, pp. 221–39; Kathy O'Donnell, 'Pit Closures in the British Coal Industry: A
Comparison of the 1960s and 1980s', *International Review of Applied Economics* 2,
1987, pp. 65–71.
100. Ashworth, *History*, pp. 235–80.

During the early period of pit closures, relocation within divisions was adequate to absorb displaced workers, though in 1962 and 1964 the introduction of inter-divisional transfer schemes was aimed at inducing migration from the worst affected divisions, notably Scotland, the north-east and Lancashire, to the west Midlands, Yorkshire and the east Midlands. Prompt housing provision, both from the NCB's stock and that heavily subsidized by local authorities, together with payment of a household grant, were important attractions in addition to piece-rate earnings. With the acceleration of pit closures after 1967, additional funds were forthcoming from government to enable the NCB to increase the levels of payment to miners aged fifty-five and over who declared themselves redundant.[101] Other factors also contributed to the relative conflict-free running down of the workforce. The close relationship between the communist, Will Paynter, NUM general secretary, and Sir William Webber, the industrial relations member of the NCB, was one; another which has been suggested is a perception within the industry that government, rather than the board, was responsible for the difficulties which necessitated closures.[102] At the same time, until the 1960s the availability of alternative employment opportunities eased the transition for an industry in which an average employment figure of 787,000 between 1951 and 1957 showed a fall of 55 per cent by 1970, even though about 480,000 new workers were recruited into the industry during that period. This reduction in the size of the workforce, and a substantial drop in production, occurred in two stages. Between 1957 and 1967, 678,000 workers left the industry, in a period when the annual rate of wastage through redundancy was less than 0.3 per cent; after 1967 that figure rose to 3 per cent, and coincided with a contraction of employment as recession hit the economy.[103] Higher redundancy levels also marked a change away from the hitherto moderate leadership and policies of the NUM. By that time, however, the mechanization programme inaugurated after the introduction of the Anderton–Shearer power loader in the early 1950s was almost completed. Between 1957 and 1963 the percentage of coal cut by power loading rose from 23 to 92 per cent.

101. Ibid.
102. G. L. Reid, Kevin Allen and D. J. Harris, *The Nationalised Fuel Industries*, London 1973, pp. 36–8.
103. Durcan *et al.*, *Strikes*, pp. 252–4; Ashworth, *History*, pp. 260–1.

Thus, despite coinciding with, and contributing to, pit closures, mechanization was not resisted by the unions.[104] A concomitant of the rapid mechanization of face operations was a greatly increased importance of machine running time and effective work organization to maximize safe and efficient production. From the filling of coal through resetting roof supports to repositioning conveyors, continuity was essential, a priority which influenced managers' definitions of a face-worker's 'stint', a standard task for completion during a specified shift time, to which piecerates related. From the workers' standpoint, therefore, effort was no longer the critical determinant of potential earnings, which meant that in practice payment by results became measured day work. This transition from piece-work incentive and local wage bargaining to day wages controlled through centralized coordination of production culminated in the National Power Loading Agreement (NPLA) of 1966.[105] Both the NCB and the NUM had long recognized the scope for reforming the wages structure, but achieved little progress except for the day wage agreement of 1955, which affected only some day workers. Ten years later, the introduction of power loading gave plausibility to the argument that in future output would be determined more by efficiency of machinery than by labour effort. For its part, under the chairman, Alfred Robens, the NCB wished to secure reform to eliminate anomalous wage differentials, to relate pay more closely to work done and to remove the pressure for local adjustments and the friction which accompanied them. Rationalization of the wages structure promised greater control over earnings, the determination of task, the composition of teams and the process of production; flexible working, continuous production and more peaceful labour relations were expected to follow. The board was seeking co-operation in return for a national guaranteed wage.[106] Fine *et al.* also point to the major incentive for the NCB to minimize total wage costs in the transition during a period when government repeatedly refused permission for the board to raise coal prices even though they were below international price levels.[107]

The NUM saw the reforms as a means of securing its long

104. McCormick, *Industrial Relations*, p. 103.
105. Joel Krieger, *Undermining Capitalism*, London 1983, pp. 76–8.
106. Ibid., pp. 16–17, 267; Ashworth, *History*, pp. 210–11, 293–5.
107. Fine *et al.*, 'Coal after Nationalization', pp. 576–7.

declared objective of negotiating national wage agreements, which would strengthen the authority of NUM officials at the centre. No such rationalization had been feasible before the introduction of national agreements, for whereas the coal-owners and the MAGB had been replaced by a single board, and the federated unions by the NUM, both inherited the intra-industry divisions stemming from differences in geology, productivity, work organization, methods and markets which had been longstanding prime sources of disunity and weakness within the MAGB and the MFGB.[108] The NPLA applied to all day wage workers hitherto receiving piece or job rates. Additions to shift rates were abolished, and each member of a coal-face team, regardless of district, received identical rates. Determination of the number of men who should make up a power loading team and of task norms was carried out at pit level by the application of work study methods. From the workers' material standpoint, the major consequence of the NPLA was the narrowing of differentials between face and other workers in the pits, so fundamental in the industry's history; from the standpoint of the NCB, the diminishing of differentials represented a departure from the historically important method of exercising a degree of managerial control through the wages structure, and specifically by monetary incentives. Method study to evaluate and equate work with pay, together with closer co-ordination of work tasks, assignment of position on the coal-face, the selection of work teams and greater supervision of underground workers now became critical elements in the exercise of managerial control. This development intensified 'the politics of productivity' by which miners responded in diverse ways to resist further encroachment of managerial prerogatives and control.[109]

The NPLA together with the 1971 day wage agreement (which embraced also those day workers excluded from the 1955 agreement) marked a further decline in the role of pit and district conciliation committees and boards. Henceforward, national bargaining assumed paramount importance and, in parallel, unity among mineworkers in pursuit of a favourable national wages policy grew; for those who lost most by the compression of differentials could feel that they had less to gain from restraint.[110] A substantial, perhaps 20 per cent, deterioration in miners' real wages

108. McCormick, *Industrial Relations*, p. 61.
109. Krieger, *Undermining Capitalism*, p. 34 and chapter 2.
110. L. J. Handy, *Wages Policy in the British Coalmining Industry: A Study of*

by comparison with other industrial workers since the late 1950s
coincided with a widespread feeling of insecurity resulting from pit
closures, which accelerated in the late 1960s, together with new
work methods involving greater supervision. The result was a
deteriorating climate of industrial relations. The election of trade
union officers expected to adopt a more agressive stance reflected
this mood. They rejected the idea of an explicit inverse trade-off
between higher wages and more jobs, a notion which may have
been implicit in the union's acquiescence in the incomes policies
and pit closures of the 1960s (though differing from the 1930s).[111]
Twin objectives of higher wages and high employment were now
pursued with vigour by a national executive in possession, since the
NPLA of 1966, of consolidated authority in national bargaining.[112]
A further strengthening of the NUM occurred in 1970 when it was
agreed that, except for all grades of supervisory officials underground, who would be represented by NACODS, all weekly paid
workers in the industry would be represented solely by the
NUM.[113]

Meanwhile, the NCB's position weakened in the late 1960s as
productivity slackened, and production costs, which had been
contained earlier, rose with inflation. The result was a conjuncture
of mounting frustration and increasing organizational power on the
side of the NUM with diminishing capacity on the side of the NCB
to meet large wage demands.[114] A national overtime ban, followed
by a national strike in 1972, ended a phase of industrial relations
characterized from the late 1950s by a period of relatively low strike
activity, and by an absence of direct government intervention in
industrial relations. By comparison with other industries, which
also experienced a decline in strikes in the 1950s, the trend in the
coal industry was less marked, which suggests that causal factors
affecting the industry generally, notably the state of the economy
and government pay policies, were at least as important as causes
specific to coal.[115] Both before 1957 and after 1971 factors peculiar

National Wage Bargaining, Cambridge 1981, pp. 172–4, 230; Ashworth, *History*, pp.
301–4.

    111. Peter J. Turnbull, 'The Economic Theory of Trade Union Behaviour: A
Critique', *British Journal of Industrial Relations* 26, 1988, p. 108.

    112. Handy, *Wages Policy*, pp. 230–1.

    113. McCormick, *Industrial Relations*, p. 63; Ashworth, *History*, p. 291.

    114. Ashworth, *History*, pp. 309, 324.

    115. J. Winterton, 'The Trend of Strikes in Coal Mining, 1949–79', *British Journal
of Industrial Relations* 13, 1981/2, pp. 15–18.

to coal appear to have assumed greater significance as explanatory factors. Coal was in excess demand to 1957, when the miners won substantial wage increases. The decline in employment in the 1960s weakened the bargaining power of those remaining in the industry, for whom alternative opportunities had deteriorated since the 1950s. An acceleration of pit closures from 1967 accompanied by a consequent sharp rise in redundancies heralded a return to firmer union postures, for which the oil crisis in the 1970s and the resulting boost to coal demand provided reinforcement. By centralizing wages bargaining, the NPLA indirectly increased the power of the union's national executive committee. Not surprisingly, the 1970s saw a rise in stoppages.[116]

The relatively peaceful interlude between 1957 and 1970 has been attributed in large measure to the effects of revisions to the wages structure, beginning with the 1955 Wages Structure Agreement affecting day workers, followed by district power loading agreements (DPLA) negotiated at local level, and culminating in the national agreement (NPLA) of 1966. Clegg has argued that by removing piece-rates as an issue, the occasion for most disputes historically, first from colliery to area management and subsequently to centralized control, power loading agreements reduced drastically the number of face-workers' stoppages. This conclusion has been challenged in a study which reveals that whereas the proportion of stoppages over pay allowances decreased from 40 per cent to 20 per cent between 1966 and 1971 and stabilized, working arrangements and manning levels took their place as the politics of productivity became the major source for dispute leading to stoppages. The fact remains, however, that while causes of disputes altered, their intensity, measured by strike-proneness, declined sharply from the mid-1960s, reinforcing the downward trend coinciding with the onset of contraction in the industry from the late 1950s. Moreover power loading agreements indirectly contributed to industrial peace by occasioning increasing use of conciliation machinery rather than the pursuit of disputes to the point of strike action, for under the NPLA additional earnings could not be negotiated at pit level.[117] In this way, argues Winterton, the PLA was an important factor in explaining the decline in strike-proneness after 1957, though especially after 1966. He also

116. Ibid.; Durcan *et al.*, *Strikes*, pp. 247–51, 267–71.
117. Clegg, *Changing System*, pp. 272–4.

suggests that by reducing the proportion of traditionally militant face-workers in the workforce, mechanization reinforced this trend.[118]

One effect of the NPLA, which was reinforced by the 1971 national wages structure agreement, was to increase the likelihood of a *national* stoppage, an eventuality which was rendered even more likely by the emergence of a less moderate NUM executive, including Laurence Daly from the traditionally militant Scottish coalfield, who became general secretary from 1969. The year 1969 brought a national strike among surface workers to secure a forty-hour week, while in 1970 Yorkshire miners mounted an unofficial strike, partly in reaction against the official national policy of supporting the government's incomes policy.[119] When in 1970/1 prices of both fuel oil and coal imports rose steeply and coal demand exceeded output for the third year in succession, miners' expectation also began to rise, and unofficial stoppages, mainly affecting South Yorkshire, South Wales, Scotland and Kent, caused a loss of 3 million tonnes. This followed a national strike threat by the annual conference of the NUM in support of large increases in minimum wages. A national strike became more likely after the next annual conference, when the two-thirds majority in a pithead ballot necessary to authorize a national stoppage was reduced to 55 per cent. In the following year the NUM's demands for wage increases, varying from 35 to 47 per cent, were rejected, as was the board's counter offer. After an overtime ban, a pithead ballot gave the NUM a 59 per cent majority in favour of further industrial action. Against a background of strikes by public sector unions reacting against the government's policy of imposing wage restraint, January 1972 saw the beginning of the first national miners' strike since 1926. The NUM refused to enlist the services of the National Reference Tribunal, and proceeded to adopt a novel form of secondary picketing, deploying 'flying pickets' at the premises of large merchants and consumers to check the movement of coal, particularly that destined for power stations, the consumers of 50

---

118. Winterton, 'The Trend of Strikes', pp. 17–18. Winterton also argues that because, typically, power loading reduced the size of teams at the face, the characteristically short stoppages traditionally mounted by face-workers became even briefer. The effect of this was to be recorded as a dispute according to the official criterion of 100 man days (p. 15).

119. Ashworth, *History*, pp. 306–9; Andrew Taylor, *The Politics of the Yorkshire Miners*, London 1984, pp. 6–7.

per cent of all coal sold on the domestic market.[120]

The combination of lost production and the cost of wage increases, at whatever level, forced the NCB into deficit, compelling the board to ask government to draft a bill to extend the board's borrowing powers and to indemnify it against exceeding the permitted cumulative deficit. Government declared a state of emergency and imposed a three-day week on industry, accompanied by daily power cuts. The Secretary of State for Employment's intervention to find a settlement failed and a court of enquiry was set up. The NUM refused either to suspend the strike pending the court's findings, or to be bound by them. Lord Wilberforce, chairman of the inquiry, accepted the arguments advanced by both sides: that under the financial conditions constraining the NCB the latter's offer was fair; but that there was justice in the miners' claim when set in the context of movements in industrial earnings in previous years. Of critical importance for the immediate, and subsequent, settlements was Wilberforce's judgement that if the NCB found it impossible, through lack of financial resources, to meet the 1971 wage demand, then 'government should provide the necessary finance, because it is unreasonable to expect miner's wages to be held down to finance uneconomic operations'. This offered a clear signal, both to the NCB and the NUM, that the taxpayer should underwrite the industry regardless of performance or ability to pay. Government was involved in tripartite discussions, and a settlement, acknowledging the miners as 'a special case', which was favourable to the miners was agreed.[121] Thus came to an end a period of a quarter of a century during which tripartism had touched industrial relations only indirectly, as Treasury and Employment ministers began to exercise greater direct influence over the board's general labour policy.[122]

Direct government intervention during the strike of 1972 and the concession represented by the settlement was repeated in 1973/4. Ministers entered into informal discussions with the NCB and the NUM in an attempt to sustain the government's incomes policy and secure the miners' co-operation in wage restraint. The rise in

120. Ashworth, *History*, pp. 302–3; Fine *et al.*, 'Coal after Nationalisation', p. 185.
121. Ashworth, *History*, pp. 302–14.
122. Ibid., pp. 243, 637.

oil prices, however, strengthened the NUM's position, prompting an overtime ban and the threat of another national strike. Again the government declared a state of emergency and the Conservative Prime Minister, Edward Heath, called a general election on the issue of 'who governs Britain?' The return of a Labour government was followed by the setting aside of incomes policy restraints to ensure a satisfactory settlement for the miners. Furthermore, the NUM, like the NCB, pressed government for more investment in the industry. While the policy which emerged from the tripartite discussions (known as the Coal Industry Examination) of the Industry Group did provide for increased investment, it was to be accompanied by an equal balance between pit closures and new capacity, and an average increase in OMS of 4 per cent.[123] These were the key features of the 1974–9 Plan for Coal.

The Decline of Tripartism

Implementation of the Plan for Coal lay in the hands of an organization which had been transformed during the late 1960s, when rapid contraction had compelled the industry to undertake a fundamental review of its functioning and to explore the scope for major economies. Simplification and centralization, with a more powerful role for the chairman, then Lord Robens, was the direction taken by these reforms, which were operational from 1967. Management structure was reorganized, involving a reduction from five to three tiers of decision, making a clearer line of responsibility between tiers. Divisional chairmen were retained but became part of the national headquarters establishment and attended national board meetings. The board held corporate responsibility, as before, for policy-making. This translated into targets set by the board, which was also charged with continuously assessing the effectiveness of the organization.

Departments at headquarters and at regional level provided the board with advice, assisted in the formulation and implementation of policy and offered guidance and services to operating units at area level. The accountability of areas, however, ran directly to the national board rather than to divisional chairmen. The task of the operating units (areas, opencast, coal products division and ancil-

123. Ibid., pp. 326–35, 357–63.

laries) was to achieve the results contained in budgets agreed with the board. Colliery management continued to be on the basis of individual mines, and only very large collieries were allowed a general manager. Another important innovation was the introduction of management by objectives, applied to area planning and control. Three years later, modification in management operating procedures resulted in changes leading towards firmer line management, allowing for greater adjustment to varying conditions of different times and places, and an unswerving commitment to economy and continuous review of the organization. The outcome by the early 1970s was an organization consisting of fewer, though more self-contained, management units, producing and marketing their product and enjoying greater authority from the board with strict accountability.[124]

Coinciding with this internal decision to maintain a constant review was the sharp rise in more direct government intervention, as the NCB's increasing financial dependence on the state after the strikes of 1972 and 1974 was followed by more frequent financial negotiation offering greater scope for interference. This was manifest in the activities of the Departments of Energy, Trade and Industry, Employment, of the Treasury, and of the Monopolies and Mergers Commission. The year 1972 also marked a return, after twenty-five years, of direct government intervention in major labour disputes. This change in the character of the tripartite relationship so far as it affected industrial relations once again revealed the reality, obscured since 1947, that in the last resort the financial terms of a settlement between employer and workers in the industry depended partly on factors other than the revenue which the NCB could earn.[125]

These pressures prompted a review of the NCB's strategy, in particular the extent to which the NPLA had improved industrial relations and raised productivity as a result of centralized managerial control and a national time-based wages structure. Faltering productivity and a resumption of industrial conflict before the end of the decade indicated failure. This was explicable not by inadequate planning on the part of the Industrial Relations Department but, in part, by the historically regionally differentiated traditions

124. Ibid., pp. 271–4.
125. Ibid., pp. 243, 644; O'Donnell, 'Pit Closures', pp. 70–4.

of mineworking, which generated responses which acted as centrifugal forces and tended to frustrate attempts at centralized administrative integration.[126] The NPLA upset the traditional basis of the politics of productivity (largely co-operative and egalitarian among Durham face-workers, for example; broadly hierarchical, individuated and competitive among their Nottinghamshire counterparts), and triggered regionally distinctive responses to the new national agreement.[127] The NPLA generated new forms of conflict between miners and management over the pace and organization of production at the coal-face, which explains why custom and working practices in the regions produced varied responses to the intrusion of managerial control through changes in manning levels, supervision and method study, the significance of which was interpreted differently according to miners' contrasting regional experience.[128]

An even more serious consequence of national wage agreements was the redirection of face-workers' bargaining power away from the coal-face towards the national negotiating arena. Denied the opportunity to increase differentials, face-workers threw their weight behind the NUM's campaign for substantial wage improvements, which intensified the financial pressures affecting the industry. Ignoring the five-year transition period, which was part of the 1966 NPLA, Yorkshire miners struck in 1970 for immediate wage parity for all regions; one year later, again initiated by the Yorkshire miners, the NUM introduced a national overtime ban, the prelude to the national strike in 1972.[129] After the strike the NCB sought to reintroduce some form of incentive payments, in effect acknowledging that the national wage agreement had constrained the earnings and the productivity of miners in those regions where the capacity for increased productivity, notably in the east Midlands and Yorkshire, was regarded to be greater than elsewhere. The NUM, however, opposed productivity incentive schemes, such a proposal being rejected by a national pithead ballot by a 62 per cent majority in 1974. Three years later, however, when the Labour government's incomes policy only allowed pay rises above 10 per cent if they were self-financing, productivity deals began to alter miners' perception of the benefits which incentive

126. Krieger, *Undermining Capitalism*, pp. 267–8.
127. Ibid., pp. 97, 270–1.
128. Ibid., pp. 95–6, 271.
129. Ibid., p. 269.

payments offered. After another narrow defeat for an incentive scheme in 1977, the NCB intensified its attempts to negotiate a scheme with the NUM.[130]

To explain why the board took this step it is necessary to examine the extent to which management had controlled the labour process in the past, and to explore the effects of changes in the managerial structure and organization. During the interwar years and until the widespread introduction of power loading machinery in the 1960s, the predominant method of getting coal was some variant of mechanized longwall, a process which, compared with the pillar and stall method, involved a degree of specialization and division of labour. It was organized in sequential shifts. After undercutting, the coal was brought down with the aid of explosives; the drilling and loading of coal occurred in the second shift; and in the third, ripping and packing preceded the movement of coal from face to conveyor. While the new machinery required particular skills in production and for maintenance, and generated a demand for electricians and fitters, mechanized longwall working has been regarded by Burns *et al.* as having displayed deskilling characteristics by comparison with hand methods of coal-getting and in this way enhanced managerial control.[131] Penn and Simpson have challenged this interpretation, stressing that, mechanized or not, safe and effective longwall working required the application of extensive knowledge of geological conditions accumulated by experience. The latter contributed a basis for skill, the importance of which suggests that the degree to which power loading introduced discontinuity and deskilling has been exaggerated. A much more important change accompanying power loading was the reorganization of work patterns, for mechanized longwall mining involved triple shift working and posed a potential threat, therefore, to miners' unity; it was a system which resulted in stoppages due to inter-shift disputes.[132]

In the context of managerial versus worker control, however, the traditional characteristics of coal mining – the difficulty and variability of production, dispersed, dimly lit and relatively inaccessible

130. Ashworth, *History*, pp. 370–5.
131. A. Burns, D. Feickert, M. Newby and J. Winterton, 'The Miners and New Technology', *Industrial Relations Journal* 14, 1983, p. 12.
132. Roger Penn and Rob Simpson, 'The Development of Skilled Work in the British Coal Mining Industry, 1870–1985', *Industrial Relations Journal* 17, 1986, pp. 345–8.

working locations underground – continued to enable face-workers
to exercise a degree of control unusual in other industries. Indeed,
Fine *et al.* saw the technological and organizational changes from
the late 1950s as having *increased* miners' control over working
arrangements, for cutting and conveying, hitherto separate tasks,
thereafter became part of a single process, while hand filling was
eliminated entirely. Furthermore, the machine skills required in the
cutting and conveying of coal and in the use of power supports
were open to all workers, rather than, as under pre-power loading
by longwall, only to the cutting shifts. This combination of a
reintegration of the work process within a single shift and the
potential enhancement of skill acquisition greatly increased cohesion by removing friction arising from sequential operations conducted by different teams.[133]

At the same time, the proportion of engineers and electricians as
maintenance craftsmen rose from 6 per cent in 1957 to 20 per cent
by 1981, which suggests that during this period not only was the
ability of miners to control the labour process at the point of
production strengthened, but the mining workforce was becoming
more skilled and strategically more powerful. This trend towards
greater control underground was strengthened by the miners'
ability, through co-ordination by elected chargehands (equivalent
to shop stewards in other industries) rather than supervisory staff,
to pace and allocate the work.[134] Another contributory factor was
the 'problematic loyalty' of first line supervisory managers under
nationalization. They were mostly ex-miners and their continuous
contact with workers at pit bottom contrasted with infrequent
encounters with more senior managers.[135] The importance of this
ambivalence stemmed from the role of the underground managers
as the conduit for information to managers above ground, a situation which enabled a strongly organized mining workforce (aver-
aging about 90 per cent unionization into the early 1970s)[136] to
continue, with considerable success, to resist increased direct super-

133. Fine *et al.*, 'Coal after Nationalization', pp. 177–9.
134. Penn and Simpson, 'The Development of Skilled Work', pp. 345–6.
135. Trevor Hopper, David Cooper, Tony Lowe, Teresa Capps and Jan Mourit-
sen, 'Management Control and Worker Resistance in the National Coal Board;
Financial Controls in the Labour Process', in D. Knights, and H. Wishart (eds),
Managing the Labour Process, Aldershot 1986, pp. 112–13.
136. Durcan *et al.*, table 8.7. The union density among those eligible for the
NUM only was well above 90 per cent. G. S. Bain and Robert Price, *Profiles of
Union Growth*, Oxford 1980, p. 45, table 2.6.

vision, control and discipline. Managerial control was also limited by the decentralizing regional tendencies which produced differential regional outcomes from the politics of productivity, reflecting customary working relationships and attitudes towards production, payment and productivity.[137] Thus the NPLA had failed to achieve the degree of greater managerial control and workforce compliance which had been expected. Neither had a system of standard costing, allied to the agreements, been successful. These failures explain why, from the 1970s, the NCB persisted with a plan to reintroduce incentive wages payments as the route to effective managerial control, now deemed to be a necessary condition for achieving the concentration of production in the high productivity pits of Yorkshire and the Midlands. This change in strategy after twenty years was not achieved until 1977/8 when, yet again, regional interests undermined a national union policy of resistance.

In considering how successful the new wages policy proved to be, it is necessary to consider the overall management of the industry of which the incentive scheme from 1977/8 was only one part. How far did the formal, hierarchical, bureaucratic structure of the NCB function as its creators had intended? Following the Fleck Report in 1953, a degree of reorganization had occurred intended to strengthen centralized control, in part by clarifying managers' roles with respect to authority and accountability at board, divisional and area levels. One effect of the reforms was to diminish the importance of divisional boards and committees. Corporate policy, finance, marketing and industrial relations were handled mainly by staff at national headquarters, whereas production and engineering became the responsibility of managers at area and colliery levels. The board concentrated on relations with government and other external bodies with respect to finance and markets, and handled industrial relations and welfare policy. This left local managers to concentrate on achieving continuity of production and target outputs in negotiations with labour. Such an arrangement had proved adequate during the 1950s, when the secular market trend was favourable. One important effect, however, was to establish the line managers, almost all of whom were professional mining engineers, as the dominant influence on industrial relations in practice; they formed a managerial cadre largely preoccupied with production,

137. See Krieger, *Undermining Capitalism*, pp. 267–77.

who regarded finance and marketing functions as of secondary importance.

When demand for coal turned down from 1959, and as pressures produced by financial stringency and competition increased, the board found that even after the increased centralization of administration in the 1960s it was difficult to overcome the deeply rooted production, as opposed to commercial, orientation of their activities.[138] Whereas the NCB had agreed jointly with the NUM and the Labour administration in the 1974 Plan for Coal that new capacity should be balanced by pit closures in a drive to increase overall productivity, by 1982 half the capacity planned in principle for closure remained in production. In the face of opposition from the NUM, and anxious to preserve good relations by observing established, slow moving, joint procedures, the NCB acquiesced in the sluggish pace of progress in the coalfields.[139] Hopper et al. concluded that the formal structure of administrative control from the centre through line managers was largely ineffective. The realities of management derived from a different operational culture, the weaknesses of which became increasingly evident. The tripartism of the early years of nationalization became more complex from the 1960s as managerial sectionalism generated conflicting objectives and competition for control within the organization.[140]

These internal contradictions intensified in the 1970s as both the Treasury and the Department of Trade and Industry (DTI) assumed an increasingly interventionist role, the former by controlling the public sector borrowing requirement through the imposition of cash limits on the industry, and the DTI by emphasizing the need for internal managerial efficiency by private enterprise criteria, an imperative conveyed through the investigations and audits conducted by the Monopolies and Mergers Commission. As the demand for coal continued to contract, marketing and financial considerations achieved increasing significance. Financial management information systems introduced in the early 1980s finally challenged the traditional supremacy of the mining department in decision-making generally, while the introduction of detailed financial performance indicators to supplement the physical measures, long preferred within the operational culture of

138. Hopper et al., 'Management Control', pp. 117–25.
139. Ashworth, History, pp. 353, 780.
140. Hopper et al., 'Management Control', p. 123.

mining engineers, affected managerial practice at area and colliery level.[141] The process of decentralization and regionalization of management coincided with a similar fragmentation within the NUM, as the financial and market criteria imposed by the board sharpened inter-regional differences and underlined latent, though persistent, divergent interests, culminating in the reintroduction of a local incentive wages element in 1977/8.

After fifteen of the twenty-two NUM areas had voted in favour of local productivity schemes – though defeated by an overall voting majority in the national ballot – the national executive committee of the NUM sanctioned area negotiations, a decision which represented a return to federalism in a constitutional as well as a material sense, for it departed from the general application of national ballots which had long been customary in the industry.[142] During the next four years, coal miners came first or second in the earnings league, while absenteeism fell from 17.6 per cent to barely 10 per cent – the lowest since nationalization.[143] Differentials between direct production and face-workers widened, as did differentials between high and low productivity coalfields. A detailed analysis of the implementation of the local incentive scheme suggests that informal practices and agreements were fundamental to its success.[144] For whereas colliery managers were instructed not to depart more than marginally from an area standard task per manshift recommended by the area work study department in negotiation with NUM lodge officials, in practice 'a considerable departure from prescribed and standard procedures' occurred. From their powerful position within the board, supported by senior line managers who largely shared the colliery managers' priorities in optimizing production even at the price of sub-optimal cost control and pay bargains, the mining engineers succeeded in resisting close supervision by the industrial relations staff. Contrasting the day wage system of the 1960s and 1970s, when the resources of the high productivity districts were deployed to benefit the rest, with the

141. Ibid., pp. 122–34.
142. Adrian Campbell and Malcolm Warner, 'Changes in the Balance of Power in the British Mineworkers' Union: An Analysis of National Top-office Elections, 1974–84', *British Journal of Industrial Relations* XXIII, 1985, p. 5.
143. Ashworth, *History*, pp. 375–80.
144. Christine Edwards and Edmund Henry, 'Formality and Informality in the Working of the National Coal Board's Incentive Scheme', *British Journal of Industrial Relations* XXIII, 1985, pp. 27–9.

incentive scheme post-1977, Edwards and Henry described the latter as 'a triumph of sectionalism and market discipline over uniformity and solidarity'.[145] A close association between average incentive earnings and a range of physical and financial performance measures offered a clear indication of the degree to which colliery managers exercised control over the scheme; and they were able to defend its higher than expected costs – one consequence of informality in the system – by pointing to a reduction in the number of disputes.

Although the NUM's national executive committee had conceded power to areas in respect of productivity deals, its negotiation with the NCB on basic wages and grading during the late 1970s occurred within the framework of the Labour government's 'social contract'. This, based on 'a rather dubious majority' on the national executive committee and most of the national membership, became official NUM policy. However, an alliance between communists and the socialist left in Yorkshire had campaigned strongly since 1967 against both pit closures and wage restraint in any form. The key figure in this shift towards militancy was Arthur Scargill, architect of the unofficial strikes in Yorkshire in 1969 and 1970, and pioneer of 'flying pickets'.[146]

The strikes of 1972 and 1974 had, after all, been successful, even though the wage increases had been rapidly nullified by inflation. Those successes had enhanced the political position of those, principally from the Yorkshire area, who called the strikes; and mass picketing which was at times both violent and illegal had proved to be an effective tactic contributing to the miners' victories. The miners were invited to believe that if the NCB could not meet the costs of financial settlements, then governments – Labour or Conservative – could be pressured to arrange a solution satisfactory to labour. That assumption underlay Scargill's strike ballot in 1980, a tactic which allowed the moderate NUM president, Joe Gormley, to secure a relaxation of the board's financial targets laid down in the Coal Industry Act of 1980.[147] Thus, the Conservative government under Margaret Thatcher, elected in 1979, capitulated to the NUM, just as the previous Conservative government under Edward Heath had surrendered in 1972 (and fallen in 1974). Why,

145. Ibid., p. 42.
146. Campbell and Warner, 'Changes in the Balance', pp. 1–6.
147. Ibid., pp. 8–10.

only three years later, when Arthur Scargill was president of the NUM did an identical tactic adopted on a national scale fail? Why did the NCB and the government resist, and defeat, the miners to win a major confrontation for the first time since 1926?

Fundamental to the outcome of the 1984/5 strike was a vastly different market situation compared with the early 1970s, at the height of the oil crisis. In the early 1980s oil prices were falling and with them the demand for coal. The rapid deterioration in the board's finances which resulted occurred in a political context heavily influenced by monetary policies, a rigid public sector borrowing requirement, massive reductions in public sector expenditure and a movement towards the self-financing of nationalized industries. Both of the Coal Industry Acts, in 1980 and 1983, contained directives to the NCB to break even in the short term.[148] Under Treasury pressure, strengthened by the influential reports from the Monopolies and Mergers Commission and the House of Commons Select Committee on Energy, the energy minister raised the target for capacity reduction to a level which implied the closure of up to fifty loss-making pits between 1981 and 1986.[149]

The publicly declared outright opposition to pit closures by the new militant NUM president, elected with an unprecedented 70 per cent majority in 1983, increased sharply the probability of a major national confrontation with the NCB and with the Thatcher government. The latter had already commenced preparations to meet such an eventuality following what was seen as a humiliating defeat by the miners in 1981.[150] The lesson learned on that occasion was that unless the NCB was compelled to adhere to operating within its budget, the NUM would continue (as in 1972, 1974 and 1981) to exercise pressure on the government itself to concede. This explains why the ground was prepared in expectation of a major crisis of politico-industrial relations, resulting in a strategy to protect the state in the event of a strike.[151]

Government phased out all grants in aid of production and required the NCB to finance a larger proportion of investment from revenue, a policy which, because it disregarded both the historical and the current economic context in which this change of

148. Ashworth, *History*, pp. 414–16.
149. O'Donnell, 'Pit Closures', p. 68.
150. Martin Adeney and John Lloyd, *The Miners' Strike 1984–85*, London 1986, pp. 2–22.
151. Ibid., pp. 79.

policy was to be introduced, quickly proved inoperable.[152] The
message concerning strict budgetary limits, however, was unam-
biguous. The 1982 Employment Act outlawed secondary picketing
affecting an industry not directly engaged in the dispute (though in
the event this legislation was not invoked during the 1984/5 strike),
and protection was removed from union funds which became
vulnerable to claims for damages. A contingent strategy was formulated by an *ad hoc* committee consisting of senior civil servants.
The removal of cash limits from the Central Electricity Generating
Board enabled coal stocks to be built up both at the power stations
and at the pithead, while the mobilization of private haulage firms
ensured continued removal of coal supplies during the strike. Some
of the power stations switched from coal to oil.[153]

Key figures in these preparations were the new chairman of the
CEGB, known to be sympathetic to the government's overall
posture, and the appointment in 1983, after a brief interim period
by the former deputy chairman, mining engineer Norman Siddall,
of a new chairman of the NCB, Ian MacGregor, whose experience
in the American mining industry, where industrial relations were
characteristically confrontational and trade unions struggled for
recognition and survival, had earned him a reputation for toughness in handling labour. In 1984 the leadership of the NUM passed
into the hands of a triumvirate on the left, led by Arthur Scargill,
superseding the traditional balance between moderate, militant and
regional representation. Scargill's presidential manifesto had included opposition to pit closures except when coal was exhausted, a
four-day week, a retirement age of fifty-five and a £100 weekly
minimum wage for all miners.[154] In reply to an offer of a 5.2 per
cent wage increase the NUM imposed an overtime ban. In response to the accelerating pace of pit closures, consequent on the
government's tight budgetary limits, the union's national executive
approved requests from Yorkshire and Scotland for strike action in
the event of specific pit closures, and signified approval for further
requests should they be forthcoming. In effect, without the national
ballot required under NUM rules, a national strike commenced.[155]

Throughout the strike senior government ministers were in-

152. Ashworth, *History*, p. 663.
153. Adeney and Lloyd, *The Miners' Strike*, pp. 70–9, 205.
154. Ibid., pp. 27, 55; Campbell and Warner, 'Changes in the Balance', pp. 15, 23.
155. Adeney and Lloyd, *The Miners' Strike*, pp. 84–9.

volved behind the scenes, though their concern was principally with the country's infrastructure and the effects of the coal dispute on it. A Cabinet committee monitored the strike, conveying information on coal stocks to the Prime Minister. Through co-ordination by the Civil Contingency Unit in Whitehall, the government was able to ensure the uninterrupted operation of power stations and manufacturing industry. This was achieved by integrating emergency action to conserve energy supplies and maintain services. In addition, improved methods of co-ordination and powers of control, notably through the National Reporting Centre and the development of support units on a large scale (100,000 in 1984), enabled the police to counteract flying pickets.[156] Thus, the balance of power, which in 1981 had produced 'victory for the miners' had been altered by government in the interim. Meanwhile, between 1980 and the beginning of the miners' strike in 1984, pit closures had caused 54,000 job losses, bringing total employment below 100,000.[157] During the strike itself the government maintained a public posture of non-involvement in industrial relations, at the same time insisting that ensuring that the conduct of the strike was legal was the responsibility of the police and the courts, leaving individuals to test the legality of the strike (which in the event led to the sequestration of union funds).[158] In 1984/5 there was to be no attempt by government to assist parties to find an 'honourable solution', and no tripartite termination of industrial unrest in the coalfields.

Two years later, the logic stemming from the self-financing independence which was the new imperative for the NCB saw a change in its name to British Coal. This was followed in 1988 by moves to privatize the electricity industry, which dominated the NCB's market, a commitment by the Conservative Party to prepare coal for privatization in 1993. The trend towards progressive decentralization of the board's activities and the corollary of region-alization was accompanied by a complementary reassertion of regional interests among trade-unionists, an extreme manifestation of this being the formation of the Union of Democratic Mineworkers (UDM) in Nottinghamshire, accounting in 1987 for about 18 per cent of all colliery workers.[159] British Coal encouraged this dual

156. Ibid.
157. O'Donnell, 'Pit Closures', p. 63.
158. Adeney and Lloyd, *The Miners' Strike*, pp. 202–4.
159. In Nottinghamshire the UDM membership density was 80 per cent. A. J.

union structure, dissolving the 1946 conciliation scheme by extending recognition to the UDM as part of a new joint scheme.[160] Nearly one hundred years after a Liberal government first intervened directly in the industrial relations of the industry, the ideology of a Conservative government espousing similarly capitalist free market philosophies decided that the state should withdraw from active participation, not only from industrial relations but from ownership and corporate policy-making.

While corporatism in any form could not be identified during or before either world war, capitalist relations did weaken, through dual control in wartime and through price and output control of the state-organized cartel in the 1930s. Fine *et al.* have interpreted developments during the early years of nationalization, when one might have expected tripartite relations to have been both formally and in practice at their most effective, as heralding a reversal of any trend towards corporatism.[161] While unification of ownership made many reforms possible to ensure welfare benefits to cushion the social effects of massive reductions in the workforce, the posture of a hierarchically-managed industry is seen as essentially capitalistic, seeking higher productivity at least cost. But governments' policies were not, overall, helpful. Through financial and commercial restrictions imposed on the board, by manipulating the market for coal and by limiting investment funds, governments hindered competitive policy. From the 1970s government policies were not only unsupportive but perverse. Ashworth concluded that 'governments did not clearly think out what they wanted from nationalization; that nevertheless they wanted a degree of control, however uncertain they were of its specific uses, which was bound

---

Taylor, 'Consultation, Conciliation and Politics in the British Coal Industry', *Industrial Relations Journal* 19, 1988, p. 224.
    160. Ibid. Opposed by the NUM, this step involved a judicial decision which declared that the 1946 scheme was not legally enforceable. The observation made by a member of the Yorkshire miners' executive, who is also an historian, reflecting on the realities of trade union experience during the past hundred years is apposite: 'When we had sliding scales we had a county union. When we had the NPLA we had a national union. Now with pit based incentives we have no union at all. The paraphernalia of the union still exists, but I wonder if the motivation of the union is still here.' Dave Douglas, quoted in Krieger, *Undermining Capitalism*, p. 276.
    161. Fine *et al.*, 'Coal after Nationalization', p. 195.

to limit the commercial autonomy of a nationalized board; and that from time to time they varied both the requirements which they sought from boards and the way they expected them to be.'[162] Government actions motivated by political expediency favouring the miners also affected relations between the NCB and its workforce, especially between 1972 and 1985 when the miners would no longer accept that the board's financial limitations could not be relaxed by government. It is because of the inconsistencies and vicissitudes revealed in the interrelations between industry, labour and government,[163] the complexities of bureaucratic organizational behaviour and the continuing centrifugal influence of regional differences[164] that tripartism – the weakest form of such quasi-corporatist involvement – is the most apposite description of the character of industrial politics which has so influenced the history of industrial relations in the British coal industry during the twentieth century.

162. Ashworth, *History*, p. 657.
163. Compare, for example, the British steel industry. Steven Tolliday, *Business, Banking and Politics: The Case of British Steel, 1918–1939*, Cambridge, Mass. 1987, p. 343.
164. Krieger, *Undermining Capitalism*, pp. 96–8.

3
Labour Relations In American Coal Mining: An Industry Perspective

David Brody

Among the forces driving the remarkable industrial growth of the United States in the late nineteenth century, none was more powerful than the surging production of coal. By 1910, the 501 million short tons mined represented nearly 40 per cent of the world's production, and far exceeded the output of America's nearest European competitors. The coal industry provided 90 per cent of the nation's energy, operated at a productivity rate per worker three times that of either Germany's or Britain's,[1] and by every measure – cheapness, quality, availability – amply met the needs of the burgeoning industrial economy.

Contempories saw coal mining differently. They considered it, as Herbert Hoover once remarked, 'the worse functioning of any industry in the country'.[2] Hoover had, of course, a standard in mind. He was measuring coal-mining practice against the managerial revolution transforming American enterprise. Of all the sectors within the central industrial order, coal mining remained least touched by the great technological and organizational advances of the early twentieth century; and so, it followed, would be its labour relations. 'It may be paradoxical . . . since without coal there would have been no industrialism,' remarked the economist Carter Goodrich in 1925, 'yet it is true that, so far as the manner of work is concerned, 'mining is still in a way a "cottage" industry.' The *indiscipline* of the mines is far out of line with the *new discipline* of the modern factories.'[3]

1. United States Coal Commission, *Report*, 5 vols, Washington, DC 1925, vol. III, p. 1659 (hereafter cited as USCC). For a recent article that argues in a similar vein as this essay, but from a different theoretical perspective, see John R. Bowman, 'When Workers Organize Capitalists: The Case of the Bituminous Coal Industry', *Politics and Society* 14, 1985, pp. 289–327.

2. A. T. Shurick, *The Coal Industry*, New York 1924, p. viii.

3. Carter Goodrich, *The Miner's Freedom: A Study of the Working Life in a Changing Industry*, Boston 1925, p. 13 (author's italics).

I

The market characteristics of coal set it apart from other sectors of heavy industry. The demand for coal was inelastic – that is to say, demand was determined much more by the level of business activity (or, in the case of the household market, by weather conditions) than by the price of coal. In theory, this meant that prices would be bid rapidly upwards in time of short supply and rapidly downwards in times of oversupply. In practice, the former condition rarely obtained – in fact, only in wartime or during national strikes. This was because of a peculiarity on the supply side of the coal trade. It was elastic on the upswing – ease of entry readily expanded production; but inelastic on the downswing – new operators preferred low or even unprofitable prices to shutting down their mines.[4] So that, despite the secular growth of coal demand, the normal condition of the industry was one of over-capacity and hence, in the absence of market controls, of extreme competition at low price levels.

In the anthracite sector, market controls did in fact come into play. Once the nation's premier industrial fuel, hard coal served mostly for home heating by the turn of the century. Produced in a geographically confined region of north-eastern Pennsylvania, and entering a fairly uniform and well-defined domestic market, anthracite was ripe for market control. The key actors were the anthracite-carrying railroads. They bought up most of the coal properties, perfected a cartel between 1898 and 1902, and thereafter set a price, f.o.b. the New York harbour, that remained virtually unchanged until the First World War.[5]

The bituminous industry could not follow that course, however. Soft-coal reserves were far richer and more extensive, and impossible to engross. The enormous Appalachian field, containing the bulk of the nation's high-grade reserves, ran from north-western Pennsylvania and Ohio all the way down to Tennessee, and there were important secondary fields in Indiana–Illinois, in Alabama, in the south-west, in the Rocky Mountains and in the north-west. The industry was made up of a large number of operators – two for

4. Richard Hannah and Garth Mangum, *The Coal Industry and Its Industrial Relations*, Salt Lake City 1985, pp. 29–31.
5. Eliot Jones, *The Anthracite Coal Combination in the United States*, Cambridge, Mass. 1914, especially chapter 3.

every three mines – totalling nearly 3,500 in 1905.[6] They constituted, as the chief mine inspector of Pennsylvania observed, 'a great
army of antagonistic elements and unorganized forces . . . [who]
continue to indulge in a cut-throat war-fare'.[7] Nor did the railroads
play a stabilizing role as they had in anthracite. On the contrary, in
bituminous they tended to stimulate inter-regional competition.
Anxious to build more traffic, and often with financial interests of
their own in the developing southern Appalachian fields, the
bituminous-carrying railroads set freight rates low enough to enable southern producers to compete in distant northern markets.

Bituminous was not wholly insulated from the consolidating
processes affecting American industry generally in this period.
Some properties were absorbed by vertically integrating steel
firms. There was also a considerable amount of merger activity by
such firms as Pittsburgh Coal, Monongahela River Consolidated
Coal and Coke, and the Consolidation Coal Company of West
Virginia. But these were large only in a relative sense, each controlling only a few per cent of the total output and even together
incapable of much affecting the level of competition. The pattern of
competition was not so much between individual operators – there
had, in fact, always been considerable associational activity within
specific fields – as between rival fields competing in a common
market. It was for this reason that the question of equitable railroad
rates was so explosive an issue for coal operators.[8] At the end of the
greatest period of business consolidation in American history, the
bituminous industry retained very much the competitive structure
with which it had begun. And it was this industry, not anthracite,
that really mattered. Bituminous output was four times larger than
anthracite by 1910, and central to the transportation and energy
needs of the nation's industrial economy.

Competition in the market-place translates into a labour policy
transfixed by labour costs. That fact bore down with special force
on coal mining for two reasons. First, mining was labour intensive,
with wages accounting for between 60 and 75 per cent of total
production costs. Second, and no less important, market pressures
acted in a wholly unmediated way on mine labour costs.

In this period, shop relations in American industry were being

6. USCC, vol. I, p. 323, vol. III, p. 1889.
7. William Graebner, 'Great Expectations: The Search for Order in Bituminous
Coal, 1890–1917', *Business History Review* 48, spring 1973, p. 50.
8. Ibid., pp. 52–3.

transformed by the managerial advances that, for convenience sake, we call Taylorism. In recent labour scholarship, Taylorism has been perceived primarily as a struggle for control of the workplace. By that definition, the coal miner remained singularly untouched. His control over the labour process went virtually unchallenged at the turn of the century and, indeed, for at least another three decades. But workers' control is only one dimension of modernizing work relations. The forces undercutting autonomous labour – mechanization, the redesign of jobs, the close supervision of work – also have the effect of insulating the wage bargain from market pressures. That is, in so far as capital can be substituted for labour, or work made more efficient by Taylorist methods, to that degree the cost pressures on the price of labour are diffused. But the obverse is equally true. In so far as the employer lacks other means of reducing labour costs, he must perforce focus on the price of labour. This was the situation that obtained in coal – much magnified in bituminous by the intensity of competition – and that peculiarly defined its labour relations.

In coal mining, as the engineering expert Hugh Archbold remarked in 1922, the production process 'still remain[s] practically the work of one man who makes a finished product'. What the miner took off the coal-face was, in effect, the product that the mine operators sold. Nor did the scale of operation compare with mass-production industry. In 1920, some 9,000 bituminous mines were in operation, on average producing 60,000 tons a year and employing seventy workers. The 702 largest mines – those whose output exceeded 200,000 tons a year – employed an average of three hundred miners. What pressures there were on mine management, moreover, did not derive primarily from production itself, but from the maintenance side of mine operations – ventilation, safety, electrically-driven haulage, etc. In the mines of Allegheny County, Pennsylvania, the proportion of day (or company) men grew from 10 per cent to 25 per cent of the underground workforce between 1890 and 1930, and it was these employees who were subject to the supervisory control characteristic of American industry generally. Even here, the US Coal Commission (1925) found management practice sadly lacking: 'More definite planning and control of transportation and of all other underground operations are needed . . .' remonstrated the Commission. 'Progress must be toward functional methods of planning and control of the work of the men and machines such as have been introduced in industries

other than coal mining.'[9]
As for the miner, he was virtually on his own. In the late
nineteenth century, a single supervisor – the mine foreman or pit
boss – directed all underground operations, and did well if in his
rounds he saw each miner once during the shift (a standard that, in
fact, came to be mandated by some state regulations). Mine practice in America relied mostly on the room-and-pillar system. The
mine was laid out from a main shaft (or, in hilly areas, from a
horizontal 'drift') from which entries (or corridors) radiated out
along the coal seam. The attack on the seam was at right angles by
cuts wide enough for one or two miners. The walls separating the
lengthening 'rooms' as the coal was taken off the face were cut
through at regular intervals for ventilating purposes – hence the
'pillars'. Along a side entry 1,200 to 1,500 feet long, there might be
twenty rooms.[10] Thus miners were widely dispersed at individual
work sites, and not susceptible to close supervision.

Not even increasing mechanization very much reduced the autonomy of mine work. By 1910 over 40 per cent of all bituminous coal
was mined by undercutting machine, removing from the pick
miner the most skilled and physically demanding phase of his job.
But machine undercutting was absorbed into existing work practice rather than transforming it. Most telling was the fact that the
machine runner and his helper were normally paid a tonnage rate,
not on a day basis as were the company men.

Payment by tonnage was the lynchpin of the entire labour
relations system in American coal mining. It is important here to
distinguish clearly between two different meanings of piece-work –
as payment for output in nineteenth-century artisan production and
as an incentive mechanism on the redesigned, timed jobs of the
modern workplace. Under the older meaning, the piece-rate translated directly into the unit cost of labour, and that was precisely
what payment by tonnage meant to the coal operator: it was the
direct labour cost on every ton of coal he sold. In its inquiries, the
US Coal Commission encountered a curious discrepancy. In the
midst of mine practice that it found scandalously slack, the Commission noticed:

9. Keith Dix, *Work Relations in the Coal Industry: The Hand-Loading Era,
1880–1930*, Morgantown, W. Va. 1977, pp. 37, 38; USCC, vol. III, 1890–1,
p. 1900.
10. See e.g. Dix, *Work Relations*, figure 3, p. 5. On US Steel's efforts to apply
steel-making supervisory techniques to its mines, see Dix, pp. 85–6.

that cost accounting and pay-roll records in coal mining are more systematic and uniform than in any other industry in the country. We question whether the plants, large and small, in any other business than mining would have had on file the unit-cost and pay-roll data which enabled coal operators without difficulty to fill out certain of the commission's questionnaires.[11]

The reason, of course, was that in mining everything turned on that information. The tonnage rate was the cardinal datum of mine operation, and from its centrality we can extrapolate the leading characteristics of coal mining labour relations.

Consider, for example, these persistent points of tension between miner and operator: first, over the weighing of the coal; second, over dockage, i.e. deductions for dirty coal; third, over screening, which separated out the undersized coal for which the miner was not credited; fourth, where payment was by carload rather than weight, over the size of cars. Certain externalities also bore on labour costs. Because of the location of mines in remote or unsettled areas, operators frequently provided company housing and operated company stores. The monopoly profits incident thereto were folded into mine cost calculation, and were taken to be a part of the more general effort at shaving the price of labour. In open acknowledgment of this fact, an 1895 labour agreement actually cut the tonnage rate by five cents for those Pittsburgh operators who abolished stores and paid in cash.[12]

Payment by tonnage meant also that the miner was, in the phrase of Carter Goodrich, 'a sort of independent petty contractor', and therefrom flowed certain other points of tension in mining labour relations. One was over 'deadwork'. It was accepted that the preparation and maintenance of his room was the miner's responsibility. But where was the precise line between deadwork and compensatable extra work caused, for example, by water in his room or by rubble in the haulways? Deadwork was always a sore point for the miner who got paid only for the coal he loaded. Payment by tonnage likewise lay behind the endemic problem of an adequate and timely supply of coal cars. If, as the US Coal Commission estimated, the miner had to wait an average of an hour and fifty minutes each shift for empty cars, the time lost was his, not the operator's.

11. USCC, vol. III, p. 1918.
12. 'Labor Question. – "Company Stores" in the Pennsylvania Mining Districts', *The Annals* 7, 1896, pp. 163–4.

This brings into focus the functional differentiation at the heart of coal mining labour relations. The term 'operator' had a quite precise meaning. The mine-owner *operated* the mine; he did not produce the coal; and, because he paid for it on tonnage basis, the labour process itself was not of primary concern to him. To be sure, low and irregular output did mean a slower recovery rate on his investment, higher overhead costs per ton and delivery problems with customers; and miners were discharged for absenteeism, for dangerous or incompetent work and for disruptive behaviour.[13] But, within broad limits, the miner was his own man. 'It seems to be very frequently the attitude of the industry,' remarked Carter Goodrich, 'that "production is in the main the look-out of the miner" . . . and that how much he works and when are more his own affair than the company's'.[14] Thus, although they came to work at a fixed time, miners habitually knocked off at their own discretion, and this notwithstanding an official eight-hour day from the late 1890s onward. At the mines observed by the US Coal Commission, half the miners left early, as did a much higher proportion of the cutters.[15] Nor did the operator supply the miner with tools or pay for powder and fuses. And because the miner invested his time in opening and developing his 'room' it remained his by a kind of proprietary right. Even after an extended absence, no one ordinarily replaced him at his designated place at the coal-face.

Finally, the miner bore a heavy responsibility for his own safety. The greatest hazard he faced – the cause of half of all mine fatalities – was from roof falls. The primary preventative measure here was the proper timbering of his room. But timbering was deadwork, and every day the miner had to balance the risks against the unpaid labour expended on timbering. In his realm as mine operator, the owner had to strike a similar balance, only with rather different stakes in mind – the costs of repair, lost production and higher insurance rates as against the costs of safer haulage and electrical systems and precautions against gas and mine explosions.

The statistics speak eloquently as to how those calculations came out. The annual fatality rate ran at roughly four per thousand full-time miners in the early twentieth century, compared to three

---

13. See e.g. the analysis of grievances in Louis Bloch, *Labor Agreements in Coal Mines . . . of Illinois*, New York 1931, part 2.
14. Goodrich, *Miner's Freedom*, pp. 30–1.
15. USCC, vol. III, pp. 1945ff.

in Germany, 1.5 in Belgium and slightly more than one per thousand in France and Britain. Coal mining was everywhere a hazardous occupation and always subject to great variations in mine conditions and practices, witness for example the highly dangerous work of 'pulling' the coal pillars in the American room-and-pillar system. Even so, a large residual can only be ascribed to the economics of the American industry. No evidence is perhaps more compelling on that score than a comparison of the adjacent states of West Virginia and Pennsylvania – the first a free-market environment, the second more highly regulated and constrained. For the years 1912–21, the fatality rate per thousand bituminous miners per year was 6.50 in West Virginia, 3.22 in Pennsylvania.[16] In some large measure, the 18,243 miners killed in those years bear witness to a labour system peculiarly defined by competitive market forces and unmediated wage/labour cost relationships.

II

In nineteenth-century coal mining, the unionizing current ran very strongly. Unvarnished industrial injustice was a lesson driven home on the unorganized miner day in and day out, with every short-weighted ton of coal, extra bit of unpaid deadwork or un-fairly distributed coal car. In hard times, slashed tonnage rates brought the sense of grievance to white heat. Then, too, a special kind of mutuality came (in the words of the miners' union consti-tution) to 'those whose lot it is to toil within the earth's recesses, surrounded by peculiar dangers and deprived of sunlight and pure air'.[17] The solidarity inside the mine was cemented above ground by the close-knit community life of the isolated mining camps. And, given the early reliance on British miners, there were always seasoned trade-unionists to serve as the nucleus for organizing activity.

Miners' unions go back at least to the 1840s, initially as local organizations, but quickly broadening in scope. The first national organization – the American Miners' Association – lasted only from 1861 to 1867, and its successors were equally short-lived, until finally, as a result of a merger of the National Federation of Miners and Laborers and National District Assembly No. 135 of

16. USCC, vol. III, pp. 1658, 1659; Dix, *Work Relations*, p. 721. Other explana-tions for Pennsylvania's better safety record would include a more skilled labour force, a more mature industry and differing geological conditions.
17. The constitution is reprinted in Bloch, *Labor Agreements*, appendix III.

the Knights of Labor, the United Mine Workers of America was formed in 1890. Although it barely survived the depression of the 1890s, the modern history of coal unionism begins with the UMWA. The miner's work was, in Carter Goodrich's phrase, 'a rough sort of craft'. His tasks – undercutting the coal face by pickwork, drilling the holes for setting the explosive powder, loading the coal after 'shooting' it from the face, timbering the roof and laying the track – involved skill and experience, but of a kind acquired on the job and picked up rather quickly by green hands.[18] With the introduction of the undercutting machine, the skill component of the job markedly fell, and fell still further as other tasks – shot-firing, for example – began to be transferred to specialists. Given payment by tonnage, moreover, the learning costs were mainly absorbed by the worker himself.

By the end of the nineteenth century, the labour supply was no longer much restricted by the scarcity of skills. On the contrary, the industry was drawing its labour force primarily from large pools of pre-industrial workers. In the north, these were mainly peasant immigrants from eastern and southern Europe, who made up 46 per cent of the workforce in Pennsylvania's anthracite fields in 1920, and 49.6 per cent in the bituminous fields. As the West Virginia and Kentucky mines developed, they built a labour force principally of mountain whites and poor blacks. The latter made up 20 per cent of the workforce there in 1920, and over 50 per cent farther south in Alabama. There was, finally, a built-in oversupply of experienced miners. Because of seasonal demand and an inability to shrink capacity, the industry matched supply to demand by irregular mine operation. The nation's bituminous mines averaged 207 days of work during the 1890s, with a low of 171 days in 1894 and a high of 234 in 1899. The more prosperous decade that followed brought the average up only to 218 days.[19] Under-employment was thus a normal condition, creating a peculiar kind of labour reserve always available to the coal operator.

In so far as trade-unionism depends on control of the labour market, in coal mining the balance of forces ran persistently against labour's cause.

In the product market, on the other hand, the balance of forces

---

18. See e.g. Dix, *Work Relations*, pp. 34–6.
19. Statistics from USCC, vol. II, p. 548, vol. III, pp. 1112, 1422.

cut rather differently. In 1868 the Workingmen's Benevolent Association, which had formed in the Pennsylvania anthracite fields, called a five-week suspension of work specifically to reduce the coal supply and force up coal prices. This remarkable action prefigured the market orientation that would become the hallmark of American coal unionism. The underlying assumption – that prices and tonnage rates were linked – was acknowledged by the WBA in the most explicit of ways: in the brief era of good feeling that followed the strike, it negotiated a sliding scale with the Anthracite Board of Trade, in which the tonnage rate moved along with coal prices.[20] There were as late as 1894 instances of direct action against an oversupplied market, but well before then union leaders had begun seeking a more orderly way of reducing the competitive pressures on the industry's labour relations. The key idea came in 1885 from Daniel McLaughlin, head of the newly formed National Federation of Miners and Laborers. At his instigation, the union issued an invitation to the nation's bituminous operators to 'a joint meeting with the board for the purpose of adjusting *market and mining prices* and give each party an increased profit from the sale of coal'.[21]

Only one operator responded to the first call, a handful to a second, but the third, at Columbus, Ohio, on 23 February 1886, produced sufficiently representative delegations from the major states (Pennsylvania, Ohio, Indiana and Illinois) as to take up the question of an interstate agreement. A wage scale for these states was agreed on, but other issues had to be left to a board of arbitration made up of equal numbers of miners and operators. Although it collapsed in 1889, this first interstate agreement prefigured to a remarkable degree the collective bargaining system that would take hold in the bituminous industry. In 1897, in a last-ditch gamble, the foundering UMWA called a national strike, and, to its astonishment, the bituminous miners came out *en masse*. The 'spontaneous uprising of an enslaved people', president Michael D. Ratchford called it.[22] Out of that three-month strike came a temporary scale and a joint conference in Chicago the following February. For the next thirty years, the interstate joint conference

---

20. Charles B. Fowler, *Collective Bargaining in the Bituminous Coal Industry*, New York 1927, p. 37.

21. Letter reprinted in Andrew Roy, *A History of the Coal Miners of the United States*, Westport, Conn. 1970, reprint of 1905 edn, pp. 248–9 (my italics).

22. Arthur E. Suffern, *Conciliation and Arbitration in the Coal Industry of America*, Boston 1915, p. 41.

would remain the basis for collective relations in the industry.

As an institution, the joint conference was emblematic of the hybrid union–management connection – at once adversarial and mutualistic – that sprang up in bituminous coal. The main business of the joint conference was of course the hard, often prolonged, bargaining over a new wage scale. The conference took the form, however, not of a bilateral negotiation, but of a kind of industry convention, with a credentials committee, committee on rules and order, and so on. By custom, the temporary chair was the UMWA president, the permanent chair a leading operator, the secretary, again, a unionist. The ritual celebration of mutual interest always accompanied the election of these officers by the entire conference – 'not an arena where foe meets foe', as one speaker asserted at the 1899 conference, but 'a friendly meeting-place of those who are interested for the benefit of all'.[23] The adoption of the interstate scale likewise departed from the form of a bilateral negotiation. The agreement, once arrived at by the scale committee (i.e. the bargaining teams of both sides), was put to the vote of the conference, with approval requiring a unanimous vote from both sides. An agreement serves to bind the contracting parties. In this case, there were multiple contracting parties – not only union and industry, but potentially competing segments within the industry. The objective of a unanimous vote by all four states was to bind these multiple interests, within the operators' ranks no less than between operators and union, to the interstate agreement. It was as much an industry as a labour agreement.

The underlying principle, implicit from the very outset but formally enunciated and accepted at the 1902 conference, was that of 'competitive equality', a term regularly used thereafter by coal men. The objective was not to standardize tonnage rates – to take wages out of competition, in the current phrase – but on the contrary to treat them as the variable factor balancing other inequalities under which competitors laboured, 'so that,' as one Illinois employer put it, 'the operators in every district might exist, notwithstanding the different conditions that prevail; and so long as we work on these lines some miners will have to accept less wages than others'.[24]

The crucial function of the interstate conference was to establish

23. Ibid., pp. 143–4.
24. Isadore Lubin, *Miners' Wages and the Cost of Coal*, New York 1924, p. 74.

the framework through which competitive equality could be implemented within the four participating states – the so-called Central Competitive Field. Four roughly comparable districts within these states – Pittsburgh (thin vein), Indiana Bituminous No. 1, Grape Creek, Illinois, and the Hocking Valley, Ohio – served as the 'basing points'. The interstate conference negotiated the standard for the basing points. This then became the basis from which the districts and subdistricts within the four states negotiated their actual tonnage rates, taking into account inequalities in transportation costs and thin vein/thick vein differentials.[25]

The interstate agreement, always a very skimpy document compared to the district contracts, had to contend with powerful centrifugal forces. In fact, it was always difficult to bring all four states into agreement on a scale. Ohio had initially resisted a uniform basing-point rate, since it had historically enjoyed somewhat lower wages than the other states. In 1906, western Pennsylvania made concessions which disrupted the interstate framework, and Illinois, the best organized district, was always troublesome. That the interstate system prevailed in the face of these immense competitive pressures testified to its genuine achievement.[26] If imperfect, it worked well enough to persuade the operators that collective bargaining could indeed serve as the mechanism for levelling the playing field in their cut-throat industry.

And what did the miners receive in return? First, of course, recognition of their union despite an inherently unfavourable labour market. Then, the benefits of collectively-determined terms and conditions of employment. If pick-mine earnings were unequal from district to district, at least they were so by consent rather than

25. The application of competitive equality was, of course, an extremely difficult and always imperfect process. Precisely what concessions a given operator needed to remain competitive was never certain. Moreover, the two principal 'inequalities' cut in quite different directions. Higher transportation costs called for a *widening* of tonnage rates below those of better situated competitors. But, from a competitive point of view, thick vein/thin vein differences actually called for a *narrowing* of tonnage rates, since the concession made (in thin-vein mines) involved a reduction of tonnage rates below what the union would have demanded if its object had been to equalize earnings between thick- and thin-vein mines. In fact, the union did aspire to get as close to equal earnings as it could, and from the first insisted on uniform wages for inside day workers across the entire Central Competitive Field. For a full analysis, see Lubin, *Miners' Wages*, chapters 4–6; and, for district rates in the Illinois agreement, Bloch, *Labor Agreements*, pp. 336–68.

26. The negotiating history is summarized in Suffern, *Conciliation and Arbitration*, chapter 5.

through the arbitrary action of the operators. With the basing-point system in place, moreover, the interstate conference served as the arena within which the pay standard would be regularly advanced (or, as in 1904, reduced) and other basic terms of work improved, beginning, in particular, with the adoption of the eight-hour day in the 1898 agreement. Most important, perhaps, was the swift eradication of the systemic forms of exploitation peculiar to coal mining. The interstate conference itself dealt directly only with the problem of coal screening, partly by mandating the standard dimensions of the screen, but then moving by 1916 to a run-of-mine basis (i.e. payment for a miner's total tonnage, but at a lower rate than for screened coal).

All other issues were left to the state and district joint conferences. The Illinois agreement regulated virtually every matter touching work in the mines – dockage for loading dirty coal, compensatable deadwork and yardage, the price for powder and blacksmithing, the distribution of coal cars (every miner to get a fair share). Contractual infractions by either side were punished by a precise system of fines, a logical extension of the monetary nexus of mine work relations. In the case of unfair discharge, reinstatement invariably included back pay for the number of days of lost work. An elaborate process of adjudication enforced the contract, beginning at the mine site and rising by steps to the executive boards of the district union and state operators' association, with arbitration as the final resort. Arbitration was not compulsory, however, and if either side rejected it, the right to strike or lockout remained.[27] Nowhere in the era before the World War was the logic of workplace contractualism pressed so far, or the jealously held prerogatives of American management so constrained by contractually-defined job rights as in the unionized mines.

But, from the outset, there had been – in the striking words of the labour historians Selig Perlman and Philip Taft – 'a gun pointed at the heart of the industrial government in the bituminous coal industry'.[28] This was West Virginia. The state's rich coal reserves, left mostly undeveloped until the 1880s, thereafter came rapidly into production. Under the theory of competitive equality, the union's job was entirely clear. The Chicago agreement of 1898,

27. See the detailed analysis in Bloch, *Labor Agreements*, part 3. The state agreement is reprinted there in appendix 2.
28. John R. Commons *et al.*, *History of Labor in the United States*, 4 vols, Boston 1918–35, vol. IV, p. 326.

indeed, specifically obligated the UMWA 'to afford all possible protection to the trade . . . against any unfair competition resulting from failure to maintain scale rates'. Because of its uncanny resemblance to an illegal conspiracy in restraint of trade, the clause was cut from subsequent agreements, but no one doubted that it remained an essential condition of the interstate system of collective bargaining. The joint conferences, as the operators remarked,

> were the schools at which the miners for years received their education in the principle that the existence in the markets of competitive non-union coal was disastrous to their movement and well-fare, and that if they were to be able to continue their union relations with the operators . . . they must at all costs both force independent operators to accept union *conditions*, and prevent any union operators from breaking away.[29]

In non-union West Virginia, the UMWA ran into a stone wall. The mountaineer-miners, only recently departed from their hard-scrabble farms and little interested in collective action, participated neither in the 1894 nor the 1897 national strikes.[30] Thereafter, the legendary anti-unionism of the West Virginia operators came powerfully into play. Many were rugged individualists of a frontier-capitalist stamp, of course. But what animated their fierce anti-unionism was a market logic specific to the coal trade – in fact, the underside of the theory of competitive equality. If some competitors accept constraints, those who do not are gratuitously advantaged.[31] The very existence of competitive equality, one might say, gave West Virginia operators an incentive to be militantly non-union. To that must be added the real locational disadvantage under which they laboured: the unionized states were significantly closer to the key markets of the industrialized Middle West.[32] If we can credit what they said, the West Virginia operators suffered from a kind of economic paranoia, convinced that the joint interstate conference was conspiring, not to force them to accept

29. A. F. Hinrichs, *The United Mine Workers and the Non-Union Coal Fields*, New York 1923, pp. 118–19.

30. David A. Corbin, *Life, Work, and Rebellion in the Coal Fields: The Southern West Virginia Miners, 1880–1922*, Urbana, Ill. 1981, chapter 2.

31. For an analysis, see Lubin, *Miners' Wages*, chapter 9.

32. From the standpoint of northern operators, West Virginia already received unduly favourable rail rates, but, of course, the West Virginia operators had no certainty that these rates would always be in place.

competitive equality, but to drive West Virginia out of the coal business.[33]

They had, in any case, ample means for keeping that threat at bay – absolute control over the coal villages (including the power to evict miners), a private army of Felts–Baldwin guards, reliable allies in the courts and state government and, not least, the stomach to see blood shed in their cause. The union may have missed its best chance by opting at the turn of the century to organize the anthracite region first.[34] Thereafter, the UMWA cannot be faulted for lack of trying, either in resources spent, or in the intrepidness of its organizers. Its hold on the Kanawha Valley, gained in 1902, could not be expanded and, after the bitterly fought Paint Creek–Cabin Creek strike of 1912–13, that too became problematic.

Year by year, the non-union threat intensified. In 1913, West Virginia coal production surpassed 70 million tons, and together with neighbouring Kentucky's 19 million tons accounted for 20 per cent of the nation's bituminous output. What this meant in competitive terms can be best understood by studying the impact on specific markets. The most important, the Great Lakes trade, supplied the nation's industrial heartland. In 1898, 86 per cent of the coal shipped up the lakes originated in the union districts of Pittsburgh and Ohio. By 1913, this had slipped to 67 per cent, while West Virginia coal, negligible in 1898, accounted for 23 per cent in that year. Eastern Kentucky, which first entered the lakes trade only in 1909, was shipping 2.6 million tons by 1921.[35]

At the outbreak of the First World War, the collective bargaining system of the Central Competitive Field stood in real and increasing peril. On their own, men on both sides of the table recognized, labour and capital lacked the capacity to bring order to the coal trade. Perforce, they turned to politics.

III

In the United States, as in Europe, coal mining has stood in special relation to the state. Even at the height of the rampant capitalism of the post Civil War era, American courts consistently held that the safety of those who worked underground was a public concern. Beginning with Pennsylvania in 1869, every producing state passed

---

33. See e.g. Hinrichs, *United Mine Workers*, pp. 122–3.
34. Donald L. Miller and Richard E. Sharpless, *The Kingdom of Coal: Work, Enterprise, and Ethnic Communities in the Mine Fields*, Philadelphia 1985, p. 282.
35. Lubin, *Miners' Wages*, p. 214.

and periodically revised and expanded its mine safety and inspection laws. But if these laws demonstrated a special public interest, their history also revealed the even greater power of the industry's market forces.

State-by-state mining legislation suffered, of course, from a crucial structural flaw: it cut across the competitive pattern of coal markets. In so far as the mine safety laws actually became effective, they did so in step with unionization: where collective bargaining expanded, so did vigorous state regulation.[36] During the Progressive era, in the wake of the Cherry mine disaster of 1909, the popular outcry for reform became intense. Led by the American Mining Congress, the industry sought agreement on a uniform code for all the mining states – in effect, to take the cost of safety out of competition. The federal government, while encouraging that development, was itself denied by the Constitution (as then understood) the power to legislate over labour conditions in industry. The US Bureau of Mines (1910), the main result of progressive agitation, carried on useful research, provided technical assistance and maintained rescue stations, but it lacked any regulatory authority over the mines. As for the industry's movement for uniform state laws, not surprisingly it collapsed within a few years. While the mine safety record improved somewhat, primarily because of new workmen's compensation laws, the acknowledged state responsibility for the well-being of miners was far out-balanced by the competitive pressures within the industry.[37]

Perceived in strategic economic and geopolitical terms, as it was in some European capitals, coal could also be invested with a quasi-public character for reasons of state. Not in industrializing America, however. Coal's development, unlike that of the railroads, did not challenge the prevailing *laissez-faire* dogma.

After the turn of the century, labour troubles in anthracite did bring the government's power into play. With winter approaching and fuel for the eastern cities dwindling, the great anthracite strike of 1902 caused a national crisis. At that point, President Theodore Roosevelt intervened decisively and forced the operators to submit to a presidential commission, whose 1903 award, among other things, set up a permanent board of conciliation to adjudicate

36. For the interaction between collective bargaining and legislation, see Bloch, *Labor Agreements*, chapter 7, 'Collective Bargaining in Legislation'.
37. The standard work is William Graebner, *Coal-Mining Safety in the Progressive Era: The Political Economy of Reform*, Lexington, Ky. 1975.

90   *David Brody*

disputes between the miners and their employers. In bituminous, the federal government lacked a comparable incentive. The industry's very fragmentation militated otherwise. Major strikes, although frequent even after the interstate system got started,[38] never closed off enough production to threaten the country's economy. Nor, alternatively, could the industry's market problems be made to inspire public intervention or, to be more precise, not the right kind of intervention. The early twentieth century saw a marked expansion of federal regulation of the business economy. This had nothing to do with the kind of unrestrained competition that was afflicting soft coal, however. The Roosevelt administration was responding to precisely the opposite problem – the sudden arresting of competition in those industries caught up by the merger movement of 1898–1903. The resulting anti-trust campaign by Roosevelt and his successors served, ironically, to discourage the coal industry's own sporadic efforts at reducing competition by means of price-fixing pools, for example, or by joint sales agencies, or, after 1910, by open-price associations. Whether or not these schemes were challenged as conspiracies in restraint of trades (as some were) did not in truth greatly matter, for, as one leading operator remarked, they were 'always made inoperative by the exigencies of business'.[39] What soft coal needed, and what operators and union leaders increasingly advocated, was structural reform that dealt with the underlying problems of excessive capacity and unrestrained market pricing.

Joining with big-business interests, coal men in 1908 lobbied for the Hepburn amendments that would have relaxed the Sherman Anti-Trust Law and permitted 'reasonable' co-operative or joint activity under the supervision of the US Bureau of Corporations. But some operators, doubting their industry's capacity to take advantage of such enabling legislation, advocated that a national mining commission be given 'universal and complete jurisdiction over the mining business', with power to restrict entry, limit production and fix prices.[40]

So sweeping a departure from American *laissez-faire* was never in the cards. When a committee of operators appealed to President William Howard Taft in 1909, he told them (so they reported) that

38. Waldo E. Fisher, 'Bituminous Coal', in Harry A. Millis (ed.), *How Collective Bargaining Works*, New York 1942, p. 250.
    39. Graebner, 'Great Expectations', p. 62.
    40. Ibid., pp. 65, 66.

he 'could see no escape from a continuance of the present system of vigorous competition'. Nor did the industry's efforts in the Senate three years later fare any better. Not even Woodrow Wilson's far-reaching anti-trust revisions of 1914 made any difference. The new Federal Trade Commission proved wholly unsympathetic: its concern was with the uncompetitive practices of big business, not the supercompetitive practices of small business.[41] In the national debate over anti-trust policy, the problems of soft coal turned out to have been beside the point.

Then the war came, and all bets were off. Overnight, soft coal became an essential industry *par excellence*. Modern war, as one Cabinet member said, had become 'an industrial game, the foundation of which is coal'. Without coal, 'the war cannot be carried on'.[42] In fashioning a strategy of industrial mobilization, the Wilson administration wanted to avoid over-centralization. It reserved for itself broad policy-making powers, while leaving the implemention of its guidelines on prices, priorities, etc. to the industries themselves. The war emergency thus opened up a remarkable chance for industrial self-regulation, but that opportunity, as it turned out, was denied to soft coal.

The very competitiveness it was seeking to master confounded the industry's hopes. After 1915, prices pushed rapidly upward, tripling in the spot market by 1917 and, in the single month after the American entry into the war in April, jumping from $3.00 to $3.72 a ton. The inelasticity of coal demand was operating wholly to the industry's advantage, perhaps for the first time. The mechanism for dealing with such problems, under the Council of National Defense (CND), was the tripartite industry group, and one was duly created for coal – the Committee on Coal Production. But by the time that body moved against the spiralling coal prices, political sentiment had turned on the industry. In its time of peril, it seemed, the nation was being gouged by a predatory 'coal trust'. 'Government officials were condemning an essentially fragmented and leaderless industry as if it were a monopoly', remarks James P. Johnson.[43]

What followed was a monumental fiasco. Eager to capitalize on

---

41. Ibid., p. 63; and, on coal's relations with the FTC in 1915–16, James P. Johnson, *The Politics of Soft Coal: The Bituminous Industry from World War I through the New Deal*, Urbana, Ill. 1979, pp. 27–31.
42. Quoting Franklin K. Lane, in Johnson, *Politics of Soft Coal*, p. 38.
43. Johnson, *Politics of Soft Coal*, p. 34.

the anti-trust enthusiasms set off by rising coal prices, the Justice
Department had brought suit against members of the Smokeless
Coal Operators Association of West Virginia for price fixing. The
charge was that they had conspired to set a price of three dollars a
ton for their coal. When the CND's Committee on Coal Production finally made its price control announcement on 28 June 1917,
the benchmark price was, lo and behold, *three dollars a ton*. Out-
raged by this apparent conspiracy, President Wilson immediately
repudiated the three-dollar proposal, arranged to have coal included in the pending Lever food control bill and then personally
fixed the maximum price for coal at two dollars a ton.[44]

Coal mining was taken over by the Fuel Administration. While
other industries gained self-regulation under the War Industries
Board, bituminous coal went through the war effectively as a
nationalized industry. Coal executives did go to work for the Fuel
Administration, and they helped to fashion a programme of ration-
alized coal marketing and transportation that resolved a nearly
castastrophic distribution problem. But at the war's end the industry, notwithstanding a newly-created National Coal Association,
was organizationally scarcely better off than it had been in 1914. On
1 February 1919, Fuel Administrator Harry A. Garfield (a college
president, not a coal operator) lifted price and zone controls,
wished the industry well, and expressed the hope that it would
mend its ways and apply the war-induced marketing structure to
peacetime operation.

The industry was not, as it turned out, so readily extricated from
the government. Back in October 1917, under the aegis of the Fuel
Administration, the joint interstate conference had granted a 15 per
cent wage increase under what became known as the Washington
Agreement. Any further relief was denied the bituminous miners
by the government despite persistently rising consumer prices and
concessions to other workers (including anthracite). With the Armistice, a peculiar and intolerable anomaly appeared. Controls over
the operators came off and coal prices jumped to record heights,
but the wage freeze remained in effect, and this despite zooming
living costs. It so happened that, under its terms, the Washington

44. As it turned out, the indicted West Virginia operators were found innocent
because, while they might indeed have conspired, the effect on the market had been
negligible: prices had risen above three dollars before their agreement had gone into
effect! Ibid., pp. 47–8. My treatment of these events relies on Johnson's trenchant
analysis.

Agreement ran until either the end of the war or 1 April 1920, whichever came first. But the stalemate over the Versailles treaty put off the formal return to peace, and the coal operators, while themselves enjoying unprecedented profits, insisted on holding the UMWA to the letter of the Washington Agreement. By the summer of 1919, the miners were up in arms, and wildcat strikes swept the coalfields.

Faced by a national coal strike on 1 November 1919, the Wilson administration intervened, resurrecting the wartime controls under the Lever Act (including the reimposition of maximum coal prices) in a futile attempt to put the genie back into the bottle. All it managed to do, however, was to pit itself against the UMWA. On 8 November, the union leadership bowed to draconian injunctions and ordered the miners back to work. It was a bitter moment for the UMWA, never forgotten. Out in the fields, the miners resisted, and only grudgingly trooped back into the mines. To settle the dispute, President Wilson appointed a special arbitration commission, which in early 1920 awarded the tonnage men an increase of 31 per cent and the day men 20 per cent. The discrepancy touched off a further wave of strikes that ultimately extracted for the day workers an increase to $7.50 a day, the union benchmark wage for the rest of the decade.

In the course of this struggle, the state's relationship to the industry fundamentally shifted. If the Republican administrations of the New Era had no stomach for the state socialism of wartime control, neither could they return to the hands-off attitude of the Progressive era. Coal had become a national problem. For one thing, the 1919 crisis had demonstrated that coal strikes could no longer be treated as private affairs. In subsequent strikes in 1922 and 1925, dwindling coal supplies and skyrocketing prices again forced public intervention, and reinforced the conclusion that the government had a stake in more harmonious labour relations in coal mining. Equally important, the health of the industry itself became a matter of public concern. The demoralized bituminous industry was too much at odds with the vision of modern industrialism celebrated by the New Era. The failure to stabilize that vital sector, remarked a trade journalist in 1927, was 'the economic crime, as well as the prime folly, of an otherwise progressive age'.[45] During

45. Ellis W. Hawley, 'Secretary Hoover and the Bituminous Coal Problem, 1921–1928', *Business History Review* 42, autumn 1968, p. 252.

the decade, there were no fewer than eight major investigations, and a plethora of proposals for rescuing the industry. Unlike in the prewar years, unrestrained competition was now acknowledged to be the industry's core difficulty, and one that demanded, in some fashion, government intercession. The key Republican advocate of coal stabilization, schooled by the wartime experience, was Herbert C. Hoover, Secretary of Commerce under Harding and Coolidge, and then President in his own right.

IV

The joint interstate system never recovered from the wartime upheaval. Competitive equality had worked under the conditions of relative economic stability that had prevailed in the prewar industry. Between 1903 and 1915, tonnage rates had advanced at an average rate of only 1 per cent a year, and coal prices had fluctuated within a very narrow range. Given evenly distributed wartime pressures, the industry's bargaining system would have been strained; with wages frozen by the Washington Agreement, it broke down entirely in the 1919 crisis.

Although rescued by the arbitration awards of 1920, the damage to the interstate system was not so easily repaired. Rank and file militancy, slow to subside under any circumstances, was fed by the internal politics of the UMWA. Weakly led since the retirement of president John Mitchell more than ten years before, the UMWA was now coming under the rule of a genuine strongman, John L. Lewis. But Lewis's position was still uncertain, he had no popular mandate, and he was challenged by a host of formidable district leaders, all to the left of him and all appealing to a restive rank and file.[46] Out of this political cauldron had come the UMWA demands of September 1919: a 60 per cent wage increase, a thirty-hour week, time-and-a-half for overtime, double pay for Sundays and holidays, and, to top things off, the nationalization and democratic management of the mines. To the heightened expectations that these demands bespoke must be added the loss of confidence in a mutualistic relationship with the operators. By their indecent behaviour when they had held the whip-hand under the Washington Agreement, they had spent the goodwill built up by years of negotiation in the joint conference.

46. Melvyn Dubofsky and Warren Van Tine, *John L. Lewis: A Biography*, New York 1977, chapter 3.

Then the postwar boom collapsed, and the underlying weakness of the interstate system re-emerged. Mine capacity had grown tremendously – by 300 million tons between 1915 and 1920 – and mostly in the southern Appalachian fields. Despite substantial union headway during the war, West Virginia and Kentucky were still imperfectly organized on Armistice Day. In the postwar reaction, the operators regained the initiative and, as in the violent Mingo County strike of 1920–1, relentlessly drove the UMWA back. So the non-union competition persisted and, with every new mine, grew more threatening. What the market collapse of 1921 revealed was an added, tactical non-union advantage: the capacity to move quickly in moments of economic crisis.[47] Unable to match southern wage cuts because of the contracts still in force, many union mines were literally swept out of the market in the terrible year of 1921.

By the time that ordeal was over, northern operators had turned massively against the interstate agreement. If they could not rid themselves of the union, at least they insisted on the freedom to bargain independently or by districts. In 1922, they refused to enter another joint conference. Ohio and Pennsylvania operators, said one of them, 'felt that they must without restriction or limitation of any character make their own agreement with their own men in their own district'.[48] It took a bitter twenty-week strike by the UMWA to enforce the interstate structure. But it had been done by might alone. The joint interstate concept no longer held the allegiance of northern operators. Any semblance of industry unity, briefly manifested in the National Coal Association, now broke down. Northern operators wanted to pursue their individual interests, they stopped thinking in industry-wide terms and they turned against government intervention or co-operation. By the mid-1920s, the competitive instinct had returned with a vengeance and the Central Competitive Field was reverting to nineteenth-century conditions.

At this moment, the burden of market reform passed to the United Mine Workers of America. But no longer on the basis of competitive equality. In certain ways, the concept had always been troublesome for the union: while accepting in principle that labour costs would have to vary among unequal competitors, the UMWA

47. Lubin, *Miners' Wages*, pp. 209ff.
48. Ibid., p. 215.

also adhered as best it could to the goal of equal earnings and
conditions for all its members. To strike a balance between these
conflicting objectives had always been at the nub of district-level
bargaining over tonnage rates.[49] Now the economic justification
for variable labour costs itself began to erode. Competitive equality
rested on the assumption of an ever-expanding market that would
always make room for all producers. That crucial assumption
broke down in the 1920s. The technology of coal utilization dramatically improved and rival fuels commanded an increasing share
of the energy market – roughly 30 per cent by 1925, and growing.
After sustained growth for half a century, the demand for coal
levelled off.

'Too many miners and too much coal.' With that phrase, John L.
Lewis signalled the union's new economic programme. 'The bituminous industry is suffering . . . the pains incidental to a long-
delayed adjustment', Lewis said in 1925. 'When it is complete,
there will be fewer mines and miners and it will be a prosperous
industry.'[50] Lewis intended to use the union's bargaining power to
speed that outcome. The key, he felt, was a uniform, high-wage
structure that would drive out 'uneconomic mines, obsolete equipment and incompetent management'. He wanted to encourage the
mechanization of the more productive mines, assure an 'American'
standard of living to the remaining miners and enable the industry
to compete on even terms with gas and oil. 'Any concession of
wage reductions will serve to delay this process of reorganization,
by enabling the unfit to hold out a little longer.'[51] Thus Lewis's
seemingly ruinous policy of 'no backward step' in the face of
non-union price competition: to abandon the $7.50 standard meant
abandoning his conception of an industry driven by high-wage
labour.

Wage uniformity was necessarily a more remote goal, given the
regional differentials and multitudinous tonnage rates built up
under the practice of competitive equality. But Lewis now rejected
its underlying principle – that the ability to pay should be a primary
criterion of wage determination – and he would as best he could
press towards the goal of an industry-wide uniform wage structure. This partly explains (if it does not justify) his ruthless sup-

49. For a detailed analysis, see ibid., chapter 3.
50. Morton S. Baratz, *The Union and the Coal Industry*, New Haven 1955, p. 60.
51. Carrie Glasser, 'Union Wage Policy in Bituminous Coal', *Industrial and Labor Relations Review* 1, July 1948, p. 608.

pression of the UMWA's proud tradition of district autonomy during the 1920s: the centralized bargaining he envisioned required a centralized union. The 1924 agreement, the final effort at holding the interstate structure together, notably invaded the bargaining rights of the districts: by its terms, they were prohibited from seeking in any way to alter their district agreements.[52] Lewis put forward his ideas in avowedly conservative, even Darwinian, terms. In 1923, he engineered the repudiation of the union's nationalization programme. In the course of the decade, he systematically purged his left-wing critics from the secondary leadership. And in his book *The Miners' Fight For American Standards* (1925) he trumpeted his credo to the world:

> The policy of the United Mine Workers of America ought to have the support of every thinking business man in the United States, because it proposes to allow natural economic laws free play in the production and distribution of coal. . . . [The aim] is not to steal mines from their owners, but to make it possible for owners of economic and properly equipped mines, operated by free labor under an American system of government, to make reasonable and continuous profits.[53]

Nevertheless, John L. Lewis's programme required the assistance of the state. Indeed, as the power of the union waned, that need mounted. Up to a point, the Republican administrations of the 1920s were amenable. Secretary of Commerce Hoover took very much the same view of the industry's problems as did Lewis. Hoover estimated in 1922 that there were 2,500 more mines and 200,000 more miners than the nation needed. After weighing and rejecting a variety of public interventions, Hoover concluded by late 1923 that 'the gradual elimination of high-cost, fly-by-night mines' could be best achieved through the industry's own efforts under conditions of industrial peace. So, at Lewis's urgent request, the Secretary of Commerce pressured the resisting operators, led by the key Pittsburgh Coal Company controlled by the brother of his Cabinet colleague Andrew Mellon, into participating in another round of interstate negotiations. The resulting Jacksonville Agreement of 1924, which continued the $7.50 day for another three

52. Section I, Joint Interstate Agreement, 19 February 1924, reprinted in Bloch, *Labor Agreements*, pp. 360–1.
53. Indianapolis, pp. 15, 186.

years, was almost as much Hoover's handiwork as Lewis's.[54]
But the Jacksonville Agreement did not hold. Under enormous
pressure from the non-union fields, it began at once to crumble.
John L. Lewis had, of course, anticipated defections, but from the
marginal operators, not from the strongest firms (such as Pittsburgh Coal and Consolidation Coal) who in fact led the exodus.
Their large holdings in non-union fields gave them powerful
leverage over the union. In a showdown, they could shift their
production to the southern Appalachian mines and starve the union
miners out.[55] What Lewis had thought would be a controlled
attrition turned instead into a deluge that threatened to sweep his
union out of the Central Competitive Field. His second miscalcula-
tion was in supposing that the Coolidge administration, as a virtual
party to the Jacksonville Agreement, would help to maintain it.
But when Lewis appealed to Hoover, the Secretary of Commerce
coolly suggested that the union take any firms violating their
contracts to court – useless advice twice over, since litigation was
too slow a process, and, given the highly dubious enforceability of
labour contracts, likely to fail in any case.[56]

When the Jacksonville Agreement expired in 1927, there were
few operators prepared to negotiate a new contract. The UMWA
called a last, futile strike, and then instructed the districts to make
whatever terms they could. The joint interstate system was dead
and so, very nearly, was the union. Half a million strong in 1922, it
had shrunk to under 100,000 by 1928 and, outside of Illinois and
Indiana, had almost disappeared from the Central Competitive
Field.

Liberated from union restraints, northern operators slashed
wages and joined the desperate scramble for business. Coal prices,
which had briefly stabilized in 1926, began to fall again, and, after
the onset of the Great Depression in 1929, plummeted. At the
depths, in 1932, prices at the pit stood at little more than a dollar a
ton, compared to two dollars in 1926 and three dollars in 1922.
From 1927 onward, the industry operated at a deficit; in 1932 over
80 per cent of the surviving firms reported net losses to the Bureau

54. Hawley, 'Secretary Hoover', p. 250; Dubofsky and Van Tine, *John L. Lewis*,
pp. 107–8; Robert Zieger, *Republicans and Labor, 1919–1929*, Lexington, Ky. 1969,
pp. 229–31.
55. John Brophy, 'Elements of a Progressive Union Policy', in J. B. S. Hardman
(ed.), *American Labor Dynamics*, New York 1928, pp. 186–91.
56. On this episode, see Zieger, *Republicans and Labor*, pp. 239ff.

of Internal Revenue.[57] As in the late nineteenth century, market pressures once more bore down directly on the wage bargain, slashing rates and restoring all those petty forms of cheating over weights, deductions and charges by which operators could squeeze another penny from labour costs. But the mitigating effect of secular growth in demand of that earlier age was now over, and from 1929 onward, the trend was absolutely downward. Between 1929 and 1932, 300,000 miners lost their jobs. For those still employed, average hourly earnings fell from 65 to 52 cents, the work week from 38 to 28 hours. Hunger and privation spread across the coalfields.

John L. Lewis excoriated the owners of the industry:

> They have practically no form of organization. They have no code of ethics. They are simply engaged in a struggle to continue their existence and remain in business. Why, the larger interests of the country are preying upon the coal industry, buying its products at less than the cost of production, and compelling the operator to sell the blood and sinew and the bone of the hundreds of thousands of men who are engaged in industry, and to sell the future of their children.[58]

When he spoke those bitter words in 1932, Lewis was no less impotent. His union was shattered, and he himself personally discredited. But bankrupt of ideas or will he was not. The core of his thinking had not changed since the mid-1920s – that a uniform, high-wage structure, achieved through free collective bargaining, would solve his industry's ills.

To get there, Lewis now recognized, required the massive intervention of the state, first, to guarantee the right of miners to collective bargaining so as to crack the hard nut of anti-unionism that had defeated the UMWA's utmost efforts, and, second, to enforce strict controls on coal competition until the full power of collective bargaining could be brought to bear. Twice, in 1928–9 and again in 1932, legislation was introduced into Congress incorporating those ideas. Strenuously resisted by both Hoover and the industry, the Watson and Kelly–Davis coal stabilization bills stood no chance of passage. But they signified what distinguished John L. Lewis from the others. He at least had a conception of what had to be done. All he needed was a law.

57. Waldo E. Fisher and Charles M. James, *Minimum Price Fixing in the Bituminous Coal Industry*, Princeton, NJ 1955, p. 18.
58. Johnson, *Politics of Soft Coal*, p. 131.

V

With the launching of the New Deal, Lewis's chance came. His initial presentation of his coal stabilization plan to the new president on 27 March 1933 fared badly. Under the pressure of the banking crisis and the nation's emergency relief needs, Franklin D. Roosevelt at first thought that a programme for industrial recovery could be put off until a later time. But as the Hundred Days proceeded, industrial recovery moved on to centre stage, and Lewis and his economic adviser W. Jett Lauck made a critical decision: they would submerge their own plans in the more general effort to stem the deflationary spiral and stabilize American industry. It was largely due to Lewis, through Lauck's participation in the drafting process of April and early May 1933, that the National Industrial Recovery bill contained section 7(a), which guaranteed to workers the right to organize and engage in collective bargaining. Lewis had less luck with the bill's approach to market regulation. He would have preferred strict controls, via industrial boards empowered to allocate production and fix prices. Instead, the administration opted for a more mixed system of industrial self-governance modelled after the War Industries Board and favoured by the progressive business interests represented by the US Chamber of Commerce.

In the resulting National Industrial Recovery Act, the key mechanisms were the codes of fair competition, each to include the mandated labour provisions of sections 7(a) and (b) (covering minimum wages, maximum hours and child labour) but otherwise tailored to the regulatory needs of the individual industries. A government agency, the National Recovery Administration (NRA), was granted broad powers of oversight and enforcement. The codes, however, were to be written and administered primarily by the industries themselves, or, more precisely, by their trade associations.

Unlike the oligopolistic sector, the fragmented bituminous industry fought this chance for market regulation tooth and nail. The National Coal Association adopted, *pro forma*, an essentially empty model code, while the various districts drew up regulations that had in common only a rejection of the labour provisions of the Recovery Act. By early August, the NRA had before it a bewildering array of coal codes. To forge from these discordant elements an acceptable coal code – that is to say, industry-wide, capable of restraining the industry's competitive impulses and in conformity with the law's labour provisions – seemed almost beyond the powers of the embattled NRA authorities, and, indeed, of President Roosevelt himself.[59] At this juncture, a revived UMWA suddenly seized the initiative. Popular uprisings are among the most intractable, the most difficult of events for historians to fathom. But in the case of America's miners in mid-1933, the ingredients at least seem reasonably clear: first, long-festering grievances brought to a boil by the Great Depression; second, an authoritarian management briefly weakened and disoriented; equally exceptional, a political environment, symbolized by section 7(a), suddenly sympathetic to labour's cause; and, finally, a dramatic organizing drive into which John L. Lewis threw his union's last remaining resources. The upshot was the swift reorganization of the mining fields, not only the traditional areas of union strength, but now also southern Appalachian and deep South territory, and even the mining properties of the steel industry. Overnight, John L. Lewis became again – to use his own phrase – 'captain of a mighty host'.

So armed, Lewis moved aggressively on two fronts in Washington. He pressed the coal industry to accept an effective code and simultaneously demanded that it enter collective bargaining with his union. In these endeavours, Lewis found an indispensable weapon in the spontaneous strikes that swept the Pennsylvania coalfields during that August. Although beyond Lewis's control, the pressure from below gave him the leverage he needed at key points in the protracted NRA negotiations. 'In the last analysis,' remarks the principal historian of these events, 'it would be a truly nationwide union that would compel the operators to agree on a single code.' On 21 September 1933, three days after President Roosevelt signed that code, John L. Lewis and James D. A. Morrow, an industry leader of long anti-union standing, signed the Appalachian Agreement that both called 'unquestionably . . . the greatest in magnitude in the history of collective bargaining in the United States'.[60]

In one decisive stroke, Lewis had set in motion the uniform, high-wage programme whose pursuit had very nearly destroyed him and the UMWA in the 1920s. The Appalachian Agreement

59. Ibid., chapter 5; Glen L. Parker, *The Coal Industry: A Study in Social Control*, Washington, DC 1940, chapter 6.
60. Johnson, *Politics of Soft Coal*, pp. 153, 163.

preserved the industry's traditional bargaining structure, with a pattern-setting central field and outlying fields that followed its lead. But now Illinois and Indiana, main components of the old Central Competitive Field, were cut out and relegated to the category of outlying fields, reflective of the long-term shift in competitive markets (and a contributing factor to the break-up of the joint conference system after 1924). The remaining parts of the old Central Competitive Field, western Pennsylvania and Ohio, were now joined to West Virginia, Virginia, western Maryland, northern Tennessee and eastern Kentucky – that is to say, the entire Appalachian region – in a new central bargaining unit that represented 70 per cent of total bituminous production and constituted the real heart of the industry.

The non-union threat that had for so long defeated the UMWA was thus finally eradicated and, while the first Appalachian Agreement did concede a north/south wage differential, it was of a scale much reduced from the past. In 1934, northern West Virginia came up to the northern standard, and in 1941, after a bitter and protracted struggle, so did the rest of the southern Appalachian field. This uniformity applied, of course, only to the day workers. Tonnage rates remained fixed in the industry's history of district and local differentials. But the significance of this rate variation rapidly declined.

As Lewis had anticipated, his wage-bargaining strategy speeded mechanization of the mines. Coal-loading machines, the major innovation of the 1930s, deprived miners of their principal remaining manual task, and ended their reign as contract/tonnage men. The relationship was entirely precise: as mechanical loading expanded, tonnage work contracted. By 1945 it was down to 25 per cent of mine labour, and shrinking, and so, by definition, was the portion of variably-compensated labour. And for the tonnage work that remained, the UMWA did all it could to reduce the historical differentials. Its most effective tactic, initiated in 1946, was to negotiate flat wage increases for all workers, tonnage as well as day men, so that, while base tonnage rates might vary, the repeated flat increments eventually narrowed the earnings differentials to practical insignificance.[61] Thus John L. Lewis came closer to the wage policy for which he had long battled – equal labour rates for all and

61. Gerald G. Somers, *Experience under National Wage Agreements: The Bituminous Coal and Flint Glass Industries of West Virginia*, Morgantown, W. Va. 1953, pp. 22–7.

no quarter to the marginal operators.

If 'competitive equality' had fostered decentralization, the campaign for wage equality brought precisely the opposite result. District bargaining did not disappear, but it was increasingly emptied of significant content. Much that had earlier been left to the districts was now absorbed into the Appalachian Agreement: the grievance procedure, key terms of employment and even the specification of tonnage rates by district.[62] District autonomy had been so undermined, in any case, that the distribution of bargaining functions scarcely mattered. Two-thirds of the districts were under virtually permanent trusteeship – Lewis's technique for seizing control over them – and hence run by his appointees, not by elected officials. And this in turn relegated the union's formal representative bodies – for collective bargaining, most importantly, the National Wage Policy Committee   to the status of rubber stamps. Although *pro forma* demands for greater autonomy were regularly heard at UMWA conventions, so commanding, even mythic, had Lewis's standing become among rank and file miners that they acceded to a concentration of power in one man's hands probably unique in American trade union annals. The only limitation on Lewis's rule, an outgrowth of the bitter internecine struggles within the UMWA before 1933, was the survival in Illinois of a breakaway union, the Progressive Mine Workers of America.

The coal producers, for their part, experienced a parallel, if much more imperfect, organizational development. The National Industrial Recovery Act, to begin with, demanded a degree of associational discipline beyond what the industry's rugged individualists had hitherto been willing to accept. The conception of an Appalachian Field was originally not the union's handiwork, in fact, but that of four key operators seeking to come to terms with the New Deal's industrial recovery programme. During the early codewriting period, these industry leaders formed two large regional organizations – the Northern Coal Control Association and the [southern] Smokeless and Appalachian Coal Association – that then combined to present a single, albeit anti-union, code covering the entire Appalachian region. Collective bargaining, implanted on this structure, cemented the unity of the Appalachian operators.

It was a fragile unity, however, with an ironic tendency to break apart at the very points when greater economic uniformity was

62. Schedule A, Appalachian Agreement, 19 June 1941.

being imposed on the industry. Thus the two signal breakthroughs on north/south wage differentials – with northern West Virginia in 1934, the southern Appalachians in 1941 – split the associations representing those regions from the industry's bargaining structure. For the remainder of the 1940s, the Southern Coal Producers Association negotiated independently, and indeed frequently took bargaining stances at odds with that of the northern Appalachian group. The outcome was, nevertheless, always substantial uniformity. The industry's organizational development had likewise come a long way, with an acknowledged industrial leadership in the north and south, and a grudging acceptance of the reality of centralized bargaining.

One fight being waged by the miners in this period carried the UMWA out on to a much larger battleground. The captive mines, so called because they were owned by and their output entirely consumed by steel companies, had never been a factor in the industry's competitive markets, and, hence, never of primary concern to the union. But the organizing explosion of 1933 engulfed these mines, and the captive miners flocked into the UMWA. On economic issues, there was no problem. Although not participants in the Appalachian negotiations, the steel companies were always prepared to apply the standards of the Appalachian agreements to the captive mines. On the question of union rights, however, they parted company with the commercial mines. After 1933, most operators accepted collective bargaining unreservedly, and only a few scattered areas – notably, Harlan County, Kentucky – put up stiff and continuing resistance. But the steel industry, bulwark of the open shop, stubbornly withheld the formal acceptance that the UMWA demanded. It took years of strife to gain full recognition and, in 1941, an embittering year-long battle to extract the union shop from the owners of the captive mines.

Their hard opposition attuned John L. Lewis to the larger struggle of the nation's industrial workers for collective bargaining. Equally important, the captive mines forged a link in his mind between the fate of the steel workers and the fate of his own union: without the captive mines, the commercial miners would not be safe, and without the steel industry, the captive miners would not be safe. From 1935 onward, Lewis committed his union's resources and his own energies very largely to the titanic battle for industrial unionism. And it was at the head of the Congress of Industrial

Organizations (CIO) in its triumphant hour that Lewis earned his enduring place in American history and contributed most signally to the American labour movement. But the rock on which he stood was always the miners' union. And, so long as the Great Depression persisted, the union's power depended on the fragile structure of price restraints put into place by the NRA. For all the swagger that John L. Lewis cultivated in this period, he was deeply fearful of any resurgence of cut-throat competition in the coal business. In late 1934, for example, as the expiration of the NRA approached, operators began to accept new contracts at below-code prices. An alarmed Lewis castigated the operators for following a 'policy of monumental stupidity'. Since they seemed 'incapable of preventing their own commercial destruction', Lewis demanded NRA action to 'meet this menacing situation'. Although his demands were satisfied, Lewis had already resumed his campaign for a more centralized price-fixing programme specifically for soft coal, and, given the power he wielded in New Deal politics, this time he succeeded. His first bill, the Guffey–Snyder Act (1935), was found, like the NRA itself in May 1935, to be unconstitutional, and the second, the Guffey–Vinson Act (1937), proved to be an administrative nightmare.[63] Nevertheless, these troubled laws worked well enough to sustain the level of coal stabilization achieved under the NRA.

There is no doubt about the effect of the New Deal intervention on coal prices; notwithstanding weak demand, they rose from $1.29 per ton at the mine in mid-1933 to $1.75 in 1934, and to $1.95 by 1938. It was this upward push that enabled the industry to grant a seven-hour day in 1934 and wage increases raising average hourly earnings from 52 cents in 1932 to 80 cents by 1939.[64] Unquestionably, the New Deal underwrote the Appalachian collective bargaining system, and, unquestionably, it did so in large measure at the earnest and unceasing behest of John L. Lewis.

VI

On 1 September 1939, war broke out in Europe, and, once again, all bets were off. In a defence-driven economy, the mineworkers did not need the state: New Deal market regulation became

63. Fisher and James, *Minimum Price Fixing, passim.*
64. Ibid., pp. 7, 17. For an economic analysis of UMWA gains, see Waldo E. Fisher, *Economic Consequences of the Seven-Hour Day and Wage Changes in the Bituminous Coal Industry*, Philadelphia 1939.

superfluous. Lewis's statist enthusiasms had, of course, always
been strictly limited and, more to the point, strictly instrumental.
By 1941, with the economy booming, Lewis was bent on exploiting the new balance of forces – hence the maximum drive for
north/south wage uniformity and, in the captive mines, for the
union shop. In that second, monumental struggle, Lewis had to
overcome not only the powerful steel interests, but a disapproving
President Roosevelt, and his National Defense Mediation Board as
well. This prolonged battle with the emergency apparatus of the
defence period reminded Lewis of what he already knew: that in
wartime the state was likely to become the mineworker's enemy.
The bitter memory of the First World War, etched in Lewis's mind,
partly explains the isolationist stance he took after the Nazi threat
began to engulf Europe. Disagreement on this issue, among other
reasons personal and political, led to Lewis's famous break with
President Roosevelt. But at bottom it was a matter of business or,
more precisely, of the marketplace economics that always dominated Lewis's thinking. He vowed that his miners would not for a
second time become 'innocent victims of an ill-advised wartime
economy'.[65]

After Pearl Harbor, Lewis's worst fears were swiftly realized.
The government moved, more decisively than in 1917–18, to
impose anti-inflationary controls over the wartime economy. The
National War Labor Board, set up to arbitrate industrial disputes
affecting war production, soon took jurisdiction over wages as
well. In July 1942, the NWLB enunciated its governing stabilization policy: the Little Steel formula, which, based on an estimated
price rise of 15 per cent from 1 January 1941 to 1 May 1942, set 15
per cent as the limit on wage increases counting back to the
beginning of 1941. Since the miners had received an 18 per cent
increase on 1 April 1941, the Little Steel formula meant that their
wage rates would effectively be frozen for the duration of the war.
Nor was there the *quid pro quo* that reconciled other industrial
unions to wartime regulation: the UMWA had no need of the
maintenance-of-membership protections held out by the NWLB as
the reward for union co-operation.

Lewis was as patriotic as the next man, and as willing to support
the war effort, but not at an intolerable cost to his coal-miners. The
Roosevelt administration had adopted 'a paradoxical policy that

65. Dubofsky and Van Tine, *John L. Lewis*, p. 390, also p. 416.

runs to the premise of rewarding and fattening industry and starving labor', he protested. It was the First World War all over again: rising living costs (inevitable, in Lewis's view, despite price controls and rationing), cost-plus contracts for business, frozen wages for workers. So far as he was concerned, Lewis told assembled Appalachian operators in March 1943, 'the Little Steel formula has outlived its usefulness. . . . It can't last because it is so viciously unfair. It seeks to deny to labor what the government gives to industry, namely, the cost of living plus a profit'. As for the War Labor Board, it had 'fouled its own nest' and was best advised 'to voluntarily resign and not cast its black shadow in the face of Americans who are merely hoping for a right to live and a right to serve in this emergency of our own country'.[66] Thus, with the Appalachian Agreement due to expire on 1 April 1943, Lewis threw down a challenge of the most fundamental kind to the power and majesty of the American government.

On the face of it, such a contest might have seemed wholly one-sided. In wartime, the President can bring sweeping emergency powers to bear. He can also, with sufficient skill, mobilize enormous political and popular pressure against a dissident like John L. Lewis. And, if he is a Roosevelt, he can even isolate such a leader and make him a pariah in his own movement. All these things Roosevelt did, but Lewis was not disarmed. Above all, Lewis was backed – in a certain sense, even instructed – by the rank and file miners, who, through the wildcat strikes that swept the coalfields from January 1943 onward, revealed to him their determination to resist on the ground the Little Steel formula.

Lewis did defy the War Labor Board. When it took jurisdiction over the stalemated negotiations in late April 1943, the UMWA refused to attend its hearings, and when it handed down a final directive specifying a new agreement on 18 June, Lewis rejected the order as an 'infamous yellow dog contract' that 'no member and no officer of the United Mine Workers would be so devoid of honor as to sign or execute'.[67] On 1 May 1943, with 75,000 miners on strike, the government seized the mines, rendering the miners public employees, and strikes illegal. The War Labor Disputes Act, passed over President Roosevelt's veto by an infuriated Congress, then

66. Colstone E. Warne, 'Coal – The First Major Test of the Little Steel Formula', in Warne (ed.), *Yearbook of American Labor: War Labor Policies*, New York 1945, pp. 282–3; Dubofsky and Van Tine, *John L. Lewis*, p. 417.
67. Warne, 'Coal', p. 291.

made such strikes criminal acts. Nevertheless, walkouts continued. Lewis proved adept at orchestrating the striking miners, and, as the government quickly discovered, there was no effective way of forcing them to work against their will. So, despite a good deal of trumpeting of its sovereign powers, the Roosevelt administration backed down. Secretary of the Interior Harold Ickes, the federal administrator of the seized mines, engaged in bargaining with Lewis and finally settled with him on 3 November 1943. By a complicated calculation involving additional worktime and portal-to-portal pay – i.e. covering travel time within the mines for which miners not hitherto been compensated – Lewis got the $1.50 a day increase he had been demanding.[68] Although he had not broken the Little Steel formula, he had in fact triumphed over the wartime state.

Ironically, Lewis faltered not when its emergency powers were at their peak, but when they were being dismantled. In the first postwar round of negotiations, Lewis successfully repeated the tactics that had worked for him in 1943. After a six-week strike that created a desperate coal shortage, Lewis forced the Truman administration to seize the mines again on 21 May 1946, and, as in 1943, he then proceeded to extract from government negotiators an historic new 'fringe' – a welfare-and-retirement fund financed by a 5 cent royalty on coal tonnage – that evaded the wage stabilization guidelines established by the 1946 steel settlement. But five months later, Lewis declared the agreement he had made with the government terminated, effective from 20 November 1946, and demanded new negotiations.

This time President Truman, pushed too far and aware of the high political stakes, held firm. 'The Administration must find out sometime whether the power of Mr. Lewis is superior to that of the Federal Government,' declared one of the President's advisers.[69] The injunction Truman obtained – in an historic break from the national policy, enunciated in the Norris–LaGuardia Act of 1932, against using that legal weapon in labour disputes – required Lewis to rescind the termination declaration as a disguised strike order that violated the War Labor Disputes Act. Once pushed into the judicial system, Lewis lost all room for manoeuvre. In contempt proceedings, the UMWA was fined a crushing $3.5 million, and

68. For an explanation, see ibid., pp. 297–8.
69. Dubofsky and Van Tine, *John L. Lewis*, p. 465.

Lewis fined $10,000 and branded by the outraged federal judge as instigator of a strike that was 'an evil demoniac, monstrous thing . . . a threat to democratic government itself'.[70] Lewis had overreached himself and, beaten and humbled, he called off the strike. If he needed any further justification for the libertarian conclusions he was drawing,[71] it was the passage of the Taft–Hartley Act the following year, for which, it must be said, Lewis had to bear some considerable responsibility.

This protracted struggle inevitably took a heavy toll on the collective bargaining system constructed back in the New Deal period. While the formalities remained – the Appalachian Agreement still served as the basic framework and negotiations did take place (and once, in the spring of 1945, came to successful conclusion) – the heart of the bargaining process had been cut out. At every critical stage, the coal operators had been shunted aside, their properties taken over and held by the government for lengthy periods. In practice, seizure proved less than draconian: only the hardy few who refused to co-operate were actually displaced in day-to-day operations by public officials.[72] But this did little to assuage the resentment of operators as they watched government negotiators trade away in 'interim' agreements precedent-setting concessions on portal-to-portal pay, safety practices and welfare benefits. Almost necessarily, the return of their properties at the expiration of the War Labor Disputes Act on 30 June 1947 ushered in a period of industrial conflict during which they tested out the uses of Taft–Hartley and the permanence of the union's wartime gains, in particular, the burdensome welfare-and-retirement fund financed by a royalty on their coal output. Yet, as it turned out, the stage was being set in that turbulent time for the completion of the system that had begun with the Appalachian Agreement back in 1933.

After a decade of war-driven demand, the coal market was by 1949 reverting to normal. The rivalry of other fuels, arrested

70. Colston E. Warne, 'Industrial Relations in Coal', in Warne (ed.), *Labor in Postwar America*, Brooklyn, NY 1949, p. 377. The union fine was subsequently reduced to $700,000.

71. In 1953, Lewis testified in favour of the repeal of the Wagner Act whose federal protections of the rights of workers to organize and bargain collectively he had so desperately desired in the early 1930s. Dubofsky and Van Tine, *John L. Lewis*, p. 476.

72. John L. Blackman, *Presidential Seizure in Labor Disputes*, Cambridge, Mass. 1967, p. 177.

during the war, resumed with a vengeance. The railroads, converting rapidly to the diesel locomotive, were about to disappear as coal's best remaining market. And operators had to contemplate, not the flat demand of the 1920s, but the dread prospect of an absolute decline. As the problem of overcapacity reasserted itself, so did the persistent question of an industrial policy for soft coal. The New Deal solution was foreclosed. There would be no returning to federally-imposed market stabilization. The second Guffey law had expired unlamented in 1943, and, with the Keynesians in the ascendancy, the economic thinking it embodied was discredited and gone from the political agenda. For their part, neither the industry nor the union had any stomach for more government intervention. It was, in fact, the prospect of Congressional legislation authorizing re-seizure of the mines that precipitated a sudden resolution of the last protracted round of struggle of 1949–50.

Intervention came from a different quarter. In basic steel, as elsewhere in the industrial economy, corporate leaders were in process of constructing their post-New Deal version of economic stabilization. For that reason, the troubles in coal had to be taken in hand. In a curious way, US Steel's Ben Fairless was returning the compliment to John L. Lewis. To protect his miners in the 1930s, Lewis had moved to force collective bargaining on US Steel. Now, seeking to perfect industry-wide bargaining in basic steel, Ben Fairless found he needed stability in soft coal. To that end, Fairless and his associates began in 1947 to participate in the bituminous negotiations, and the captive mines for the first time came under the national agreement. To this was added a sudden spurt of consolidation within soft coal. In 1945, George Love merged three of the largest firms into the giant Pittsburgh Consolidation Coal Company, and he too had an interest in stabilizing coal's collective bargaining. Between George Love, who became the dominant force among the commercial operators, and US Steel's Harry Moses, who exerted comparable influence over the captive-mine group, it became possible to bring virtually the entire industry, north and south, into the National Bituminous Wage Agreement of 1950.[73] Shortly thereafter, Love and Moses established the Bituminous Coal Operators Association (BCOA) which, in alliance with

73. I am relying here on Dubofsky and Van Tine, *John L. Lewis*, chapter 20 and pp. 494–7.

the Southern Coal Producers Association, effectively united the bituminous industry for purposes of collective bargaining.

Thus, almost improbably, what John L. Lewis had hoped for since the 1920s became a reality. The confrontational bargaining of the 1940s suddenly gave way to private, informal discussions between John L. Lewis and Harry Moses (who had resigned from US Steel to become head of the BCOA). Expiration dates no longer applied: after one year, either side was free at any time to reopen the contract and the outcome took the form of 'amendments' to the basic National Bituminous Wage Agreement. When Lewis and Moses came to a decision, the terms were announced with a flourish and automatically approved by their respective organizations.[74] National strikes, almost endemic in the previous decade, became virtually extinct. The only significant walk-out occurred in 1952 during the Korean War and was directed not at the industry but at Truman's Wage Stabilization Board for scaling back a wage increase to which the BCOA had agreed.

Lewis's doctrine of high-wage uniformity had carried the day. Average earnings rose from $14.75 a day in 1950 to $24.25 a day in 1958, and royalty payments for the welfare-and-retirement fund added another forty cents of labour cost to every ton of coal. Most telling was the conclusion drawn by economists studying the wage impact of trade unions: in soft coal, over 30 per cent of earnings in the mid-1950s could be credited to the UMWA, 'the largest effect . . . estimated for any industrial union'.[75] That the industry shared Lewis's market reasoning was evident in the contractual provisions aimed at enforcing uniformity on the entire industry. Of these schemes, the most notorious was the 'protective wage clause' of 1958, which committed all signatories not to buy or process coal from, nor lease coal lands to, firms not abiding by the national agreement.[76] Not surprisingly, the UMWA soon faced an anti-trust suit charging 'a conspiracy with the large operators to impose the agreed-upon wage and royalty scales upon the smaller, non-union operators, irrespective of their ability to pay'.[77]

In the face of a production decline of 20 per cent between 1950 and 1960, coal prices remained almost stable. Labour costs per ton,

74. For a description, see Charles R. Perry, *Collective Bargaining and the Decline of the United Mine Workers*, Philadelphia 1984, pp. 188–95.

75. Albert Rees, *The Economics of Trade Unions*, Chicago 1977, p. 73.

76. C. L. Christenson, *Economic Redevelopment in Bituminous Coal*, Cambridge, Mass. 1962, pp. 268–9.

77. *Pennington v. UMWA* (1965), in Perry, *Collective Bargaining*, pp. 80–2.

despite a 64 per cent rise in wages, actually fell by 8 per cent as increasing mechanization pushed output per miner up from 6.77 to 12.83 tons a day. Jobs, of course, fell off just as dramatically. The industry, which had employed 415,582 miners in 1950, employed 141,646 in 1963. It was a trade-off which Lewis, if not jobless miners, had always welcomed. And if final proof of the UMWA's 'co-operative' stance was needed, it was forthcoming when the industry entered a cyclical downturn in 1958. The union refrained from reopening the agreement, and for the next five years wage rates remained unchanged.

VII

At his retirement in 1960, John L. Lewis must have been well pleased by his handiwork. He had accomplished what America's miners had been struggling for ever since that effort by the Workingmen's Benevolent Association back in 1868 to push up anthracite coal prices by going on strike: that is, to master by their own collective effort the market forces that bore down so heavily on them.

But those market forces, after persisting for a century, were about to change. Three events, all dating from about the time of John L. Lewis's departure from the scene, can serve as markers of a remarkable transformation of the American coal business:

First In 1961–2, coal burned by the nation's electric utilities reached 50 per cent of total coal consumption (compared to 18.6 per cent in 1950). At that point, reversing a fifty-year trend, the coal industry began a sustained expansion, driven entirely by America's insatiable appetite for electricity. Between 1960 and 1980, coal output doubled, 80 per cent of it now absorbed by the utilities. The impact on the coal market, although little noticed, was far-reaching. For one thing, the utilities, as coal consumers, had no interest in the price stabilization fostered by the BCOA–UMWA bargaining system. Hence the sudden challenge to the UMWA from the unlikeliest of places – that triumph of New Deal reform, the Tennessee Valley Authority. With its increasing reliance on steam-generating plants in the 1950s, the TVA adopted a policy of buying coal from the lowest bidders, no questions asked. The result was a proliferation of low-wage truck mines and the renewal of a non-union threat from Kentucky and West Virginia.[78] On the other hand, the nature of utility demand very

78. Christenson, *Economic Redevelopment*, pp. 257–68.

much moderated the market pressures that had called forth the
BCOA–UMWA structure in the first place. For electric power
companies, fuel was a major cost item, generally over 60 per cent of
production expenses, and price became correspondingly important
as a determinant of their coal-market behaviour. Because their
modern furnaces were multiple fuel burners, moreover, the utilities
could switch to coal when its price was falling. Thus, unlike major
industrial consumers of the past, the demand for coal by the utilities
was quite elastic and served as a counterforce to the extreme
pressure on prices that had always marked a declining coal market
in the past.[79]

Second In 1961, Pennsylvania passed the Surface Mining Recla-
mation Act. This measure, the first even modestly effective
environmental response to coal strip-mining, signalled the commercial success of a radically different extractive process in the industry.
Suited to the low-grade requirements of steam-powered generators,
surface mining took hold in some eastern districts, and then expanded
dramatically in the west. A very minor factor in Lewis's day, the
Rocky Mountain and High Plains states were producing a quarter of
the nation's coal in 1980, almost exclusively as suppliers of steam coal
for the utilities. With its giant steam shovels and dragline excavators,
surface mining could scarcely even be classified operationally in the
same category as underground mining. Resistant to the UMWA
(hence its support for the 1961 Pennsylvania environmental legislation), many of the surface mines remained non-union, or, especially
in the west, were organized by other unions. Surface mining confronted the UMWA with a different and more formidable kind of
regional threat, for while the south's advantage had been cheaper
labour, the west's was higher efficiency.[80] And in so far as the
surface-mine sector could not be accommodated within the
BCOA–UMWA structure, its uses were thereby further eroded.

Third In 1959, the General Dynamics Corporation purchased
the Freeman Coal Company. This marked the start of a wave of
acquisitions that in little more than a decade essentially brought to

79. I am following the economic analysis in Reed Moyer, *Competition in the Midwestern Coal Industry*, Cambridge, Mass. 1964, chapters 3, 4, 6.
80. Richard H. E. Vietor, *Environmental Politics and the Coal Coalition*, College Station, Tex. 1980, chapter 3, and appendix A; Perry, *Collective Bargaining*, pp. 36–8, 65–72; and for regional and surface/deep mining shifts, Curtis Seltzer, *Fire in the Hole: Miners and Managers in the American Coal Industry*, Lexington, Ky. 1985, table 5, p. 212.

an end the existence of coal as an independently-run industry. The
need for steady, high-volume sources of supply had already
speeded the process of consolidation under way in soft coal. In the
mid-western region, for example, the four largest firms moved
from 26.8 per cent of total production in 1949 to 54.6 per cent in
1962, and the eight largest controlled 74.2 per cent by then. On top
of this, the rapid depletion of the nation's cheap oil and gas reserves
now led to a revaluation of coal as an energy source, not only for
burgeoning utility needs but also (so it was hoped) for other uses
through new liquefication and gasification technologies. Oil companies, utilities and other resource-oriented firms moved quickly to
establish a stake in the coal business. By 1976, only three of the
forty largest coal producers remained under independent management. Concentration of control by the energy conglomerates,
together with the predominating utility demand, meant a coal
market no longer subject to the severe downward price pressures of
the past.[81]

As an economic regulator, the BCOA–UMWA structure had
effectively been superseded.

Inside the underground mines, meanwhile, a workplace crisis
was building up. Traditionally, mining had been an autonomous
labour process, with miners left largely unsupervised and pay fixed
by tonnage output. In the era of hand-loading, which prevailed up
to the 1940s, the essential thrust of union work rules had been to
regulate mine practices that cut into tonnage earnings or increased
unremunerated work. The spread of coal loading machines after
1930 changed all that, drastically reorganizing the labour process,
replacing the individual miner with teams of specialists and the
tonnage pay system with day work. Now the operator discovered
that labour productivity did matter, and that supervision was a
managerial function. Nor did innovation abate. A second wave of
mechanization, initiated in 1948, began to replace undercutting and
loading machines with continuous mining equipment, which by
1970 accounted for half of all deep-mine output, and in the 1980s
even more sophisticated longwall technology seemed in the
offing.[82]

As workplace control disappeared, miners began to demand a

81. Vietor, *Environmental Politics*, chapter 2; Robert Stobaugh and Daniel Yergin
(eds), *Energy Future: Report of the Energy Project at the Harvard Business School*, New
York 1979; Moyer, *Competition*, table 19, p. 68.
82. Perry, *Collective Bargaining*, p. 26.

different brand of work rules. Unlike the men paid by tonnage and claiming proprietary places at the coal-face, specialized day workers wanted explicit rights governing the allocation of job opportunities – hence the inclusion for the first time of seniority rights in the 1941 and 1952 agreements. With the new division of labour, too, pay equity and job classification became problems that increasingly occupied the grievance system.[83] Mine safety was likewise cast in a different light. So long as the cost had in considerable measure been borne by the contract miners (through labour they considered deadwork), the union had shied away from safety as a bargaining issue. But under the new system, health and safety became strictly management costs, and, as new machinery sped operations up, these became matters of increasing concern to the miners. So the 1941 agreement provided for mine safety committees, the first step in an aggressive collective bargaining intrusion into the realm of health and safety.[84] As mine foremen gained supervisory functions, finally, the power they wielded over workers entered union calculations – hence, among other reasons, the organizing drive directed at them by the UMWA in the 1940s.

Between the rationalizing impulses of the modern operator, and the work-rule defences of the modern miner, tensions necessarily existed, just as they did across the entire mass production sector. In mining, however, this conflict deepened into a severe crisis. There were, first of all, the traditions of the workers' control era to surmount – for operators no less than miners. The speed of technological change, too, kept things unsettled. As mining became capital intensive, operating schedules, in particular, became hotly contested. Union practices going back to the hand-loading era – no Sunday work, no night shifts, vacation shutdowns – came up against insistent employer demands for continuous mine operation so as to maximize the return on expensive equipment. In the early 1970s, finally, twenty years of sustained productivity growth came to an abrupt end, evidently triggered by the Coal Mine Health and Safety Act of 1969, but then fed by widespread workplace strife over the next decade. From a record 15.61 tons in 1969, output per worker in underground mines plunged to 8.25 tons a day in 1978. At the opening of the 1981 negotiations, the BCOA

83. For a detailed treatment of seniority and job classification issues, see Gerald G. Somers, *Grievance Settlement in Coal Mining*, Morgantown, W. Va. 1956, chapters 4, 5.
84. Perry, *Collective Bargaining*, pp. 29–36.

complained bitterly about 'the heavy cost burden that low productivity imposes on coal produced under the UMWA–National
Agreement'.[85] Whatever the real productivity advantage of the
non-union competition – for deep mines, the BCOA put it at 39
per cent[86] – it was a far cry from Lewis's conception of unionized
operations at the cutting edge of the industry's efficiency.

Converging over an extended period, these developments – the
market impact of utility demand, the takeover of the industry by
the energy conglomerates, an expanding non-union sector, the
workplace crisis – came to an altogether predictable end. The
Bituminous Coal Operators Association, with 130 members in
1980, spoke for fewer than twenty companies in the 1988 negotiations. A larger number bargained independently with the UMWA.
And more bargained not at all, for by then the industry was
becoming increasingly non-union.[87] The BCOA–UMWA structure that had been the life's work of John L. Lewis served for
roughly twenty-five years; after the early 1970s, it began to weaken
and then rapidly disintegrated.

Perhaps, in this post-industrial age of American capitalism,
twenty-five years ought to be taken as the normal life span of any
collective bargaining system. Across the economic spectrum, from
steel to trucking, postwar labour–management settlements have
come apart more or less on the same schedule. But no union has
handled its troubles so badly; or, between 1947 and 1985, lost so
large a share of its organized territory;[88] or today seems so incapable of mounting a fresh attack on the transformed coal industry. To encapsulate the UMWA's breakdown in a single event: in
1986 this once-mighty union announced that it was ready to seek
shelter in merger with a larger organization.

In some measure, the sad fate of the UMWA has to be laid at the
feet of John L. Lewis. To prevail over the market forces in soft coal,
he always insisted, required that the union be run as a 'business
proposition'.[89] During the long years of struggle, that notion
mostly manifested itself in Lewis's iron-handed control over the

85. Ibid., p. 29.
86. Ibid., p. 231.
87. *New York Times*, 31 January 1988.
88. Leo Troy, 'The Rise and Fall of American Trade Unions', in Seymour M.
Lipset (ed.), *Unions in Transition: Entering the Second Century*, San Francisco 1987,
table 7, p. 87.
89. Dubofsky and Van Tine, *John L. Lewis*, pp. 384, 385.

union's affairs. But after the 1950 agreement, when the need for militancy and solidarity subsided, the UMWA experienced in full measure what it meant to be run as a 'business proposition': in organizing – strong-arm tactics and money talked; on collective bargaining – a happy community of interest with the industry; towards the rank and file – the demand only for dues, for passive loyalty, and submission to the contract's no-strike clause. Two decades of this regime were enough utterly to drain the UMWA of its vitality as a workers' organization. The history of failed leadership, rebellious rank and file and collective bargaining disarray that followed Lewis's departure is beyond the limits of this essay; and beyond its limits, too, is any attempt at assessing the degree to which the UMWA's own troubles contributed to the breakdown of the BCOA–UMWA structure.[90]

But there is the future to think of. It would be the crowning irony if, by virtue of his brief triumph, John L. Lewis had exhausted the chances of succeeding generations of America's miners to control their own industrial destinies.[91]

90. In addition to Perry, *Collective Bargaining*, *passim*, see e.g. Peter Navarro, 'Union Bargaining Power in the Coal Industry, 1945–1981', *Industrial and Labor Relations Review* 36, January 1983, pp. 214–19; William H. Miernyk, 'Coal', in Gerald G. Somers (ed.), *Collective Bargaining: Contemporary American Experience*, Madison, Wis. 1980, pp. 1–48; and, on the internal union history, Brit Hume, *Death and the Mines: Rebellion and Murder in the United Mine Workers*, New York 1971; Joseph E. Finley, *The Corrupt Kingdom: The Rise and Fall of the United Mine Workers*, New York 1972; Paul F. Clark, *The Miners' Fight for Democracy: Arnold Miller and the Reform of the United Mine Workers*, Ithaca, NY 1981.

91. For a highly favourable assessment of current UMWA president Richard L. Trumka, however, see e.g. *Business Week*, 15 February 1988, pp. 65–6.

4
Entrepreneurial Politics and Industrial Relations in Mining in the Ruhr Region: From Managerial Absolutism to Co-determination

Bernd Weisbrod

Introduction

Coal is the stuff from which the industrial revolution was made. It supplied the key resource for the steel age and advanced the revolution of transportation both on land and water. In Germany, the soaring demand for coal in the steel industry and in the forced expansion of the railroad system caused the 'take-off' of the early 1870s that eventually pulled the entire economy along with it.[1] The deceleration of the growth rate at a high level in the 1880s and the nearly complete cartelization of the coal industry in the 1890s revealed even then the structural problems that would evolve into a constant source of concern for national economic policy in the 1920s following the surge in demand created by the war. Once the flywheel of economic growth, coal was to become a politically privileged 'old' industry and a structural burden for the modernization capacities of the German economy. Thanks to the two world wars and the energy crises of the postwar period, coal had attained the political status of a national resource industry, the organization of which could no longer be left to the marketplace or to self-regulation by employers and unions. Until coal was replaced by oil as the most important energy source, it attracted the interest of the general public; even up to the present this has not only regulated or where necessary increased the pressure on the industry to adapt to economic realities by way of intervention, but also allowed indus-

---

1. Cf. Carl-Ludwig Holtfrerich, *Quantitative Wirtschaftsgeschichte des Ruhrkohlenbergbaus im 19. Jahrhundert. Eine Führungssektoranalyse*, Dortmund 1973. Judged by the growth rate trend and by the forward and backward linkages, the railroad represented the leading sector of the greatly interdependent 'core of growth'.

trial relations in mining to become a political test case of the relationship between employers and unions in general.

Therefore, mining was an industry of strategic importance in many respects:

- Being a key industry in industrialization, its organization of production and of the market supplied the model for capitalistic, economic organization in the age of 'organized capitalism': it was characterized in Germany even before the First World War by large-scale industrial organization, consolidated structures, and monopolized control of the market.
- Being a national resource industry, it inevitably attracted state supervision and intervention. Before the First World War, such state involvement was modelled on the tradition of the state management in the early nineteenth century, the so called *Direktionsprinzip*. Later, supervision and intervention would act as a substitute for profitability and economic peace in an 'old' industry and would finally provide the structural framework for crisis management in the private sector following the Second World War.
- In this economic and political setting, the organization of industrial relations in mining attained an importance that greatly exceeded the narrow scope of the coal and steel industries themselves. Due to the special status of coal, mining-specific labour disputes and industrial relations became the test case for the validity of entrepreneurial power claims in large-scale capitalist enterprises, or more specifically, for the effectiveness of union representation of workers' interests.

Industrial relations in German mining, especially in what was its most important sector by far, the hard coal mining in the Ruhr region, undoubtedly corresponded in many respects to the particularities specific to mining in other countries. However, nowhere else was the organization of industrial relations so greatly charged by the effects of major changes in the political constellation; nowhere else were such changes in the political system so directly linked back to the corresponding reorientation of industrial relations as in Germany. This study primarily deals with the entrepreneurial policy strategies affecting the labour market and the workplace in Ruhr mining. It is a topic requiring a look into not only how economically motivated such strategies were, but also how far they were politically determined and relevant. This context appears to be less evident in the situation of other European mining nations; in the history of industrial relations in German mining, this perspective is unavoidable. Neither the *Herr im Haus* position prior to the First World War, to which the mining industrialists stubbornly held despite growing opposition even from the state, nor the acceptance and rejection of a system of collective wage contracts during the revolution and postwar period can be sufficiently explained solely by the economic rise and decline of the coal industry. Neither does this explain the forced implementation of a new plant social policy of *Werksgemeinschaften* in the union-free power structure of the Third Reich, nor the offer of co-determination made to the unions following the Second World War and its later utilization in overcoming the coal crisis. Despite all the similarities to the strategies of problem-solving in comparable countries, there is still a significant margin which can only be explained by investigating the changing political framework and the national traditions of a specific industrial culture.

These assumptions require a broader understanding of the field of industrial relations than is generally used by industrial sociologists or, in other contexts, by union historians.[2] It is not solely concerned with the increasingly efficient regulation of power struggles within a company as an optimal management strategy, or with the inevitable domestication of class conflict as a result of the legal regulation of industrial relations and the equality in status of the organized labour force. Rather, the development of industrial relations should be understood as a part of that historical process of interest aggregation in which claims of economic power and political participation are incorporated into the political system by corporative arrangements.[3] Contrary to the fundamental conviction of mining entrepreneurs, neither the individual company nor the economic self-organization of such a key branch of industry as coal mining once was existed in a political vacuum. During the various political stages of German development, industrial relations both

2. On this point, cf. Jonathan Zeitlin, 'From Labour History to the History of Industrial Relations', *Economic History Review* 60, 1987, pp. 159–84.
3. See the contributions based on the corporative model by Philippe C. Schmitter in: *Comparative Political Studies* 10, 1977; Gerhard Lehmbruch and Philippe C. Schmitter (eds), *Trends Towards Corporatist Intermediation*, London 1979; and Suzanne Berger (ed.), *Organizing Interests in Western Europe. Pluralism, Corporatism, and the Transformation of Politics*, Cambridge 1981.

on a company and branch level presented a picture puzzle of a national political culture which set it apart from all the other mining countries despite comparable conditions of production and the step-by-step advancement towards the modern and supposedly more rational model of union-organized and state-regulated labour disputes.[4]

As the trend towards state intervention increased, industrial relations appeared to the mining employer to be one continual conflict with the labour force and the unions over workplace and labour market control. The 'tripartism' that eventually evolved between the employers, the unions and the state in the modern industrial state may prove to be a national characteristic of Germany's economic organizational structure, making it the first 'post-liberal' nation.[5] Or it may prove to be a particularity specific to mining and found in all mining countries. However, the example of Germany shows more clearly than do others that tripartite corporatism did not represent a stable mechanism for regulating labour disputes; rather, these were themselves part and parcel of a continual process of political negotiation, the parameters of which were determined by the feedback which the unions, employers and the state received from the labour force, management and the public's political convictions as a whole.

When industrial relations are studied more closely, the corporative model inevitably loses many of its systemic traits. Although the existence, simultaneously and successively, of pluralist, liberal-corporative, and state-corporative forms of interest organization may serve a macrohistorical form of analysis, it fails when applied as an analytical category of the process of change in individual cases of industrial relations.[6] Instead of referring to a corporative system, it would be more appropriate in the field of industrial

4. See Gerald D. Feldman's judgement of 'the ambivalent modernization of industrial relations' in the Weimar Republic: 'Obviously, the basic foundations of any industrial relations system are political and derive their legitimation not only from the acceptance of the participants but also from the acquiescence of the nation as embodied in its system of representation and its administration.' 'The Weimar Republic: A Problem of Modernization?', *Archiv für Sozialgeschichte* (*AfS*) 26, 1986, p. 15.

5. Werner Abelshauser, 'The First Post-Liberal Nation: Stages in the Development of Modern Corporatism in Germany', *European History Quarterly* 14, 1985, pp. 285–318.

6. See the interpretation by Ulrich Nocken, 'Korporatistische Theorien und Strukturen in der deutschen Geschichte des 19. und frühen 20. Jahrhunderts', in Ulrich von Alemann (ed.), *Neokorporatismus*, Frankfurt 1981, pp. 17–39.

relations to speak of a 'corporate bias',[7] which indeed helped to establish a trend towards the 'institutionalization of class antagonism' (Theodor Geiger) but was in no way irreversible as the history of labour relations in mining shows. In the tripartite form so typical for mining, it was linked to certain political and economic constellations in which the state's interest in industrial co-operation made itself felt. The state's influence essentially consisted of a political compensation transaction for organized labour during war, inflation and the postwar period; it was not able, however, to establish a lasting balance of forces.[8]

The history of industrial relations in mining was politically determined in Germany simply because the unions were excluded from any sort of corporatist arrangement both in the period before the First World War and during the Third Reich. Fully developed tripartite corporatism only evolved during the periods of reconstruction that followed the defeats in both world wars, in which the political integration of the labour movement was then the order of the day. Thus, in addition to the 'corporate bias', allowances must also be made for a 'political bias', namely the political price to be paid for refusing, accepting or abandoning corporative industrial relations. The employers in mining proved by way of their strategy that they were very aware indeed of this dimension of their action: in the 'bargained corporatism' of German mining, it was not only the unions, as the employers never tired of pointing out, but also the employers themselves who advanced their political position, be it that of an opponent in class struggle or of a partner in wage contract negotiations.[9]

7. See Keith Middlemas, *Politics in Industrial Society. The Experience of the British System*, London 1979, pp. 371ff. Using Britain as an example, Middlemas meanwhile emphasizes even the uniqueness of the war-related corporatism that collapsed following the 'competitive equilibrium' of the postwar period. K. Middlemas, *Power, Competition and the State*, vol. I: *Britain in Search of Balance 1940–1961*, London 1986.

8. For a criticism of a harmonizing view of a corporatism that is related to the pluralist model, see Leo Panitsch, 'Recent Theoretizations of Corporatism: Reflections on a Growth Industry', *British Journal of Sociology* 31, 1980, pp. 159–87. See also Klaus von Beyme, 'Neo-corporativism: A New Nut in an Old Shell?', *International Political Science Review* 4, 1983, pp. 173–96.

9. On the concept of 'bargained corporatism', see Colin Crouch, *The Politics of Industrial Relations*, Manchester 1979.

'Herr im Haus': The Long Shadow of State Management in Mining

The development of mining in the Ruhr region in the nineteenth century is characterized by two major innovative phases, which overshadowed both the economic and organizational picture of a stormy and steady rise in production. Until the middle of the century, mining was subordinated to the governmental Direktions-prinzip, the state management policy, and thus under the immediate supervision of the Prussian mining authorities. The liberalization of the Prussian mining law in the second half of the century led, after a transition period, to the complete freedom from such supervision of the now privately-run mining industry. This lasted until the great May strike of 1889, after which the industry was once again increasingly restricted by outside interference on the part of the state and the organizations of both the employees and the employers.

The three phases of mining law, namely that of state management, liberalization, and political, that is, legal and organizational fencing in of Ruhr mining, coincided roughly with the long-term economic cycles of development. However, a causal connection between legal and economic development should not be automatically derived from this, even though the outstanding role of the state in Germany is irrefutable during both the phase of early industrialization and the protective tariff period of full industrialization. In the tradition of the historical school of German political economy, the fact is quickly overlooked that the components of state capitalism in German industrialization were not only the cause but also the consequence of market-economy processes that were sanctioned and advanced, for example in Ruhr mining, by the liberalization of the Prussian mining laws.

The acceleration of the growth rate in the liberalization phase from an annual 5.2 per cent (1824–1850) to 9 per cent (1851–1874)[10] was certainly due to the expansion of underground mining beginning in the pre-liberalization 1840s, which resulted in the founding of a second generation of mines in the middle belt of the later Ruhr region in the vicinity of the Hellweg cities. These new mining operations employed about 400 or 500 men and were responsible for the quantitative growth during the liberalization period in Ruhr

10. Holtfrerich, *Wirtschaftsgeschichte*, p. 24.

124

Bernd Weisbrod

mining that increased the average pit output tenfold between 1851 and 1880 and the average manpower level fivefold.[11] The legal release of miners from required membership in guild-like organizations (*Knappschaften*) sanctioned the growing need for free wage workers and enabled the second generation of mines, by contrast with the traditional Ruhr valley mines of the first generation, increasingly to recruit non-mining labour from the local regions of migration with seasonally fluctuating labour markets. The enormous financial requirements of this phase of expansion could be secured thanks to the reform of company and corporate law, which reduced the investment risk of mining shares by limiting the personal liability of the shareholders in the traditional mining companies and which especially mobilized the necessary investment capital by permitting the establishment of stock companies.

The strong fluctuations in the business cycles of the seventies revealed even then the structural dilemma of a branch of industry that tended to build up excess capacity with long investment cycles and a great price elasticity.[12] During the economic upswings both before and especially after the Great Depression – the latter being characterized by curbed growth combined with considerable fluctuations in prices and wages in Ruhr mining as late as the 1880s – it was the still unorganized workers who asserted their demands in mass strikes. These increasingly forced the state, but not the employers, to intervene. The deceleration of the growth rate after the 1870s to 4.7 per cent (1875–1913) is in a certain sense a statistical illusion because expansion continued with giant strides at a high level. Yet the exploitation of labour, which had been made possible by the one-sided legal privilege of the employers, and the improved yield of more valuable types of coal in easily accessible coal seams ran up against the limitations of labour organization and production technology. As the productivity curve began to flatten out, it could only be compensated by the quantitative increase due to a third generation of mines on the northern border of the Ruhr

11. See Gerhard Adelmann, *Quellensammlung zur Geschichte der sozialen Betriebsverfassung. Ruhrindustrie unter besonderer Berücksichtigung des Industrie- und Handelkammerbezirks Essen*, vol. 1: *Überbetriebliche Einwirkungen auf die soziale Betriebsverfassung der Ruhrindustrie*, Bonn 1960, no. 97, pp. 143ff.
12. See Holtfrerich, *Wirtschaftsgeschichte*, pp. 162f. Since coal in the nineteenth century was not considered to be storable, the pressure was intensified by the given cost structure to produce with excess capacity, even when prices were falling, and to reduce the relatively high labour costs in order to cover at least a portion of the high fixed costs besides the variable costs.

region. The workforce of these new mines was on the average twice the size of the others and was increasingly recruited from the eastern provinces of the Reich.

The employers found themselves confronted, on the one hand, with a labour force which had become more self-confident since the May strike of 1889 and, on the other hand, with politically motivated government intervention. However, the consolidation of Socialist, Christian, Polish, and Liberal miners' unions did not occur until the mass strike of 1905, in which three-quarters of the labour force held out for three weeks without being able to force the employers into concessions. Promises made by the government to the workers led in both cases to the end of the strikes; in 1912 the third major strike before the First World War failed due to political disunity among the miners' unions. Such promises and the diminishing number of economic options forced the employers finally to protect their powerful position, in the first place, by creating a Rhenish-Westphalian coal syndicate in 1893 as a check on prices, and, in the second place, by establishing a mining association (*Zechenverband*) themselves in 1908 to counter the unions.

Against the backdrop of the developments roughly outlined here, the behaviour of the mining entrepreneurs with regard to industrial relations must be seen as a confrontation with the changing economic and political conditions in the three different stages in the development of Ruhr mining. It would certainly be wrong to try to attribute these stages directly to upheavals in industrial relations. Yet it seems appropriate to ask how far employers' strategies for market and workplace control – later in interaction with union strategies – not only reflected but also effected the great changes in the economic and political constellation of the mining industry by deliberately pressing home their version of industrial relations. It especially needs to be explained how the mining entrepreneurs were able to maintain their relative autonomy in managerial control despite the increasing politicization of industrial relations, and how they managed to refuse to acknowledge the unions until the middle of the First World War by sticking faithfully to this Herr-im-Haus position, thus warding off the collective wage contract in this key sector of German industrialization.

Traditionally, industrial relations in Ruhr mining were based on the state management principle, according to which the Prussian state assumed control not only of the *Bergregal* (mining royalties) but also of the actual direction of the mining operations up until the middle of the nineteenth century. For those miners registered in the lists of the Knappschaft, the state-controlled benefit fund and fraternity of the miners, this meant a degree of security in their living standard that was unattainable not only for other wage labour groups in the early phase of industrialization, but also for the less privileged day labourers in mining. Hiring and firing practices, wage payment and working hours, and the provision for sickness, disability, and old age for these people were all regulated by the authorities. The direct state protection of the legally privileged mining industry also justified, however, the demand for a special loyalty to the employer and an absolute subordination to state disciplinary power, similar to that required of civil servants. This did not stand in the way of the self-confident pride that the Knappschaft comrades took in their profession.[13]

In this traditionally corporative organization of industry and labour, the shareholders in the old cost-book mining companies were only of secondary importance and only gradually grew accustomed to their role as entrepreneurs. In all cases of dispute, the mining authorities ensured that the mining and Knappschaft regulations that they had issued were observed, in which, for example, the working day was stipulated to be eight hours long excluding travel time to and from the pit face – a regulation which would later be heavily contested – and the procedure for additional shifts or penalty measures was expressly stated. Whereas the mining authorities could resort to drastic measures in dealing with Knappschaft members – arbitrary departure was considered to be an act of serious disobedience and was punished with expulsion from the Knappschaft – they had to accept the fact that the state management principle was being undermined to a certain degree by the shareholders, despite repeated affirmations of the state's vested authority. It is true that the members of the Knappschaft also pressed for a relaxation of their corporative wage agreements in view of the good money to be earned in 'free' mining, but only the stockholders were eventually successful in directly taking advantage of the increased demand for coal by recruiting additional day workers for the mines and by dodging the state price settings. Thus, the state management principle was already being disre-

---

13. On this point, see extensively Klaus Tenfelde, *Sozialgeschichte der Bergarbeiter an der Ruhr im 19. Jahrhundert*, Bonn-Bad Godesberg 1977, pp. 63ff.

garded in the marketplace before it was finally abandoned in industrial relations.[14]

The withdrawal of the state from its entrepreneurial function began with the Miteigentümergesetz (co-ownership law) of 12 May 1851, in which the technical and economic management of the pit operations was transferred to the owners. It thus sanctioned the undermining of the state management principle but at first left a degree of governmental protection intact in the regulation of industrial relations. The mining authorities retained the mining rights and thus remained responsible for the enforcement of the safety regulations. As part of the inspection principle (*Inspektionsprinzip*), they also retained the right to set the rate of 'normal pay' as a minimum wage at least for the Knappschaft members, and to supervise their hiring, firing, or transfer. Not until the Knappschaftsgesetz of 1854 and the Freizügigkeitsgesetz (the law guaranteeing the free movement of labour) of 1860 were these relicts of the state management principle in industrial relations also revoked.

Despite the sweeping liberalization of the Prussian mining laws by the General Mining Act (Allgemeines Berggesetz) of 24 June 1865, which instituted the virtual economic independence of the industry until the First World War, these last remnants of the state management principle represented a sort of moral reserve of the labour conditions in Ruhr mining, which could be partially reactivated during the major labour disputes of 1889 and 1905.[15]

With regard to labour law, the position of the mining entrepreneurs towards the labour force in the Freizügigkeitsgesetz was originally modelled on the master–servant relationship as laid down in the regulations for domestic service. When a contract was broken, there was no such conventional penalty as existed in a free contractual relationship; instead disciplinary measures to punish 'disobedience' or 'recalcitrance' were available to the employer. On the other hand, the employer was required by the Knappschaftsgesetz to contribute to a mandatory miners' insurance in the form of a self-administered corporation and no longer as an institutional fund, which forced him to acknowledge the state's interest in the

---

14. See Holtfrerich, *Wirtschaftsgeschichte*, pp. 26ff.
15. See the definitive work by Wolfram Fischer, who emphasizes the obligatory nature of the Prussian mining laws in comparison with the Napoleonic mining laws: W. Fischer, *Wirtschaft und Gesellschaft im Zeitalter der Industrialisierung. Aufsätze, Studien, Vorträge*, Göttingen 1972, pp. 139ff., as well as Tenfelde, *Sozialgeschichte*, pp. 163ff.

social and political protection of the miner against accident, illness and death. Within the framework of self-administration as provided in this law, he was also forced to co-operate on a practical basis with the elected representatives of the Knappschaft.[16] The final break with the traditional corporative labour regulations, which the miners had also urged during the economic upswing of the fifties, eventually came about due to the pressure exerted by an association of mining interests founded in 1858 in the chief mine district of Dortmund as the first indications of crisis became apparent. The Bergbau Verein, one of the very first entrepreneurial associations to be founded, did not merely pursue general economic and political goals early in its history, such as the reduction of freight rates, but also assumed the rudimentary functions of an employers' organization when it called for legislation ensuring 'increased performance and fewer demands' from the labour force.[17]

As a result of the influence of the market, the fronts formed anew during the transition phase, just as they had towards the end of the state management period. The employers considered the one-sided right to give notice while maintaining the work guarantee secured by the Knappschaft, as demanded by the miners, to be unreasonable, and were able to press for the complete abandonment of the pre-modern organization of labour, in view of the 'self-cost crisis', and claim unlimited disciplinary power over their labour force. Theoretically, the 'free' labour contract might even have been in the interests of both the employee and the employer during the economic upswing, especially since the authorities were still responsible for mediation in labour disputes and the confirmation of work regulations (*Arbeitsordnungen*). In the General Mining Act of 1865, these last few provisional safeguards were also finally dropped. The linchpin of the power relations in a company lay in the work regulations, which were drafted solely by the employer and merely had to be presented pro forma to the mining authorities, and which provided a 'general contractual offer' which according to legal fiction was the basis of each concluded contractual agreement.[18]

---

16. See the survey by Gerhard Adelmann, *Die soziale Betriebsverfassung des Ruhrbergbaus vom Anfang des 19. Jahrhunderts bis zum Ersten Weltkrieg*, Bonn 1962, pp. 51ff.

17. See ibid., pp. 53f., cf., for example, the 'Eingabe des Vereins für die bergbaulichen Interessen an das Oberbergamt Dortmund, 22. Juni 1859', in Adelmann, *Quellensammlung*, vol. 1, no. 87, pp. 125ff.

18. See Tenfelde, *Sozialgeschichte*, pp. 261ff.

With this the miner was indeed left defenceless and at the mercy of the arbitrary handling of his employment situation. The employer acquired the disciplinary rights that had once belonged to the mining authorities during the period of state management without being compelled to guarantee the corresponding protection of the earlier miners' laws. The free labour contract was not in a position to replace this protection, as the unsuccessful petition to the king by the miners of Essen in 1867 would show. A quarter of the entire labour force signed the petition against employers who planned arbitrarily to prolong shift hours, force extra shifts and set unfavourable piece-work rates under the threat of firing especially the older miners. The miners' demand to be treated in a humane and dignified way no longer found any support among the authorities: 'They [the mine-owning shareholders] see us only as machines and work instruments without wills, whose labour they can exploit wherever possible to their advantage . . .'[19]

As a result of the Freizügigkeitsgesetz, employers extended the work hours by about an hour for the same pay and made working conditions worse in a variety of ways. One such way was the extensive practice of the much despised *Wagennullen*, in which miners were not paid for loads of coal mixed with a quantity of rock that the employers found unacceptable.[20] Because the authorities were strictly forbidden to prescribe the setting of the total amount of work hours in the work regulations, the employers used every opportunity to prevent the ever longer descent into the mines from subtracting from the work hours at the face. The miners, for their part, demanded (for the first time in the strike demands of the Essen miners in 1872) that the eight-hour day included the descent and ascent into the pits in order to counter the slow but steady extension of the time underground. Despite the favourable economic development, which in view of sky-rocketing coal prices made a wage increase of 25 per cent appear to be reasonable, the initiative of the Essen negotiating commission remained unsuccessful. The Bergbau Verein, which declared itself not to be the competent body, denounced the protest as a 'Jesuit strike' attribut-

19. See 'Eingabe der Essener Bergleute vom 29. Juni 1867', reprinted in Otto Hue, *Die Bergarbeiter. Historische Darstellung der Bergarbeiter-Verhältnisse von der ältesten bis in die neueste Zeit*, vol. 2, Stuttgart 1913, pp. 169f., as well as the 'Bericht des Oberbergamts an das Handelsministerium vom 20. Sept. 1867', in Adelmann, *Quellensammlung*, vol. 1, no. 121, pp. 187ff.
20. See Tenfelde, *Sozialgeschichte*, p. 266.

able to the activities of the Christian Socialist chaplains and praised
the unyielding position of the mine-owners for their patriotic
resistance to the 'pathological labour movement' and the 'terrible
agitation of the Social Democratic Party and in part of the clerical
party'. The latter, in the words of the chairman, Dr Friedrich
Hammacher, could only be dealt with by way of a 'long cultural
battle *(Kulturkampf)*'.[21]

In view of the support that the employers received from the
authorities, who classified local work stoppages as acts of disobedience endangering the state, mining entrepreneurs even felt
themselves to be justified in nearly provoking such strikes in
agreement with neighbouring mines during poor economic times.
In 1877 the larger mines in Dortmund presented a joint, stricter set
of labour ordinances and prevailed in enforcing them against the
striking miners, in part with the help of mass firings and new
hirings.[22] The strike at the mine *Germania* in Dortmund in 1883,
which caused a sensation because of the forceful occupation of the
mine, the police intervention and a spectacular 'insurgent trial', was
also caused by the provocative stance of a manager whose actions
directly confirmed the abusive, dictatorial practices in mining that
had been publicly denounced the year before: the increased usage of
Wagennullen, the subtle changes in piece-work rates by installing
larger mine cars, and the immediate extension of the work hours
without complying with proper notification requirements. Such
gross breaches of contract belonged to the characteristics of the new
'managerial absolutism' that was to supply the moral legitimation
for the first strike of Ruhr miners from all the coalfields during the
economic upswing of 1889.

The mining entrepreneurs had used the opportunity that the state
had offered them. They usurped the authority of the state management principle to legitimize their control over the business operations without assuming patriarchial duties as the state had done.
The Herr-im-Haus standpoint did not grow out of the initial phases
of an autonomous, patriarchial business management with company

21. See 'Bericht über die Dreizehnte ordentliche Generalversammlung des Vereins für die bergbaulichen Interessen vom 9. Juli 1872', pp. 4f., in *Die Entwickelung
des Niederrheinisch-Westfälischen Steinkohlen-Bergbaus in der zweiten Hälfte des 19.
Jahrhunderts*, ed. by the Verein für die bergbaulichen Interessen, Berlin 1904, vol. 12,
pp. 218f. Cf. also Hue, *Bergarbeiter*, vol. 2, pp. 310f.
22. See Tenfelde, *Sozialgeschichte*, pp. 504f. At the mine *Margaretha* the entire
labour force was replaced within a period of six weeks, cf. *Die Entwickelung*, vol. 12,
pp. 220f.

insurance, company housing, and other ways of bonding workers to the company; these were hardly to be found beyond the first generation of the family mines in the Ruhr valley and the phase of economic upswing. The Herr-im-Haus standpoint emerged from the more hierarchically organized mines of the second generation in the 'strike belt' between Essen and Dortmund, in which most of the mine managers who were recruited from state service attempted to enforce their military view of service in agreement with neighbouring mines under the pressure of rising costs.

As early as 1867, the Essen miners complained that the pit foremen were being selected only on their merits as slave drivers and that the mining managers, being 'exclusively theoretical people', had no idea of the oppressive situation experienced by the miner since they themselves were most unlikely to have been in a similar situation. The mine inspectorate concurred with the first point but argued that the second point was unjustified since many of the mining managers, being former Prussian civil servants, had begun their careers by 'learning practical handwork'.[23] The integration of former mining assessors from the Prussian civil service into private mining management undoubtedly led to a close political and societal accord between the state supervisory authorities and the private mining industry, which was in a position to offer quite different perspectives for advancement than could the limited state administration. However, it is questionable whether the impact of the self-image of state civil servants was decisive for the emergence of the Herr-im-Haus standpoint. It is more likely that the backing of the mining authorities enabled the mining assessors employed in private industry to undergo a process of adaptation that ended in the development of a primarily market-oriented, authoritarian style of management in hierarchically-organized, large-scale industry.

In any case, this modern justification of the Herr-im-Haus standpoint became decisive for the labour force even though it was tempered by a patriarchal management policy. This is supported by the fact that, firstly, repeated appeals were made to the supervisory authorities in the expectation of aid true to the patriarchal tradition of state-managed mining, whereas the 'retired' mining assessors were never reminded of their self-image as former civil servants. Secondly, strike behaviour also revealed a certain dis-

23. See Hue, *Bergarbeiter*, vol. 2, p. 171, and Adelmann, *Quellensammlung*, vol. 1, p. 198.

tinguishing factor: of two neighbouring and comparable mines in Essen, one, whose mining director himself came from the ranks of the miners, joined neither the strike of 1872 nor that of 1889; whereas the other mine, which was directed by the authoritarian management of a mining assessor with patriarchal ambitions, entered both strikes. The trade mark of this mining assessor, Krabler, who would later advance to chairman of the board of the Bergbau Verein, did not consist of a management style based on conciliation and mutualism as would have corresponded to the patriarchal tradition. His style of management was based on subordinating the miners to his authority as business manager through institutionalizing hierarchy, bureaucracy and efficiency controls.[24]

Bismarck's turn towards a policy of protective tariffs (1879) corresponded with the pressing ideas of the 'Verein zur Wahrung der gemeinsamen wirtschaftlichen Interessen' (association for the protection of common economic interests) or Langamverein of coal and steel industrialists in the Ruhr district as had the Anti-Socialist Act of 1878. This created an atmosphere in which any sort of insubordination in a company appeared to be a sign of Socialist agitation. The Bergbau Verein prided itself on its co-operation with the police authorities in order to 'eradicate dubious elements'.[25] Even the 'Christian-Social demagoguery' of the miners' unions, which relied on the Bochum Centre Party parliamentarian von Schoreemer-Alst publicly to voice accusations against the mine-owners, was viewed suspiciously, and some employers even thought nothing of controlling as closely as possible the way miners voted in the elections. Due to the discrimination built into the restricted franchise of the Prussian three-class vote, the mine-owners had nothing to fear in the local and state elections. In the Reichstag elections, where universal male suffrage gave them more to worry about, the wave of persecution had an impact: the Social Democrats, who were confined to simple election campaigning, were definitely weakened in the eastern districts of the Ruhr region where they had received up to 15 per cent of the vote in 1877. However, in the first Reichstag elections (1890) following the decision not to prolong the Anti-Socialist Act and following the great May strike, they bounced back to capture 26 per cent of the

24. See Adelmann, *Betriebsverfassung*, pp. 90ff., and Adelmann, *Quellensammlung*, vol. 2, p. 39.
25. Cited in Tenfelde, *Sozialgeschichte*, p. 524.

vote and became a political heavyweight with which the mining entrepreneurs would long have to contend in addition to the ever powerful Centre Party in the Ruhr region.[26]

The first general miners' strike of major proportions, that of May 1889, supported by broadbased Knappschaft agitation, arose from the spontaneous work stoppage of haulers and horse-drivers and included up to 80 per cent of the workforce. For the mining employers, it represented a double challenge. First, it confirmed the end of the fiction of a form of individualized labour relations. The unions consolidated themselves and, following initial disagreements, finally concurred on the general strike demands of 1905. Second, it announced the end of the state's abstinence in sociopolitical affairs within the framework of the so-called New Course (*Neuer Kurs*). The state once again revived at least part of its obligations from the period of its direct management of the mines.[27] The employers reacted especially grudgingly as the Emperor engaged himself, in a theatrical manner, on behalf of his subjects' complaints. He ordered the subordinate authorities to ensure that the mining companies made adequate concessions on the justified interests of the labour force, who had demanded from the Bergbau Verein a 15 per cent wage increase, the eight-hour day including travel time to and from the pit face, and the elimination of extra shifts. In addition to this, the Emperor gave three labour delegates an audience, not without warning them against Social Democratic subversive activities.

Since the chairman of the Bergbau Verein, Dr Hammacher, found himself in a position in which he could evade neither the direct pressure of the head of the Reich nor public opinion, he negotiated a protocol with the labour representatives in Berlin. This protocol evoked extreme disapproval from his colleagues in the Ruhr district not only because of its content, but also because of its conciliatory form as such. In an 'Essen protocol', the Bergbau Verein retracted several of the concessions that had been made,

26. See Tenfelde, *Sozialgeschichte*, pp. 566f.
27. For the 1889 strike see Tenfelde, *Sozialgeschichte*, pp. 573ff.; a view opposing this thesis on the 'rationalization of the labour dispute' can be found in Franz-Josef Brüggemeier, *Leben vor Ort. Ruhrbergleute und Ruhrbergbau 1889–1919*, Munich 1983, pp. 182ff. See also Max Jürgen Koch, *Die Bergarbeiterbewegung im Ruhrgebiet zur Zeit Wilhelm II. (1889–1914)*, Dusseldorf 1954, pp. 33f., and more recently Rudolf Tschirbs, *Die Ruhrunternehmer und der große Streik von 1889,* in Karl Ditt and Dagmar Kift (eds), *1889. Bergarbeiterstreik und Wilhelminische Gesellschaft,* Hagen 1989, pp. 87–112.

especially that of the creation of a workers' committee. (Already in Berlin the labour representatives had given up their demand that travel time in and out of the mines be included in the eight-hour shift.) Instead it proposed negotiating wage concessions individually at each mine. A resumption of the strike, which, especially in the eastern parts of the Ruhr region, had earlier encountered heavy-handed and provocative police and military action, fizzled out due to the exhaustion of the workforce and the unrelenting intervention of the authorities, who incorrectly suspected that the Social Democrats were attempting a takeover.[28]

In the meantime, the Bergbau Verein had organized the forces of the 'counter-revolution' (Kealy): the long-time chairman Hammacher was replaced by the Krupp director Jencke, who shared Krupp's uncompromising stance, and the association itself was undergoing a changing of the guard. The younger mine directors detached themselves from the old mine-owner elite with its connections to bankers and political circles in Berlin and – contrary to the assurances made in the 'Essen protocol' – locally locked out workers who had breached contract. Finally, a strike insurance association, the Ausstands-Sicherungsverband, was formed under the leadership of the young director Emil Kirdorf from the Gelsenkirchener Bergwerks AG. This association linked the compensation claims of the mines affected by strikes to their unwillingness to negotiate. Although this aggressive regulation meant it was not officially recognized, this association was able to sustain itself as effective, local strike protection until the Zechenverband was founded in 1908.[29]

It was not the labour organizations with their tendency to split into fractions who learned a lesson from the May strike, but the employers. They reacted to the threatening public intervention by intensifying their managerial authority and by denying any sort of recognition to collective interest articulation from either the unions or the state. They freed the Bergbau Verein from its chamber-like

28. On the audience and the protocols, cf. Ernst Jüngst, *Festschrift zur Feier des Fünfzigjährigen Bestehens des Vereins für die Bergbaulichen Interessen im Oberbergamtsbezirk Dortmund in Essen*, Essen 1908, pp. 140ff.; as well as *Entwickelung*, vol. 12, pp. 225ff.

29. See *Jahresbericht des Vereins für die bergbaulichen Interessen 1889*, Essen 1890 with the contract of the Ausstands-Sicherungsverband in appendix 5. For the early history of the Zechenverband, cf. Paul Osthold, *Die Geschichte des Zechenverbandes 1908–1933*, Berlin 1934.

relations to government authorities by implementing a network of modern interest organizations, including the cartelization of the entire production within the Rhenish-Westphalian coal syndicate (RWKS), thanks again to the lengthy attempts under Kirdorf's aegis.[30]

Kirdorf had been a trained textile merchant from a bankrupt family business before he switched to the mining industry as an architect of the Gelsenkirchener Bergwerks AG where he was long dependent on the Disconto-Gesellschaft, a Berlin bank. It was precisely this 'dependent' position of his that made him into the most ardent advocate of the rejuvenated Herr-im-Haus standpoint. Unlike in Krabler's case, it cannot be argued that Kirdorf was motivated by an authoritarian tick stemming from the title of mining assessor.[31] This power claim was expressly directed against the tendency of the Prussian Minister of Commerce von Berlepsch to intervene in labour affairs. In mining, only the rudiments ever actually materialized from the state's obligation, as declared in Wilhelm II's February decree, to regulate working conditions in order 'to ensure the preservation of the health, morality and economic needs of workers and their right to equal treatment before the law', and from its admission that legal labour representatives were necessary to 'maintain the peace'.[32] The newly created industrial court for mining in Dortmund, in which employer and employee were each represented by one lay judge, turned out to do little else than initiate the arbitration of labour disputes, while the amendment to the mining law of 24 June 1892 did not produce anything more than the mandatory regulation stipulating that detailed labour ordinances were to be issued which – rather ineffectively – had to be shown to the workers for comment.[33]

Thus, the position of the Bergbau Verein won out on the issue which the new chairman of the association, Director Jencke, had so

30. See Maura Kealy, 'Kampfstrategien der Unternehmerschaft im Ruhrbergbau seit dem Bergarbeiterstreik von 1889', in Hans Mommsen and Ulrich Borsdorf (eds), *Glück auf Kameraden! Die Bergarbeiter und ihre Organisationen in Deutschland*, Cologne 1979, pp. 175–97. On the founding of the syndicate, see Volkmar Muthesius, *Ruhrkohle 1893–1943. Aus der Geschichte des Rheinisch-Westfälische Kohlen-Syndikats*, Essen 1943.
31. See Helmut Böhme, 'Emil Kirdorf, Überlegungen zu einer Unternehmerbiographie', *Tradition* 13, 1968, pp. 282–300, and 14, 1969, pp. 21–48, especially p. 28. On the 'revolt' of the young directors see now Tschirbs, *Ruhrunternehmer*, p. 105.
32. See Adelmann, *Quellensammlung*, vol. 1, no. 233, p. 144.
33. See the justification of the amendment of the mining law in ibid., no. 181, p. 306. On the industrial courts for mining, see *Entwickelung*, vol. 12, pp. 58ff.

vehemently propagated during the consultations of the Central
Association of German Industrialists (Centralverband deutscher
Industrieller) on the trade-regulation amendment. Jencke argued
that participation of the workforce in stipulating work regulations
would be out of the question just as much as state regulation would
be. There could be no 'equality' between 'supervisors' and 'subor-
dinates' because the work regulations were 'the outcome of the
sovereign will of the employer' based on his property rights: 'I
acknowledge neither any right of the state to dictate to the employer the content of the work regulations, nor any right of the
worker to comment on the work regulations in such a way that
needs to be heard before they are issued.' The worker could, of
course, quit should the work regulations be altered to his disadvantage, and 'naturally' sufficient notice would be given; there
would be no changes made from one day to the next.[34]

Further strikes would be necessary before the employers would
be willing to abandon this unconditional version of the Herr-im-
Haus standpoint and would be forced to accept a legal regulation of
work hours and the creation of labour representation in mining.
The strike of 1905 was supported by up to three-quarters of the
labour force, which had doubled since 1890, and lasted a total of
nearly six weeks. The demands made in this strike by the four
miners' unions included the eight-hour shift with travel time in and
out of the mines, worker committees, mine inspectors, minimum
wages, the end of Wagennullen and 'humane treatment' as such.[35]
Characteristically, the strike was prompted by exactly such arbitrary, immediate changes of the work regulations as Jencke had
denied would occur. It took place at second-generation mines
which had since become less profitable and where young directors
– none of them mining assessors – were attempting to balance out
stagnating productivity by making working conditions more stringent. Especially in the case of the arbitrary extension of travel time
to and from the pitface beyond the shift time stipulated in the
normal work regulations of the Bergbau Verein – namely, eight
hours plus half an hour for each descent and ascent – the mine
managements did not always observe the notification rules for

34. See *Verhandlungen, Mittheilungen und Berichte des Centralverbandes deutscher Industrieller*, no. 50, June 1890, pp. 127f.
35. See Koch, *Bergarbeiterbewegung*, pp. 77ff.; see also Hue, *Bergarbeiter*, vol. 2. pp. 575ff., and Heinrich Imbusch, *Arbeitsverhältnis und Arbeitsorganisation im deutschen Bergbau. Eine geschichtliche Darstellung*, Essen, n.d. (1908), pp. 566ff.

extending work hours or adhere to the stipulated period for appeal.[36] As Director Gustav Knepper from the Stinnes mine *Bruchstraße* in Langendreer forced through an extension of the work hours, disregarding both the appeal period and the protest of the workforce, and gaining support from Stinnes for his actions to maintain authority in the mine, a strike broke out from such pent-up aggression that even the leadership of the unions was no longer able to contain it regionally.[37]

The argument with which Krupp had confronted his miners in 1889, namely that negotiations could not be conducted with those in breach of contract,[38] was once again brought up; but it had since lost much of its credibility. It was undeniable that there was something to be said for the reasonableness of other complaints as well: wages lagged behind the rate of economic growth; due to a widespread worm disease, many had been forced to accept a loss of earnings through no fault of their own; pit closures in the Ruhr valley gave a taste of the social consequences of the competition for syndicate quotas, in which by now even the steelmills and the state itself participated.[39] In view of the widespread disapproval of cartels and of the growing importance of mediation and collective wage contracts in other branches of industry, the mine-owners and directors advocating their *Herrenstandpunkt* found themselves confronted with opposition not only from middle-class public opinion but also from the ranks of industry.[40]

Although the state had always equated the protection of the non-striking worker with the protection of public order, now it came out openly on the side of the striking workers. It did this

---

36. See 'Beratung über eine Normal-Arbeitsordnung für Bergwerke', in *Bericht über die 33. Generalversammlung des Vereins für die bergbaulichen Interessen vom 30. Dez. 1891*, pp. 12ff. and 'Fassung vom 28. November 1892', *Glückauf*, 1892, no. 97, pp. 1098ff.

37. See Adelmann, *Betriebsverfassung*, pp. 130ff.; see also Elaine Glovka Spencer, *Management and Labour in Imperial Germany. Ruhr Industrialists as Employers 1896–1914*, New Brunswick 1984, pp. 94ff.

38. See 'Friedrich Krupp an die Arbeiter der Zeche Hannover, 10. Mai 1889', in Adelmann, *Quellensammlung*, vol. 1, no. 128, pp. 213f.

39. On the first efforts towards nationalization, cf. Hans Georg Kirchhoff, *Die staatliche Sozialpolitik im Ruhrbergbau 1871–1914*, Cologne 1958, pp. 131ff.

40. See e.g. Günter Brakelmann, 'Evangelische Pfarrer im Konfliktfeld des Ruhrbergarbeiterstreiks', in Jürgen Reuleucke (ed.), *Fabrik, Familie, Feierabend. Beiträge zur Sozialgeschichte des Alltags im Industriezeitalter*, Wuppertal 1978, pp. 297–314. For the position of the 'syndicate group' in industry generally, cf. Hartmut Kaelble, *Industrielle Interessenpolitik in der Wilhelminischen Gesellschaft. Centralverband deutscher Industrieller 1895–1914*, Berlin 1967, pp. 68ff.

partly from tactical reasons and partly because it was forced to do
so since the employees refused to comply with the government's
insistence on direct negotiations. They argued that it was in principle not possible to negotiate with contract breakers who could
not even command authority within their own ranks.[41] Instead,
they attempted to provoke a military occupation of the Ruhr region
by exaggerating reports of the situation in order to get the government on their side. However, Kirdorf had to submit to being told
by Chancellor von Bülow that such a struggle against the Social
Democrats, which the Bergbau Verein was alleged to be leading in
practice on behalf of the government, was not in the interests of the
'general welfare'.[42] The threat of possible intervention by the Reich
government prompted the Prussian government quickly to an-
nounce an amendment to the mining laws which promised finally
to fulfil the most important demands of the 1889 strike: to prohibit
Wagennullen, to set a normal shift at eight hours with no more
than an additional half-hour for descent and ascent and to establish
obligatory worker committees. Thus, the strike came to a 'happy
end', without having been able to force the mine-owners themselves to make any concessions, let alone to negotiate with the
unions' Commission of Seven (*Siebenerkommission*).

As the collapse of the third major Ruhr miners' strike in 1912
would show (the Christian union (*Gewerkverein*) did not participate, thus making it easier to denounce the strike as a political one)
the government concessions of 1905 represented anything but a
green light for the Social Democrats, from whose mass-strike
agitation the *Alte Verband* completely distanced itself.[43] Nevertheless, the mine-owners portrayed their fight as being a national,
political struggle to defend not only 'authority within the industry'
but also the existence of the Reich itself. Yet the way in which the
workers' committees were set up prompted the Alte Verband to
boycott the elections at first. Voter participation was thus only 11
per cent, and most of the seats went to Christian union representatives. Even the employers' apprehensions of a politicization of the
elected safety controllers – legally established following a mine
disaster in 1909 at Radbod in which 344 died – remained unwar-

41. See Adelmann, *Quellensammlung*, vol. 1, no. 156, pp. 255f.
42. See Klaus Saul, 'Staatsintervention und Arbeitskampf im Wilhelminischen
Reich 1904–1914', in Hans-Ulrich Wehler (ed.), *Sozialgeschichte heute. Festschrift für
Hans Rosenberg zum 70. Geburtstag*, Göttingen 1974, pp. 479–94, pp. 488f.
43. See Koch, *Bergarbeiterbewegung*, pp. 101ff.

ranted. The majority of the labour force at first participated in the elections, in which the Alte Verband won about a third of the seats, but quickly lost interest in the controllers, who were directly dependent on those they were to control.[44]

The lack of public support during the 1905 strike bitterly disappointed the employers. Since they denied each and every abuse, the strike, which came on them like 'a thief in the night', must have appeared to them to be an epidemic spread by 'rabble-rousers'. And it remained a total enigma to them as to why the government, under pressure from public opinion, conceded to these 'rabble-rousers', thus arousing everywhere the 'greediness of the workforces'.[45] They proceeded on two fronts to strengthen their position against recognizing the unions, and this position became even more entrenched during the strike. First, they increased their control of working practices with the help of bureaucratic procedures. Even the autonomous, influential position of the pit foremen, which had been occasion for repeated confrontations, became subjected to the more impersonal criteria of economic efficiency. Second, they attempted with the help of the Zechenverband and its labour exchange to control the labour market better in order to master the great degree of fluctuation in personnel – perhaps the most effective form of 'individual bargaining' before unionization. This strategy was reinforced from within by the intensification of company welfare policy and from without by a consequent opposition to collective wage contracts in the mining industry.

Thus, industrial relations in mining finally stepped out of the long shadow of the state management principle. The inherited Herr-im-Haus standpoint had to be legitimized anew to the degree to which the labour force also turned away from the traditional forms of interest articulation. It is true that the filing of complaints and the petitioning, which still reflected the old expectations of state protection for conciliatory industrial relations, were to mobilize the new form of state intervention well into the twentieth

44. See Adelmann, *Betriebsverfassung*, pp. 134ff. and pp. 145ff.; the supervising authorities were also sceptical, cf. Adelmann, *Quellensammlung*, vol. 1, no. 253, pp. 407ff. See the apprehensions of the Zechenverband (*Jahresbericht* 1910, pp. 4ff.) and the rather modest outcome, in Evely Kroker, 'Arbeiterausschüsse im Ruhrbergbau zwischen 1906 und 1914', *Der Anschnitt* 29/30, 1977/78, pp. 204–15.
45. See the position of the chairman of the Bergbau Verein, Krabler, in *47. ordentliche Generalversammlung des Vereins für die bergbaulichen Interessen*, 2 June 1905, p. 6ff. and the *Jahresbericht für das Jahr 1905*, Essen 1906, pp. 5ff., for the consequences.

century.[46] But the modern miners' movement that emerged from
the mass strikes of the Wilhemine period did not gain its strength
from this traditional rendering of emerging class consciousness,
which even the Socialist Alte Verband could not fully do without;
instead it was rooted primarily in the experience of relatively
autonomous working practices and a considerably self-determined
working-class culture, a fact that the mining employers indirectly
acknowledged with their new strategy.[47]

The labour force, working in small crews at piece-rates, was
bound into an intricate network of mutual dependence despite the
continually growing size of the operational units with an ever more
heterogeneous composition. The distance of the miner from the
mining management increased by giant strides: the average workforce doubled each decade from about 400 men in 1880 to over
2,000 men before the First World War.[48] The miner's work underground during this period remained primarily semi-skilled manual
labour and relatively unaffected by the changes in the methods of
extraction. The replacement of the pillar and stall system by the
longwall system, beginning in the eighties, with the corresponding
concentration of operations and the mechanization of transportation did, however, prepare the ground in the long term for the
disbanding of crews and the deskilling of hewers. Over the medium term, this reorganization at least helped to improve the
supervisory function of the pit foremen.[49] Although these men
were increasingly and deliberately distanced from the mine management as in the case of the Stinnes mines, and although they were
supervised by offices above ground and occasionally sought refuge
in union-like alliances, they were constantly in conflict with the
crews underground. The cost-conscious integration of the pit

46. See Tenfelde, *Sozialgeschichte*, pp. 397ff. and 608ff. See also the documentation of the old tradition of petitioning in mining in Klaus Tenfelde and Helmuth
Trischler (eds), *Bis vor die Stufen des Throns. Bittschriften und Beschwerden von Bergar-
beitern*, Munich 1986.
47. On this point, see especially Brüggemeier, *Leben vor Ort*.
48. Cf. the figures in Adelmann, *Quellensammlung*, vol. 1, no. 97, pp. 143ff.
49. See Klaus Tenfelde, 'Der bergmännische Arbeitsplatz während der Hochin-
dustrialisierung (1890–1914)', in Werner Conze and Ulrich Engelhardt (eds), *Arbeiter im Industrialisierungsprozeß. Herkunft, Lage und Verhalten*, Stuttgart 1979, pp.
283–335, as well as K. Tenfelde and Imgard Steinisch, 'Technischer Wandel und
soziale Anpassung in der deutschen Schwerindustrie während des 19. und 20.
Jahrhunderts', *AfS* 28, 1988, pp. 27–74, 31ff.; see also Wolfhard Weber, 'Der
Arbeitsplatz in einem expandierenden Berufszweig: Der Bergmann', in Reulecke
(ed.), *Fabrik, Familie, Feierabend*, pp. 89–113.

foreman in company 'quota systems' led inevitably to a worsening
of the work atmosphere, which the crews at the face could counter
only by maintaining their collective autonomy in working practices
and wage negotiations.[50]

Hence, the work experience in the mines created its own culture
of self-protection and self-confidence, independent of class tradition, and it was reflected in the 'informal structures of solidarity'
(Brüggemeier) of daily community life. What appeared to the
supervising authorities to be threatening indications of the chaos
and immorality of a self-made 'Wild West' – namely the high
degree of mobility, the 'half-open family structure', the weakness
of the bonds of traditional means of socialization in a workforce
often recruited from rural labour conditions – was in reality the
breeding ground for a relatively autonomous working-class culture. It was not, however, automatically dominated by socialist
ideas. Instead it reflected a multitude of cultural backgrounds
differentiated by national, regional, religious, generational and
political patterns. The employers may have been at odds with the
authorities about the desirability of military intervention in strikes,
but in the public policing of the industrial labour force they could
feel thoroughly committed to a broad 'anti-democratic alliance'
with the local authorities.[51]

Within the company itself, however, it required exertions of a
different kind to improve social control over the workforce, which
was increasingly backed by the consolidating miners' unions. The
cumulative company welfare policy that was established following
the 1905 strike was meant not only to set off and supplement the
intensified disciplinary intervention of the unions in the labour
process, but also to legitimate a general and moral claim to leadership using a counterplan of a 'works culture'. This form of pa-

50. See Spencer, *Management*, p. 86ff., as well as Spencer, 'Between Capital and
Labour: Supervisory Personnel in Ruhr Heavy Industry before 1914', *Journal of
Social History* 9, 1975, pp. 178–92; see also S. H. F. Hickey, *Workers in Imperial
Germany. The Miners in the Ruhr*, Oxford 1985, pp. 156ff. On disciplinary measures
and the functional differentiation of the pit foremen according to the 'Stinnes-
system', cf. now in detail Helmuth Trischler, *Steiger im deutschen Bergbau. Zur
Sozialgeschichte der technischen Angestellten 1815–1945*, Munich 1988, pp. 42ff., 72ff.,
and Trischler, 'Steiger und Bergleute. Betriebliches Sozialverhalten während der
Industrialisierung', in Gustav Schmidt (ed.), *Bergbau in Großbritannien und im Ruhrgebiet. Studien zur vergleichenden Geschichte des Bergbaus 1850–1930*, Bochum 1985, pp.
147–98.

51. See H. G. Spencer, 'Businessmen, Bureaucrats and Social Control in the
Ruhr 1896–1914', in Wehler (ed.), *Sozialgeschichte*, pp. 452–66, p. 463.

ternalism did not grow out of the corporative mining tradition as
had the authoritarian paternalism of the state administration.
Rather, it corresponded to a model of a 'derived', 'secondary'
paternalism whose 'managerial absolutism' and 'company welfare
policy' – unlike the traditional Herr-im-Haus position – were in
keeping, on the one hand, with the stipulations of a bureaucratic,
authoritarian organization in a large-scale enterprise, but, on the
other hand, were also suited to the necessities of an increasingly
critical crisis of legitimation.[52]

The efforts of company welfare policy, which created a social
bureaucracy in all large-scale industry, concentrated on developing
company and family insurance schemes, company housing, company training and careers, and company retirement benefits. At the
same time it also often included efforts to encourage loyalty to the
company at the expense of the unions by supporting moves to
establish factory clubs and associations.[53] In this way, the employers used the company family health insurance and factory relief
fund, for example, into which the money from fines was channelled
following the end of the Wagennullen, as a means of counteracting at
factory level the increased influence of the unions caused by the
amendment of the Knappschaft law in 1906. This amendment – valid
as of 1912 – guaranteed secret elections for the Knappschaft aldermen,
labour's freedom of movement, and stronger governmental control
of standards.[54] The intentions of the employers became most obvious
in company housing, even if the direct threat to throw out those
breaching contract was actually rarely carried out. By way of an
accelerated development of mine settlements, it was possible to
house 11.9 per cent, 21.2 per cent and 34.9 per cent of the workers
in company housing in the years 1893, 1901 and 1914 respectively.
Due to their greater need for migrant workers, the northern pits

52. On this point, cf. Ludwig H. A. Geck, *Die sozialen Arbeitsverhältnisse im
Wandel der Zeit. Eine geschichtliche Einführung in die Betriebssoziologie*, Berlin 1931, pp.
59ff., 130ff., and in connection with this, see the example of Siemens: Jürgen Kocka,
'Management und Angestellte im Unternehmen der Industriellen Revolution', in
Rudolf Braun et al. (eds), *Gesellschaft in der Industriellen Revolution*, Cologne 1973,
pp. 162–201. See also Tenfelde, *Sozialgeschichte*, pp. 336f., and Tenfelde, *Technischer
Wandel*, pp. 42f.
53. On this point, cf. in detail Rudolf Schwenger, *Die betriebliche Sozialpolitik im
Ruhrkohlebergbau*, Munich 1932.
54. See Martin H. Geyer, 'Staatliche Sozialpolitik und Knappschaftsreformen in
Deutschland 1880–1910', in Schmidt (ed.), *Bergbau*, pp. 96–146; see also M. H.
Geyer, *Die Reichsknappschaft. Versicherungsreformen und Sozialpolitik im Bergbau
1900–1945*, Munich 1987, pp. 23–69.

proved to have a rate in the self-provision of housing nearly twice as high as their southern counterparts.[55] These measures eventually led to a slowdown in the high rate of turnover, at least within the core of the labour force. For example, the personnel turnover at the Essen mine *Prosper* was more than 100 per cent in the year 1900, whereas among the tenants of company houses and flats (nearly 22 per cent of those employed), the rate was only about 10 per cent.[56] Until the First World War, this form of 'mine running' continued. Beginning in the 1890s, the annual average number of job changes clearly surpassed – in the northern district more than in the southern one – that of the manpower level of the respective mines.[57] Even the labour exchange of the Zechenverband, established in 1909 and opposed by the unions due to its obvious use for disciplinary purposes, could not change this situation in any way. 'Black lists' were declared to be in conflict with national policy and public morals and had to be withdrawn following the fierce debates in the Reichstag in 1909 and a test case trial. The waiting period for those breaching contract was reduced from six months to two weeks, and it was thoroughly in the interests of the member mines not to withhold the permits required for job changes from the masses of the workforce even without granting parity status to the unions in labour exchanges.[58] According to the Zechenverband, the number of the successful job changes registered by its labour exchanges for the entire labour force in Ruhr mining rose from 46.7 per cent in 1910 to 74.4 per cent in 1913. The registered number of breaches of contract also rose in this period from 3.1 per cent to 6.2 per cent.[59] Control of the labour market would prove to be just as unattainable a goal for entrepreneurial ambitions as was the control of the workplace.

The founding of the company unions, which united nationwide

---

55. See Brüggemeir, *Leben vor Ort*, table 4, p. 274. The degree of self-provision corresponded roughly to the percentage of migrants from the east; cf. Christoph Kleßmann, *Polnische Bergarbeiter im Ruhrgebiet 1870–1945. Soziale Integration und nationale Subkultur einer Minderheit in der deutschen Industriegesellschaft*, Göttingen 1978, table 11, p. 270.

56. See Adelmann, *Betriebsverfassung*, p. 173, for further examples. According to another source, 53.8 per cent of the miners left their positions in 1902, but only 7.9 per cent of those renting company housing left; cf. Spencer, *Management*, p. 74.

57. Adelmann, *Die soziale Betriebsverfassung*, p. 155, and Adelmann, *Quellensammlung*, vol. 1, p. 145f. See also *Entwickelung*, vol. 12, pp. 52f.

58. See Osthold, *Zechenverband*, pp. 64ff.

59. Calculated on the basis of the annual reports of the Zechenverband for the years 1910 and 1913.

in non-strike, 'yellow' umbrella organizations, did not fulfil its expectations. The efforts of the Krupp director, Alfred Hugenberg, who took over the direction of the Zechenverband in 1912, only produced a certain degree of success after the double shock of 1912. The first shock was the 'three-league strike' (*Dreibund-Streik*) which, however, ran into immediate trouble because the Christian miners' union refused to join in. In this case the employers reacted for the first time by making drastic deductions in strikers' pay. The second shock was the outcome of the 1912 Reichstag elections in which the Social Democrats came head of the polls. In the period between these events and 1913, about 9 per cent of the mining labour force were incorporated into the company unions, especially from among the ranks of the white-collar employees.[60] But the majority of the workforce, of which 36.7 per cent – in the northern district even more than 50 per cent – were by now recruited from the eastern provinces,[61] could not be bought off from using either the individual form or the collective form of interest articulation. The membership in the Socialist, Christian and Polish miners' unions steadily increased in the years preceding the First World War, except following the disappointing strike of 1912.[62] However, the Alte Verband could not maintain the high level of membership that it had enjoyed in 1905, namely 29.4 per cent of the labour force; by 1912 it had fallen to 17.3 per cent. Combined with the 30,000 organized Polish mine-workers and the 40,000 members of the Christian union, the 70,000 members of the Alte Verband still comprised about 36 per cent of the entire labour force and somewhat more than half of those working underground.[63]

Thus, it was all the more important for the employers in heavy industry to stick to the most vital part of their defence strategy, namely to refuse to recognize the unions and to oppose collective wage contracts. In the meantime, a rapid process of concentration and consolidation within the industry in a race for the syndicate quotas had begun, favoured by advantageous yield. To compete

60. See Klaus Mattheier, *Die Gelben. Nationale Arbeiter zwischen Wirtschaftsfrieden und Streik*, Düsseldorf 1973, pp. 201f.

61. On the difficult process of socially integrating the 'Poles' and the political consequences of the national subculture in the Ruhr region, see Kleßmann, *Polnische Bergarbeiter*, passim and table 5, p. 265, and table 11, p. 270; see also Koch, *Bergarbeiterbewegung*, table 13, p. 147.

62. See Koch, *Bergarbeiterbewegung*, table 7, pp. 142f., and Brüggemeier, *Leben vor Ort*, table 11a and b, pp. 282f.

63. See Hickey, *Miners*, table 11, pp. 234, 240.

with the steel works, the large mines invested in large coking plants
and facilities for by-products in the hope of increasing their profit
through diversification. The steel producers obtained their own
mines in order to secure a source of coal independent of the
syndicate. Even the 'pure' mining companies such as Kirdorf's
Gelsenkirchener Bergwerks AG sought ways to expand within the
coal and steel industry, in order to prevent the syndicate from being
destroyed by the differences between foundry mines and 'pure'
mines.[64] The four largest mining companies, Gelsenkirchen, Harpen, Hibernia and Phoenix, supplied the market with about a third
of the entire coal production in 1907; the foundry mines produced a
percentage just as large.[65] In this way, the question of collective
wage contracts actually affected the entire complex of the coal and
steel industry, yet it was still argued that such contracts were not
possible due to particularities specific to mining.

Based on the fact that the average net income in neighbouring
mines differed from one another by 10 per cent to 15 per cent – one
of the central reasons for 'mine running' – and that, on the whole,
wages reacted to price changes, the Zechenverband came to the
conclusion that wages would resist any 'arbitrary manipulation' in
the long run.[66] In order to justify the infeasibility of collective wage
contracts in Ruhr mining, it was pointed out that the coal seams
were supposedly irregular in comparison to England and that
indeed the entire piece-work set-up did not adapt itself to such
contracts.[67] There was certainly more at stake as the Zechenverband declared itself not to be the party responsible for discussing
wage issues. The young business manager of the Zechenverband,
the mining assessor von und zu Loewenstein, who would run the
mining associations well into the Third Reich, warned of the
danger that the collective wage contract would bring. In his view,
it was no more than a 'fixed idea', a legalistic 'soap bubble', a

64. See Muthesius, *Ruhrkohle*, pp. 78ff. See also Winfried Feldenkirchen, *Die Eisen- und Stahlindustrie des Ruhrgebiets 1879–1914. Wachstum, Finanzierung und Struktur ihrer Großunternehmen*, Wiesbaden 1982, pp. 129ff.
65. See Jüngst, *Festschrift*, pp. 17f., and Adelmann, *Betriebsverfassung*, p. 151.
66. See Osthold, *Zechenverband*, pp. 93ff.
67. See mining assessor Hilgenstock, 'Lohntarife im britischen und rheinisch-westfälischen Steinkohlenbergbau', *Glückauf* 43, 1907, pp. 1625ff., 1677ff., 1705ff., 1741ff. For another view, see Director Brauns, 'Die Möglichkeit von Tarifverträgen im Ruhrbergbau', *Soziale Praxis* 17, 1908, 1. 593ff. and 617ff., as well as the reply by mining assessor Hilger, 'Die inneren Grenzen des Tarifvertrags unter besonderer Berücksichtigung des Bergbaus', *Glückauf* 44, 1909, pp. 1396ff., 1432ff., 1466ff., 1492ff.

'power factor' of the Social Democrats to 'oppress and ravage national and monarchical elements' and to advance the decline of the entire institution of the state.[68] Negotiation with labour organizations, explained Emil Kirdorf in his defence of the cartels to the Verein für Sozialpolitik (Association for Social Policy), was out of the question not only because collective wage contracts were detrimental to production yields, but also because their real goal was 'to fight for control of, or more specifically for the destruction of, the entire economic prosperity of our industry'.[69]

The position of mining entrepreneurs on collective wage contracts represented one of their most fundamental political convictions.[70] Although they had to resign themselves to the fact that the unions existed, they attempted to find pragmatic means of 'containing' them by way of company welfare policy, and they were indeed rather successful with regard to the Christian union, which since 1909 had been indirectly linked to the government through its association with the Centre Party. These efforts then paid off during the Dreibund-streik of 1912.[71] On the question of collective wage contracts, however, they remained firm. Together with their colleagues from the iron and steel industries, they unequivocally defined the position of the leading industrial organizations that they dominated, although in 1910 the association members from medium-sized enterprises in the general association of German metal industrialists (Gesamtverband deutscher Metallindustrieller) were forced to conclude a collective wage contract for the shipyards. Unlike its free-trade rival, the Bund der Industriellen (the Industrialists' League), the Centralverband deutscher Industrieller (Central Association of German Industrialists) remained

68. See *Geschäftsbericht auf der 3. ordentlichen Hauptversammlung des Zechen-Verbandes 1910*, pp. 7ff.

69. See Kirdorf's commentary on Gustav Schmoller's paper on the same topic: 'Das Verhältnis der Kartelle zum Staate', in *Verhandlungen des Vereins für Socialpolitik in Mannheim 1905*, Leipzig 1906, pp. 272ff., especially p. 289.

70. Therefore the statement by Gustav Schmoller is not very convincing when he said of Emil Kirdorf that in May 1905 he was on the verge of negotiating a collective wage contract for miners with the Social Democrats. Schmoller probably wanted to signal Kirdorf's willingness to compromise to Lujo Brentano, who advocated a state system of collective contracts with offices of arbitration. However, before the conference of the Verein für Socialpolitik, Kirdorf showed no signs of such a willingness. Cf. Lujo Brentano, *Mein Leben im Kampf um die soziale Entwicklung Deutschlands*, Jena 1931, pp. 252f.

71. See Elaine G. Spencer, 'Employer Response to Unionism: Ruhr Coal Industrialists before 1914', *Journal of Modern History* 48, 1976, pp. 397–412.

firm in its opposition to the 'parity moralism of class struggle' up until the First World War. It condemned any initiative supporting the idea of collective wage contracts, be it from academic social reformers (*Kathedersozialisten*) or from the Bavarian government. It also engaged itself on the political level to promote a front against the Socialists by way of a *Sammlungspolitik*. Although the directors of the syndicate kept a low profile in this matter, because of their independent power in labour struggles and the anti-cartel stance of the Bund der Landwirte (Agriculturalists' League) among other factors, their political strategy against collective wage contracts was thoroughly successful until the First World War. Whereas collective wage contracts existed for about half of the workers in the construction and printing industries, and for about a fifth of those in other mid-sized branches of business, labour relations in heavy industry remained completely free from any such contracts.[72]

The syndicate protection also paid off financially, although it had to be defended against growing public criticism. One especially important effort was to block the Prussian attempt to take over the Hibernia AG, an attempt which was linked not only to the welfare traditon of state intervention in mining, but was also supposed to counter trends towards creating trusts similar to those in America on the grounds of national security interests.[73] What resulted was a 'spite-trust' of private owners that thwarted the takeover, prevented the state from exercising any influence on the management of Hibernia despite its 46 per cent participating interest, and left it to develop new fields, where it could not be dangerous to the syndicate as an outsider. (Its share of Ruhr production rose from 0.5 per cent in 1903 to 4.5 per cent in 1913.) Not until the threat of a government-enforced syndicate cast its shadow in the war economy of 1916 was it possible for the state to impose its share of influence.[74]

Despite the ever-smouldering conflict with the foundry mines, the syndicate had paid off for the mining entrepreneurs. Generally, the accusation of 'profiteering' at the expense of both consumers and workers was refuted, the price increases were justified by the

72. See Peter Ullman, *Tarifverträge und Tarifpolitik in Deutschland bis 1914*, Frankfurt 1977, pp. 171ff. and table 7, p. 228. Cf. also Kaelble, *Interessenpolitik*, pp. 76ff.

73. See Charles Medalen, 'State Monopoly Capitalism in Germany: The Hibernia Affair', *Past and Present* 78, 1978, pp. 82–112.

74. See Friedrich Schunder, *Tradition und Fortschritt. Hundert Jahre Gemeinschaftsarbeit im Ruhrbergbau*, Stuttgart 1959, pp. 219f.

continually increasing wages in the Ruhr region as in other mining districts, and it was pointed out that dividends had remained approximately the same measured against the proceeds of sales and corporate capital. In the case of the Gelsenkirchener Bergwerks AG, it was claimed that dividends had dropped since the founding of the syndicate to a level that was nearly 30 per cent per ton below that of the twenty-year period preceding the syndicate, and the net profit had even fallen by 40 per cent.[75] Yet, in addition to the average dividend of about 12 per cent, in exceptional cases up to 80 per cent, considerable capital increases should also be calculated into this; so should the increased amount of depreciation allowance, which in Kirdorf's estimation gave the mining companies the equity capitalization they needed to become more or less independent of the banks for the first time. Between 1893 and 1911, production doubled while the gross profit and issued dividends of the mining stock companies increased nearly fivefold; the dividend paid on each ton produced increased from 0.59 marks in 1893 to 1.34 marks in 1911.[76]

The assertion that Kirdorf made to the Verein für Sozialpolitik that the syndicate issue and the labour issue had 'nothing to do with one another as such'[77] was in any case hardly credible. The RWKS, which controlled over 97 per cent of the Ruhr production and about 60 per cent of the Prussian hard coal in 1905,[78] functioned as the price setter in the 'uncontested' area, as well as for the inferior Saar coal, of which 75 per cent was in state hands. It could also react to the English and Silesian competition in the 'contested' areas along the coast and east of the Elbe by price reductions. The extent of its monopoly power was evident, among other things, during periods of recession when, in order to keep prices up, it cut back production (1901) or reduced wages and the number of shifts (1908). It was therefore the topic of repeated debates in the Reichstag, in which a broad coalition ranging from the Social Democrats to the Conservatives unsuccessfully demanded, up to the First

75. See Ernst Jüngst, 'Arbeitslohn und Unternehmensgewinn im rheinischwestfälischen Steinkohlenbergbau', Essen, n.d. (1906), as a special reprint in *Glückauf* 42, 1906.
76. See Brüggemeier, *Leben vor Ort*, p. 90.
77. *Verhandlungen des Vereins für Socialpolitik 1905*, p. 285. The main lectures were held by Brentano and Schmoller deliberately on these two topics: 'Das Arbeitsverhältnis in den privaten Riesenbetrieben' and 'Das Verhältnis der Kartelle zum Staat'.
78. The upper Silesian share equalled 24 per cent, that of the Saar coal was 9.5 per cent. Cf. *Statistisches Heft des Bergbauvereins*, 1912, no. 4, p. 4.

Germany

149

World War, that state cartel supervision be instituted.[79] The political masterminds of the Social Democrats drew the conclusion from the miners' strike of 1905 that a fight against the syndicated mine-owners could not be won by union means alone, and that because of this situation, each strike had to become a political strike – independent of the revolutionary character of the mass strike.[80] One could also conclude, as had Max Weber, that it would be in the profit interests of the syndicate if the authoritarian standpoint of the industrialists were to encourage a Social Democratic radicalization which would in turn cripple the initiatives of social reform. In any case, for the employers in mining the syndicate was on both counts not only an economic facility, but also the expression and guarantee of a specific form of industrial relations. For Max Weber, it made no difference whether state or private monopoly reflected those 'German traditions' that he saw at work in the giant mining enterprises, namely those of 'the thrills of philistine overlords' (spießbürgerliche Herrenkitzel) and the 'policeman jargon' (Schutzmannsjargon), seeking not only power but 'the showing off of power'. As opposed to foreign entrepreneurs who were contemplating the demands for collective wage contracts, in 'these gentlemen' beat 'the heart of a policeman'.[81]

It might be that these 'German traditions' in mining were more strongly pronounced as a result of the after-effects of the state management principle. When making international comparisons, it is especially easy to underscore the role of the state in the process of delayed industrialization.[82] Certainly the great social homogeneity and bureaucratic mentality of the primarily Protestant caste of mining assessors contributed to the establishment of the authoritarian style of management in mining. A growing percentage of this elite, who were schooled in the tradition of the Prussian civil

79. See Fritz Blaich, *Kartell- und Monopolpolitik im kaiserlichen Deutschland. Das Problem der Marktmacht im deutschen Reichstag zwischen 1879 und 1914*, Düsseldorf 1973, pp. 92ff.
80. See Karl Kautsky, 'Die Lehren des Bergarbeiterstreiks', *Die Neue Zeit* 23, 1904/5, no. 24, pp. 772–82.
81. See Max Weber's contributions to this debate at the conference of the Verein für Socialpolitik in 1905, in M. Weber, *Gesammelte Aufsätze zur Soziologie und Sozialpolitik*, Tübingen 1924, pp. 394ff., especially pp. 396f., 404f. Max Weber, who had been referring to the conditions in mining in the Saar region, also then replied to Kirdorf's defence of the cartels.
82. See Werner Berg, *Wirtschaft und Gesellschaft in Deutschland und Großbritannien im Übergang zum 'organisierten Kapitalismus'. Unternehmer, Angestellte, Arbeiter und Staat im Steinkohlenbergbau des Ruhrgebiets und von Südwales 1850–1914*, Berlin 1984.

service, went directly into the private sector of mining (in the 1880s, this was true for 10 per cent of the trainees, in the 1890s for 13 per cent, and in the 1900s for 21 per cent) and assumed there leading positions.[83] However, they had to share their privileged position with mining directors, such as Gustav Knepper or Ernst Tengelmann, who had been recruited from the level of business management and who often proved to be the driving force behind confrontations like those in 1905. At the same time, in some exceptional cases, certain individual mining assessors also took on conciliatory tones, much to the aggravation of their colleagues; two examples were Friedrich Trippe, who advocated negotiations following the 1905 strike, and Otto Krawehl, who made the concession in the 1912 strike of granting his workers paid holiday for the first time.[84]

The exceptional severity of the industrial relations in mining prior to the First World War thus appears to be due less to the pre-industrial residue of the mentality of mining assessors than it was to the general need for the 'ascertainment of advancement' of a professional elite of business managers in the modern, large-scale firm.[85] In any case, Kirdorf's vision of a large-scale industry free of unions came true in the iron and steel industry as well, where the academically-trained engineers in metallurgy also began to create a self-confident, professional elite. Academic training was still less widespread in the leading positions of the iron industry than it was in mining, and it lacked a certain governmental aura, but it confirmed the general trend towards the professionalization of business management.[86] In addition, the mining assessors forfeited a portion of their prominent entrepreneurial position as a result of the cartelization and the tie-in with steel, even though they continued to act quite independently on the 'coal-side' of the combines. This differentiating Ruhr elite drew the legitimation of its fundamental authoritarian stance from a wide range of societal models: more

83. See Bernd Faulenbach, 'Die Preußischen Bergassessoren im Ruhrbergbau. Unternehmermentalität zwischen Obrigkeitsstaat und Privatindustrie', in *Mentalitäten und Lebensverhältnisse. Beispiele aus der Sozialgeschichte der Neuzeit (Festschrift für Rudolf Vierhaus zum 60. Geburtstag)*, Göttingen 1982, pp. 225–42; see also Helmuth Croon, 'Die wirtschaftlichen Führungsschichten des Ruhrgebietes in der Zeit von 1890 bis 1933', *Blätter für deutsche Landesgeschichte* 108, 1972, pp. 144–59.
84. See Spencer, *Response*.
85. See the ambivalent attitude of Berg, *Wirtschaft*, p. 413.
86. See Toni Pierenkemper, *Die westfälischen Schwerindustriellen 1852–1913. Soziale Struktur und unternehmerischer Erfolg*, Göttingen 1979, pp. 54ff.

than by the specific socialization experienced by mining assessors, it was motivated by the generational outlook of those elites who had seen the 'Great Depression' and the period of the founding of the Reich, the political identification with the honorary member of the Bergbau Verein, von Bismarck, against the 'folk without a fatherland', and the professional ethics of a new generation of bureaucratic managers in the large-scale industrial enterprises.[87] In any case the relative affinity with the state and the conception of service of the mining leadership cadre did not lead the mining assessors to accept the state's efforts at mediation during the great miners' strikes. On the contrary, industrial relations in mining had finally stepped out of the long shadow of the state management principle. The arbitrary, legalized privileges of the employers in the actual stipulation of labour conditions, the substantial core of the Herr-im-Haus standpoint, by which the major labour disputes in Ruhr mining had been sparked, now had to be protected not only from the demands of the young union movement for the regulation of industrial relations through collective wage contracts, but also from the threats of state intervention in the social order of the company and in the monopolized order of the market.

On both counts, the mining employers remained undefeated until the First World War. Their strength was the weakness of the unions, which, because they were politically divided and not always sure of the highly mobile and autonomous workforce, could only expect concessions from the state. In this regard, the groundwork for the 'tripartite structure of industrial relations' so typical of mining was not prepared by the traditional 'pit militarism' (Götz Briefs), but by the modern adaption of the Herr-im-Haus standpoint based on secondary paternalism and monopolized market control. The unions were compelled in the long run on to that specific 'German path' of politically and legally securing their position in industrial relations less by the pre-liberal tradition of the state than by the deliberate ostracism on the part of the mining employers. The employers were successful in their refusal to acknowledge these consequences of the system they themselves had induced. Yet war conditions left them without a choice.[88]

87. See Elaine G. Spencer, 'Rulers of the Ruhr: Leadership and Authority in German Big Business before 1914', *Business History Review* 53, 1979, pp. 40–64.
88. See Hickey, *Workers*, pp. 288.ff.

The Forced Partnership: From Conditional Co-operation to the Break-up of Collective Bargaining

Mining employers were forced to abandon two basic positions during World War I. First, their refusal to recognize the unions could no longer be maintained because there had been a party truce, the *Burgfrieden*, and the unions had fallen in line behind the national war effort. Second, the government could no longer be kept out of the syndicate since private industry's control of the disposition of such a national resource as coal had, in principle, been called into question by war economy measures. The leaders of Ruhr mining considered neither of these concessions to be binding in the long run. Thus, they bowed unwillingly to the supposedly blackmailing pressure of political forces and, as it was to be feared that the war would end badly, even began to blame the coming defeat on these political concessions. Their inclusion into the network of war economy agreements with the unions and the state, which would become the basis of the special status of the coal industry and of industrial relations on the whole in the Weimar Republic, was thus linked from the very beginning to one condition: the return to the supposedly apolitical shaping of labour and market relations in mining. This appeared to the employers not only to be an economic necessity, but a national duty.

Of all the conditions taken to secure markets in the war economy, the participation of the Prussian state in the syndicate was the easiest to bear. In Upper Silesia the state was already participating (about 15 per cent) in the production quota agreement dominated by the local 'noble families'. In the Saar district, which was continually losing ground to the Ruhr region, the state – with its monopoly – took over the leading role in price-setting based on the RWKS. Thanks to the defensive measures taken by private industry, that is, by the shareholders and the syndicate itself, the state was blocked from obtaining the 4 per cent needed to take over the Hibernia AG, the third largest 'pure' mining company after the Harpener Bergbau AG and the Hanielsche Zechen, and thus from establishing itself in the heart of the syndicate.[89] The RWKS had also to face the fact that the foundry mines had a further advantage.

89. See Medalen, 'State Monopoly Capitalism' and Muthesius, *Ruhrkohle*, pp. 133ff.

They could expand a plant's own consumption, not included in the production quota, by buying mining operations in addition and by incorporating processing plants and could fix quotas on the basis of the positive business cycle of 1907. Their share of syndicate production rose from 19.5 per cent in 1904 to 30.5 per cent in 1912.[90] As the issue of continuing the syndicate was being debated, the 'spite-trust' finally found itself forced to give up its Hibernia shares in 1915 since, as a result of a Reich ordinance, a compulsory syndicate could be averted only if 97 per cent of all the producers approved, that is, the approval was needed of the state-owned mines (4 per cent) in Recklinghausen that had been taken over from Thyssen in 1902. With about 11 per cent of production under its control, the state joined the syndicate. It thus laid the basis – later together with the Recklinghausen mines and those on the left bank region of the Rhine, and the shipping and smelting works investments and oil concessions of Preußag and Bergag under the holding of Veba – at least theoretically, for control of the power of private capitalism in this primary industry during the Weimar Republic.[91]

Actually, the participation of the government-controlled Hibernia in the syndicate did not have any influence on business policy. The managerial control of private industry was threatened instead by exigencies of the coal and labour markets in the war economy. The drop in production as a result of the military call-up of 28 per cent of the labour force at the beginning of the war was counteracted primarily by reorganization of the labour force above ground, the exemption from military duty for certain age groups and the recall from military service of underground mining personnel, the request for 'voluntary' Belgian labourers and the demands for prisoners of war. In the long run, however, the mining associations were not able to succeed completely in getting the military to fulfil their demand either to press-gang unemployed Belgians within the framework of the Hindenburg Programme or to provide military disciplinary authority over the up to 70,000 prisoners of war.[92] In certain mines, as much as two-thirds of the workforce

90. See the review by Gebhardt, *Ruhrbergbau*, pp. 31f., and Friedrich Schunder, *Tradition und Fortschritt. Hundert Jahre Gemeinschaftsarbeit im Ruhrbergbau*, Stuttgart 1959, pp. 223ff.

91. See Hans-Joachim Winkler, *Preußen als Unternehmer 1923–1932. Staatliche Erwerbsunternehmen im Spannungsfeld der Politik am Beispiel der Preußag, Hibernia und Veba*, Berlin 1965. See also Gebhardt, *Ruhrbergbau*, pp. 38ff.

92. See Osthold, *Zechenverband*, pp. 138ff, 174ff., 197ff. On the inherent logic of the deployment of foreigners as a whole, see Ulrich Herbert, *Geschichte der*

consisted in the end of prisoners of war and foreign workers and only roughly a third were skilled German miners. In February 1918, the 'enemy foreigners' equalled 5.6 per cent and prisoners of war 12.9 per cent of the entire labour force in Ruhr mining, so that the proportion of foreigners made up about a fifth, and in exceptional cases a quarter, of the labour force.[93]

The dependency on government endeavours had a greater effect in the area of the control of raw materials. Even with daily production yields equal to those of the prewar period, an effort that could be achieved only through the destructive exploitation of resources, the increased demand could not be satisfied. After the failure of the Kohlenausgleichstelle (coal compensation bureau) in the newly founded Kriegsamt (War Office) under General Groener, which had been run for all practical purposes by the RWKS, mining was forced to submit itself in February 1917 to a national commissioner for the distribution of coal, similar to that in England.[94] This application of Wichard von Möllendorff's co-operative concept, forced on mining by the High Command, had been agreed upon by Stinnes, who hoped to be able to reduce excess demand by closing 25 to 30 per cent of the operations not vital to the war effort. However, the principle of a controlled economy threatened to strike back at the mining industry itself. The consumers, the southern German states and the unions demanded, although unsuccessfully, not only an effective control of prices but also of the profits of the privileged producers of raw materials. A corresponding move by the War Office against the exorbitant profits ensured, in August 1917, the political demise of General Groener, whose inability to check the inflationary wageprice spiral and the emerging strike movement by taking more heavy-handed measures against the unions, aroused the indignation of heavy industry and the High Command.[95]

Here the hands of the heavy industry entrepreneurs themselves were tied by the Auxiliary Service Act (Vaterländisches Hilfsdienst-

Ausländerbeschäftigung in Deutschland 1880–1980. Saisonarbeiter, Zwangsarbeiter, Gastarbeiter, Berlin 1986, pp. 82ff.

93. See Osthold, *Zechenverband*, pp. 134, 190, and Herbert, *Ausländerbeschäftigung*, p. 85.

94. On this point, see extensively Gerald D. Feldman, *Army, Industry and Labor in Germany, 1914–1918*, Princeton 1966, pp. 257ff.

95. Ibid., pp. 385ff. On the booming profits evident even in the early years of the war, see Jürgen Kocka, *Klassengesellschaft im Krieg. Deutsche Sozialgeschichte 1914–1918*, Göttingen 1973, pp. 25ff.

gesetz) of 5 December 1916. Hugenberg had boasted in August 1915 that the Zechenverband had not permitted 'this English form of "constitutional regime" in the factory in the first place', and in view of the negotiations being held in conjunction with the War Munitions Act, he saw no reason to consider the possibility of recognizing the unions as a 'reward for the German worker for his patriotic behaviour' once the war was over: if one did that, one was giving away the most important weapon needed 'should there again be war' since entrepreneurial resolve and flexibility were based 'first and foremost on the unconditional authority of the entrepreneur in his firm . . . just as the striking power of the army is based on military obedience'. From such a 'German' point of view, conciliation boards and wage demands during times of war must not only have possessed a 'blackmailing character' but even have 'bordered on treason'.[96]

The Rhenish-Westphalian heavy industry found itself alone in its stance as the High Command commenced actually to realize its demand for the forced recruitment of manpower within the framework of the Hindenburg Programme. Instead of the desired 'militarization' of labour relations, the government, the southern federal states and those processing industries in which joint war committees already existed preferred instead to incorporate the unions in order to stabilize the war effort. The Reichstag negotiations about the auxiliary service, which were conceived as a national demonstration, ensured the unions precisely those compensations that heavy industry so feared. Aside from their declared intent, these compensations were to become in the long run the basis for mutual recognition that was to uphold the tripartite state of industrial relations during the Weimar Republic. They were joint arbitration committees chaired by the military and mandatory worker committees in firms with more than fifty employees which – unlike those in mining up to that point – were also responsible for wage issues and could file collective complaints with the arbitration committees.[97]

96. See 'Denkschrift des Zechenverbands vom Aug. 1915', reprinted as a retrospective justification as from 1934 in Osthold, *Zechenverband*, pp. 225ff.
97. On the negotiations, see Feldman, *Army*, pp. 197ff; on the behaviour of the unions, see Hans-Joachim Bieber, *Gewerkschaften in Krieg und Revolution. Arbeiterbewegung, Industrie, Staat, Militär in Deutschland 1914–1920*, vol. 1, Hamburg 1981, pp. 296ff.

The governmental recognition of the parity principle and the necessity of negotiating on the parliamentary level was a milestone on the way towards consolidating the unions and simultaneously establishing parliamentary procedures in Reich politics. The employers realized what was occurring. Heavy industry, in particular, was angered by the regulation under which a licence to quit work in favour of an 'appropriate improvement of working conditions in the auxiliary service to the country' could not be denied (section 9). The Zechenverband strove immediately to have exempted from this regulation at least those who had been recalled from military service, and to prevent the migration to other firms which was theoretically possible for any worker following a fourteen-day waiting period. On the whole, the annual rate of job turnover in Ruhr mining did not fall under 100 per cent in either 1917 or 1918.[98] In essence, the mining employers were concerned rather with the 'corrosive effect on labour relations' which – just as during the first strikes by Krupp miners in February – would force them, 'contrary to all precedents, to capitulate before the danger of a strike'.[99] Following the wave of strikes that shook Silesian mining in April 1917, the local military commanders began to ignore the Auxiliary Service Act to the benefit of employer demands, as well as quietly collaborating with the unions that had become unnerved by the wildcat strikes. The declaration of a state of siege, not the Auxiliary Service Act, proved to be the actual prerequisite for the militarization of the factories.[100]

In the war year 1917 not only did the war aim aspirations of heavy industry fade away in the wake of the parliamentary peace initiatives – Hugenberg, like Stinnes, had pushed strongly for the annexation and Germanization of the Belgian and French coal mining regions, among other things[101] – but so did the chances for

98. See 'Denkschrift des Zechenverbandes vom 5. Jan. 1917', in Osthold, *Zechenverband*, p. 245. In 1914 and 1915, the job turnover rate equalled 171 per cent and 120 per cent, respectively, due to both call-ups and releases from military service. In 1919 this figure was 118 per cent. Afterwards, it fell until 1923 when it reached 48 per cent. Adelmann, *Quellensammlung*, vol. 1, no. 99, pp. 145f.

99. See the petition of the Bergbau Verein (Hugenberg and Stinnes) to von Hindenburg, 23 February 1917, in Alfred Hugenberg, *Streiflichter aus Vergangenheit und Gegenwart*, Berlin 1927, pp. 202f.

100. See Bieber, *Gewerkschaften*, vol. 1, pp. 455f. On constructive union activities during the war, see also *Material zur Lage der Bergarbeiter während des Weltkrieges. Eine Sammlung von Eingaben der Bergarbeiterverbände*, n.d. (printed in Bochum, 1919).

101. On Hugenberg's role as a founding member of the Alldeutscher Verband in the discussion on war goals, see Fritz Fischer, *Griff nach der Weltmacht. Die Kriegsziel-*

a return to the status quo in industrial relations. The Auxiliary Service Act finally opened the door for the unions to the large-scale enterprises without having curtailed the syndicalist undercurrents in sections of the labour force. Just the contrary: the latent protest potential among miners was strengthened by the worsening of the supply of provisions and the cuts in wages due to the use of unskilled manpower, about which the worker committees repeatedly complained. It was also strengthened by the incorporation of the unions as a disciplinary authority in the war economy, a development which the Independent Socialist Party (USPD) publicly denounced. With a drop in real weekly wages to about 62 per cent of the prewar level for cutters and drivers (1917), the primarily young workers who had migrated from the east to the more recently developed northern regions of the Ruhr became a reservoir for the syndicalist-influenced forms of protest at the local and factory levels that would lead to the major miners' strikes in January and in the summer of 1918.[102] Hugenberg and Stinnes reacted to this increase in tension with different means. As chairman of both the Zechenverband and the Bergbau Verein, the 'mining outsider' and Krupp director Hugenberg had occupied since 1912 the most important positions among the mining associations and had personally assumed administrative control over the political mining funds. With these funds, he laid the basis for the co-ordinated support of numerous 'patriotic' organizations, such as the anti-parliamentary Deutsche Vaterlandspartei, founded with his help in 1917, and for his press empire with the help of which he would later be able to make good his claim to the leadership of the Deutschnationale Volkspartei (DNVP).[103] While

politik des kaiserlichen Deutschland 1914/18, first published 1961, Düsseldorf, 1977, pp. 141ff. See also Peter Wulf, *Hugo Stinnes, Wirtschaft und Politik 1918–1924*, Stuttgart 1979, pp. 29ff.

102. See Klaus Tenfelde, 'Linksradikale Strömungen in der Ruhrbergarbeiterschaft 1905 bis 1919', in Mommsen and Borsdorf, *Glück auf*, pp. 199–223; see also Eberhard Lucas, 'Ursachen und Verlauf der Bergarbeiterbewegung in Hamborn und im westlichen Ruhrgebiet 1918/19. Zum Syndikalismus der Novemberrevolution', *Duisburger Forschungen* 15, 1971, pp. 1–119. On the general background, see also Jürgen Tampke, *The Ruhr and Revolution. The Revolutionary Movement in the Rheinish-Westphalian Industrial Region 1912–1919*, London 1979, pp. 33ff. and Barrington Moore, *Injustice. The Social Bases for Obedience and Revolt*, London 1979, pp. 227ff.

103. See Dankward Guratzsch, *Macht durch Organisation. Die Grundlegung des Hugenbergschen Presseimperiums*, Düsseldorf 1973; see also Heindrun Holzbach, *Das 'System Hugenberg'. Die Organisation bürgerlicher Sammlungspolitik vor dem Aufstieg der*

Hugenberg's political publicity work aimed to block reform to the end, work which his vice-chairman on the board of the Zechenverband, Hugo Stinnes, thoroughly supported, the latter followed a more flexible policy towards the end of the war. Officially, the Zechenverband was still rejecting all negotiation with labour representatives as 'unnatural' in the summer of 1918; otherwise there would be 'no barring ' the way to a unionized 'state within a state'.[104] Following the initial, unsuccessful contact within the framework of the war aims debate of 1917, Stinnes actually advanced, relatively independently of any influence from the association, an agreement between heavy industry and the unions in October 1918 in order to solve pragmatically the question of recognition. The way was clear on the national level for the agreement that had already been prepared in Berlin by representatives from the processing industry with the unions on the establishment of a *Zentralarbeitsgemeinschaft*. This agreement, the Stinnes–Legien agreement of 15 November 1918, would not only lay a new basis for industrial relations in heavy industry but also for the social order of the Weimar Republic in general.[105]

For Stinnes, the recognition of the unions as a guarantee for a demobilization free from war economy bureaucracy was merely the consequence of his 'reinforcement practice' (Wulf). As was openly admitted, the bourgeois alliance partners of industry were no longer of much use for keeping industry intact, and the unions were demanding a modest political price considering the revolutionary situation: namely, their recognition as a negotiating partner, the eight-hour day and collective bargaining.[106] The heavy industry entrepreneurs in the Ruhr region were even forced to sacrifice their favourite child – their support of the 'yellow' com-

NSDAP, Stuttgart 1981, pp. 42ff., and John A. Leopold, *Alfred Hugenberg. The Radical Nationalist Campaign against the Weimar Republic*, New Haven 1977, pp. 6ff.
   104. See Osthold, *Zechenverband*, p. 260.
   105. See Wulf, *Hugo Stinnes*, pp. 87ff.; on the history of the establishment of the ZAG, see in detail Gerald D. Feldman, 'Das deutsche Unternehmertum zwischen Krieg und Revolution: Die Entstehung des Stinnes-Legien-Abkommens', in Feldman (ed.), *Vom Weltkrieg zur Weltwirtschaftskrise. Studien zur deutschen Wirtschafts- und Sozialgeschichte 1914–1932* Göttingen 1984, pp. 100–27; G. D. Feldman, 'The Origins of the Stinnes–Legien Agreement: A Documentation', *IWK* 19/29, 1973, pp. 45–104.
   106. See the justification made by Jakob Reichert, the chief business manager of the VDESI, to the association of Ruhr chambers of commerce on 30 December 1918: *Entstehung, Bedeutung und Ziel der Arbeitsgemeinschaft*, Berlin 1919.

pany unions. Just a few days after the revolutionary events in Berlin and against the backdrop of the Hamborn strike movement, they conceded additionally in direct negotiations with the four miners' unions (Alter Verband, Gewerkverein Christlicher Bergarbeiter, Polnische Berufsvereinigung and the Hirsch-Dunckerscher Gewerkverein) the eight-hour shift, including descent and ascent time, a 25 per cent wage bonus for overtime and back-to-back shifts, a minimum wage for piece-work labourers and the cancellation of the black lists.[107] They left no doubt by the end of the year, however, that all this could not be upheld in the hour of defeat since, according to Hugenberg, the political significance of such concessions as the unions eventually could have had from the war booty had already been gambled away by the Auxiliary Service Act.[108] As a 'document of German weakness' this law was said to have prepared the way for the revolutionary stab in the back. In view of the confusion over 'the terms of yours and mine' as expressed in the call for socialization, of the 'political government' which now had to be contended with, and of the general compliance towards the 'ever pushing and shoving masses', the agreements on the Arbeitsgemeinschaft essentially bore the label of being 'coercion'.[109]

In fact, neither the legislative takeover of the central core of the Stinnes–Legien agreement by the Council of the People's Representatives (Rat der Volksbeauftragten), nor the negotiated solution with the miners' unions were able to calm down the situation. With the strike movement in the Ruhr in the spring of 1919, the German revolution entered into its phase of disappointed hopes. The cry for socialization released long pent-up expectations of finally being free of the inhumane Herr-im-Haus system forever. Expressions of this were often to be found in symbolic acts of dismissing supervisory personnel and the occupation of the office

107. See Rudolf Tschirbs, *Tarifpolitik im Ruhrbergbau 1918–1933*, Berlin 1986, pp. 37ff. See also Osthold, *Zechenverband*, pp. 265f. It was also agreed to eliminate female labour and to avoid intervention in the conditions at the company by dismissing supervisory personnel, committee members or security officials.
108. See Hugenberg's speech given before the Vereinigung der Handelskammern (Association of Chambers of Commerce) on 30 December 1918. Quote cited in Wulf, *Stinnes*, pp. 87ff.
109. Business manager von Löwenstein at the twelfth convention of the Zechenverband, 30 June 1919. Under the motto 'work and honesty', the 'desire for a strong personality' was expressed, one 'who would not only show the German people the right path to take, but would also be able to influence the people to the degree that they would follow his will' (p. 15).

buildings of the Zechenverband (and the RWKS) by the Essen
workers' and soldiers' council.[110] Neither the first government-
appointed commission on socialization nor the revived commission
that followed the Kapp putsch came to any clear conclusions in
their reports of February 1919 and July 1920, respectively.

Beyond a common commitment to socialization in principle, the proposed
contradictory strategies ranged from plans for immediate and complete socialization with compensation (Karl Kautsky), to step-by-
step expropriation with help from a state investment fund supplied
by profits (the government's guidelines), to a public-benefit, co-
operative control concept (Walther Rathenau). The opposition in
the mining industry, which did not take part in the negotiations,
centred on referring to the danger of a reduction in performance
and wages as an inevitable result of the amount of red tape involved
in running a state-led integrated coal industry. At the same time,
the purpose of such opposition was to stall for time politically.[111]
The immediate danger of socialization was finally lifted by the
defeat of the Weimar governing coalition in June 1920 and the
heightening tensions over the problem of reparations at the Spa
conference in July of that year. At this conference a monthly coal
delivery of two million tons was set – under the conditions of an
acute coal crisis – despite the effective publicity of Stinnes's opposition: he had always argued that socialization would give the Allies
easy access to the German mines for reparation purposes.[112]

In the light of the government's unkept promise to socialize,
collective bargaining, which had been granted to the unions as a
welcoming bonus into the new republic, appeared to the radicalized
labour force to be a poor instalment on the road to freedom.

110. See Hans Mommsen, 'Die Bergarbeiterbewegung an der Ruhr 1918–1933',
in Jürgen Reulecke (ed.), *Arbeiterbewegung an Rhein und Ruhr. Beiträge zur Geschichte
der Arbeiterbewegung in Rheinland-Westfalen*, Wuppertal 1974, pp. 275–314. See also
Jürgen Tampke, 'Die Sozialisierungsbewegung im Steinkohlenbergbau an der
Ruhr', in Mommsen and Borsdorf, *Glück auf*, pp. 225–48.
111. See *Die Sozialisierung des Bergbaus*, ed. by the Bergbau Verein, Essen, n.d.
See also *Verhandlungen der Sozialisierungskommission über den Kohlenbergbau im Jahre
1920*, 2 vols, Berlin 1920, as well as the *Bericht der Sozialisierungskommission über die
Frage der Sozialisierung des Kohlenbergbaus vom 31. Juli 1920*, Berlin 1920. Summar-
ized in Peter Wulf, 'Die Auseinandersetzungen um die Sozialisierung der Kohle in
Deutschland 1920/1921', *Vierteljahrsheft für Zeitgeschichte (VfZ)* 25, 1977, pp. 46–98.
112. See Charles S. Maier, 'Coal and Economic Power in the Weimar Republic:
The Effects of the Coal Crisis of 1920' in Hans Mommsen, Dietmar Petzina, Bernd
Weisbrod (eds), *Industrielles System und politische Entwicklung in der Weimarer Repub-
lik*, Düsseldorf 1974, pp. 530–42.

Whereas the first collective wage contract from October 1919 ensured minimum wages, social bonuses and the seven-hour shift or eight-hour day, the unions were forced to accept wage reductions in the less profitable mines in the southern district and – in compliance with the collective contract ordinance of December 1918 – *Allgemeinverbindlichkeit*, the principle of general validity of contracts for all workers. Because the employers wanted to prevent the grant to the unions of a closed shop by way of a collective wage contract in any case, the members of the no-contract syndicalist Free Workers Union, who demanded inflation supplements and a further shortening of work hours, thus came to enjoy the benefits of wage agreements.[113] Those miners' associations which were officially acknowledged as a party to wage contracts, and which were only capable of 'superficially treating social-economic differences in wage policy' (Mommsen), became more and more isolated from the workforce. This weighed all the more heavily as the Social Democrats in government put down both the strike movement in the spring of 1919 and the armed general strike against the Kapp putsch in March 1920, using white terror by deploying an armed force that was in no way loyal to the Republic. It was the revolutionary threat to the union position from below, not the Arbeitsgemeinschaft set up under certain reservations, that moved the employers to back the unions, at least in negotiating wages policy. They did this the more readily since the unions attempted to find refuge in the unsustainable accusation that the strike movement was being manipulated by the Bolsheviks. However, the miners' unions were to discover what was to be expected from the Arbeitsgemeinschaft as a political alliance as the employer representatives in the Arbeitsgemeinschaft for Ruhr mining refused openly to support the elected government against the Kapp putschists or, by contrast with other industries, to compensate financially for the wage losses of the strikers.[114]

Like the movement supporting socialization, the workers' council movement could also be reined in. What remained of the elaborate council ideas of the Left following the consultations on the Weimar Constitution was merely a politically powerless 'tem-

113. See Tschirbs, *Tarifpolitik*, pp. 79ff.
114. See Gerald D. Feldman, 'Big business and the Kapp-Putsch', *Central European History* 4, 1971 pp. 95–130, and the documentation in Feldman and Irmgard Steinisch, *Industrie und Gewerkschaften 1918–1924. Die überforderte Zentralarbeitsgemeinschaft*, Stuttgart 1985, pp. 158ff.

porary national economic council' and the constitutional stipulation
that workers and white-collar employees were to participate
'equally in co-operation with the entrepreneur on fixing wages and
working conditions as well as in the entire economic development
of productive forces'. Organizations on both sides and their agreements were to be recognized (Article 165 of the Weimar Constitution). This constitutional stipulation of collective labour rights was
to be secured at the company level by the Company Council Act
(Betriebsrätegesetz) of February 1920 against which the employers
strongly protested even to the point of threatening to close down
operations.[115] They were less afraid of socialization from below
than of the consequent politicalization of the labour force in the
annual elections and the organizational consolidation of unions in
the plant. Following violent demonstrations infront of the Reichstag in which forty-two people were shot, the company councils
retained merely a right, limited by the obligation to keep the peace
of the company, to supervise wage and social policy issues, without
any right to economic co-determination. Later on the employers
were even successful in legally containing the 'administrative control of contracted wages' of the company councils further and
avoiding for all practical purposes the budget control exercised by
union representatives on the supervisory boards of the companies.[116] The recruitment of union functionaries from within the
plant with access to the facilities of company welfare forced the
heavy industry entrepreneurs in particular to expand their social
policy further in order to counteract union support within the plant
with welfare and training opportunities.[117]

As early as 1920, the situation for mining employers had thus
become a rather agreeable one as opposed to the earlier revolutionary threat, all the more so since the co-operative control in the Coal
Administration Act (Kohlenwirtschaftsgesetz) of March 1919,
which constituted the last remaining substitute for socialization,
proved to be far from effective under inflationary conditions. The

115. See Kurt Briegl-Mathias, *Das Betriebsräteproblem*, Berlin 1926; see also
Hans-Otto Hemmer, 'Betriebsrätegesetz und Betriebsrätepraxis in der Weimarer
Republik', in Ulrich Borsdorf et al. (eds), *Gewerkschaftliche Politik: Reform aus
Solidarität. Zum 60. Geburtstag von Heinz Oskar Vetter*, Cologne 1977, pp. 241–69.
116. On the escalation of conflicts on labour law, see Martin Martiny, *Integration
oder Konfrontation? Studien zur Geschichte der sozialdemokratischen Rechts- und Verfas-
sungspolitik*, Bonn–Bad Godesberg 1976, pp. 99ff.
117. This side-effect of the Betriebsrätegesetz is also overlooked by Schwenger,
Sozialpolitik.

simple majority of the RWKS dominated the National Coal Association (Reichskohlenverband) as a clearing-house and was usually able to push through its price proposals with the help of the unions in the National Coal Council (Reichskohlenrat) even against government pressure.[118] The 'monetarization' concept of Hugo Stinnes began to pay off: the Arbeitsgemeinschaft policy was for him only a part of a larger plan in which the wage–price deal that had been established during the war would be expanded and the demands for wage increases would be passed on to the consumers in the form of higher prices. The foundry mines were also not in a position to speak out against such a move since the necessary increase in personnel and performance was to be expected only from such wage incentives.[119] By the end of the war, Stinnes had already implored the labour representatives to work together to improve the position of German coal in the expected coal crisis of the postwar period, so that an international position could once again be achieved when the time came to economize.[120]

On the national level, the idea of the Zentralarbeitsgemeinschaft as a form of industrial self-administration had already proved itself in 1919 to be an 'organizational fantasy' (Feldman) once the parliamentary system was established and the grassroots democratic workers' and soldiers' council movement had lost out in the battle for legitimacy. Yet for the heavy industry entrepreneurs, this opened up the possibility of creating a 'producer alliance' (Feldman), at least as long as the inflation lasted.[121] This was especially advantageous for them in that they could react to the strike movement by way of a two-tier strategy. At one level, they worked out an economic understanding with the unions, whereby the worldwide coal shortage and the blockade of socialization in reparations policy were like aces dealt into their hands. At another, they reacted to the strike movement by striving to defend the company against the politicalization of industrial relations with the help of concessions in wage policy.[122]

118. See Wulf, *Sozialisierung*, pp. 48ff.
119. See Tschirbs, *Tarifpolitik*, pp. 50ff.
120. See Wulf, *Stinnes*, pp. 106f.
121. See Feldman and Steinisch, *Industrie und Gewerkschaften*; for an extensive presentation on the idea of the 'producer alliance', see Feldman and Heidrun Homburg, *Industrie und Inflation, Studien und Dokumente zur Politik der deutschen Unternehmer 1916–1923*, Hamburg 1977, and Feldman, *Iron and Steel in the German Inflation, 1916–1923*, Princeton 1977.
122. See Maier, *Coal Crisis*.

Under these circumstances, the linch pin of industrial relations in mining became the issue of hours. Here the utopian demands for a six-hour shift by a workforce that had become disenchanted during the course of the revolution directly confronted the employers' production demands, which had been accepted in principle by the unions.[123] Compared to the prewar period, German coal production had fallen by about half in 1919. Roughly a quarter could be attributed to the territorial losses stipulated by the peace treaty. The other quarter was due to cutbacks in work hours and exhaustion of the mining sites and crews. Within five months the length of a shift had been reduced by an hour-and-a-half to seven hours, and in a special law of 17 June 1922 this was finally stipulated to be the standard length of work hours in mining. Those working underground, who bore the brunt of extra production for reparations, thus lost some of their advantage in shorter hours, as against those working above ground who enjoyed a two-hour reduction of work hours thanks to the legal eight-hour day.[124] Under these conditions the issue of additional shifts was to become the litmus test for the new collective industrial relations.

In view of the smouldering threat from the political Left, Hugenberg had already secured in February 1920 the massive support of the Social Democratic National Commissioner Severing against any further shortening of work hours. He had proposed military force and governmental 'arbitration' in order to compel an agreement on overtime on a 'voluntary' basis in the interest of national policy. Subsequently the mining employers were indeed dependent on negotiated solutions which, however, in the end secured wage supplements of up to 100 per cent for overtime. The syndicalist workers who had won an impressive 27 per cent of the seats in the first elections for the company councils in Ruhr mining (1920) were still not satisfied and kept pushing for a further reduction of work hours.[125] Under these increasingly unfavourable conditions for the

123. See Tschirbs, *Tarifpolitik*, pp. 59f., and Gerald D. Feldman, 'Arbeitskonflikte im Ruhrbergbau 1919–1922. Zur Politik von Zechenverband und Gewerkschaften in der Überschichtenfrage', *VfZ* 28, 1980, pp. 168–223. See also Osthold, *Zechenverband*, pp. 338ff.
124. See the review in Osthold, *Zechenverband*, pp. 338ff.
125. They even surpassed the Alte Verband (32 per cent) in 1924 with 42 per cent. As the renewed strength of the Left in the Great Depression shows (1931: 27 per cent), the two-thirds leadership position of the Alte Verband in the mid-Weimar years was only temporary, whereas the Christian union continually increased its share from 18 per cent (1920) to nearly 27 per cent (1931). See Bernd Weisbrod,

unions, a plan was developed under the auspices of Labour Minister Brauns in the course of 1921–2 which would become typical for industrial relations in mining during the entire Weimar Republic. With the help of government-led negotiations or arbitration, the wage increases demanded by the unions were linked to the overtime sought by the Zechenverband in a method whereby the state provided both direct and indirect benefits in the form of price supplements or wage subsidies in order to secure the legislative protection of the seven-hour shift.[126] It was this principle of government-supported, compensatory wage policy that would be sanctioned at the end of 1923 by arbitration decree and would become the basis of industrial relations, especially in heavy industry, during the stabilization period of the Weimar Republic.

Stinnes was first to realize that the need for overtime in mining could eventually be used as a lever to overcome the general principle of the eight-hour day. Under his auspices, industry demanded the suspension of the eight-hour day in return for an industrial guarantee for a foreign loan of the national government. Such a guarantee was urgently required for reparation purposes but was eventually thwarted by heavy industry.[127] Stinnes's complete cunning in conceptionally linking 'monetarization', international vertical integration and a rollback in social policy was revealed finally in his autonomous negotiations on reparations with the Mission Interalliée de Contrôle des Mines et des Usines (MICUM) in the occupied Ruhr region in 1923. The Zechenverband sought to use the favourable situation in October 1923, first, to reintroduce prewar work hours (eight-and-a-half hours including ascent and descent time) in a unilateral and illegal dictate, preferably with the backing of the occupational forces, and second, to eliminate the co-operative supervision of prices.[128]

Now for the first time, the state-owned mines, which were only second in size to the Stinnes mines in the syndicate (1924: 14.3 per

---

Schwerindustrie in der Weimarer Republik. Interessenpolitik zwischen Stabilisierung und Krise, Wuppertal 1978, pp. 128f.
126. On this compensatory governmental social policy in mining, see Hans Mommsen, 'Sozialpolitik im Ruhrbergbau', in *Industrielles System*, pp. 303–21.
127. See Gerald D. Feldman and Irmgard Steinisch, 'Die Weimarer Republik zwischen Sozial – und Wirtschaftsstaat. Die Entscheidung gegen den Achtstundentag', *AfS* 18, 1978, pp. 353–439.
128. See Feldman and Homburg, *Industrie*, pp. 146ff; Wulf, *Stinnes*, pp. 425ff.; and from the viewpoint of the employer, Hans Spethmann, *Zwölf Jahre Ruhrbergbau*, vol. 3: *Der Ruhrkampf 1923 bis 1925 in seinen Leitlinien*, Berlin 1929, pp. 181ff.

cent), broke ranks. Some other mines forced through plant agreements with their workers, who waived their contractual rights for wage bonuses and further employment. The Zechenverband attempted instead to bring the unions to heel on the work hours issue by threatening them with equally illegal mass firings. It was above all due to the intervention of the Labour Minister Brauns that, in the face of this 'return of the old "Herr-im-Haus standpoint"' (Preller), the plan of a state-protected overtime agreement was legally established by means of decrees on work hours and shifts; only this policy saved the collective contract as such in heavy industry. For all practical purposes, this meant a return to the eight-hour shift in mining, and for industry as a whole it meant a renewed extension of the regular working day beyond eight hours – in the continuous operations of heavy industry even up to twelve hours. However, overtime was linked to contractual stipulations which could be enforced by state arbitration and could be declared binding in the public interest. State arbitration, on the functioning of which the conclusion of collective contracts in heavy industry was going to depend, was less of a delegitimizing burden to the autonomous bargaining position of both employees and workers than the political prerequisite as such for the acceptance of collective bargaining on the part of the mining employers.[129]

The mining employers had striven for an open break with contractual stipulations for hours and wages in order, first, to get rid of the minimum wage in piece-work, the social wage, and other 'production-stifling' elements of the contracted wage; second, to eliminate the levelling of the wage spread conditioned by inflation; and third, generally to exert wage pressure as a 'camouflaged means of labour struggle' in order to push through their shift schedules.[130] This switch in the industrial relations strategy in mining which went hand in hand with the national task of bringing to an end the passive resistance against French and Belgian occupation in the Ruhr region and with the stabilization of the currency still did not go far enough for the employers. Governmental intervention prevented them, however, from being able to take full advantage of the weaknesses of the unions. The latter had been bled financially dry by the inflation, they were politically discredited and organiza-

---

129. In this sense, see the classic study by Ludwig Preller, *Sozialpolitik in der Weimarer Republik*, Stuttgart 1949, pp. 313ff. See also Weisbrod, *Schwerindustrie*, pp. 301ff, and Tschirbs, *Tarifpolitik*, pp. 190ff.
130. Tschirbs, *Tarifpolitik*, p. 224.

tionally weakened by the Arbeitsgemeinschaft policy and the drop in real weekly wages to as much as 30 per cent of the prewar level. In 1923 only 19 per cent of the labour force in the Ruhr region belonged to the Alter Verband (1918: 46 per cent, 1924: 15 per cent).[131] Compared with this, the mining companies were experiencing a financial consolidation. The opening balance sheet on the basis of the new gold-based currency permitted a consolidation of debts. The rapid paying off of the advanced MICUM payments by the national government also included compensation for losses of interest and money value, although Ruhr mining had already switched to accounting in gold and the emergency taxation ordinances of the bourgeois governments, bypassing the Reichstag for the most part, brought noticeable tax relief.[132] In any case, the mining industry was in a better position to survive the general credit crisis, which only eased following the Dawes Plan, than other industries because of its special status in reparations policy. It even came back for extra cash from the government when the wage concessions that were necessary in order to end the great labour dispute of May 1924 had to be financed. The four-week dispute began as a lockout of the workers who quit work after seven hours despite arbitration because the unions imagined themselves to be under no contractual obligation with regard to overtime. It ended as a strike with broad public support against the confrontation course of the Zechenverband, which in face of a radicalized labour force, demanded the extension of standard work hours and the abolition of wage protection clauses. Wage bonuses for overtime and thus the safeguarding of the regular seven-hour shift below ground (plus one hour of overtime) as well as the defence of contractual minimum vested rights were, however, only to be achieved with help from state arbitration and credit aid.[133]

For the consolidation of mining under the conditions of the international market, it was not primarily the return to prewar work hours continually sought by employers that was decisive, but

131. Ibid., p. 193.
132. See Klaus-Dieter Krohn, *Stabilisierung und ökonomische Interessen. Die Finanzpolitik des Deutschen Reiches 1923–1927*, Düsseldorf 1974.
133. See Hans Spethmann, *Der Maistreik 1924 im Ruhrbergbau. Ein grundsätzlicher Arbeitskampf*, Berlin 1932; for an opposing view, see *Unternehmer und Kommunisten während der Bergarbeiterkämpfe im Mai 1924*, ed. by Verband der Bergarbeiter Deutschlands, Bochum 1924.

the improvement of productivity. And this meant first and foremost the dismantling of inflationary manning levels, a concentration of the mining operations, the enforcement of company rationalization, and the securing of markets. As opposed to conditions before the war, miners also now had to face unemployment. In connection with the extension of shift hours, nearly 30 per cent of the miners were laid off. Until the Great Depression, 100,000 dismissed miners disappeared from the job market in Ruhr coal mining in part due to the alleged lack of qualification. Long-term mass unemployment, such as existed in the winter of 1925–6, could be avoided thanks only to the positive effects of the British miners' strike.[134] The Ruhr labour force, which was reduced by nearly a quarter between 1924 and 1929, achieved a production level a quarter greater than before. The production yields in 1929 per man and shift surpassed those of prewar Ruhr mining by more than a third and would climb to 178 per cent of prewar production levels during the Great Depression thanks to further cutbacks.[135]

Unquestionably, the most important factor in this development was the 'negative rationalization'.[136] Pits were acquired for their transferable syndicate quotas – and then shut down – and less productive working points were phased out. Mechanization of coal-getting had already been well advanced during the inflation period by the introduction of pneumatic picks and drill-hammers; now the hauling and handling of coal was greatly improved by the use of conveyors in flat seams and 'major' working points. The long-term effects this was to have on the qualification of hewers, the intensification of surveyors' control and the dissolution of the work crews were, however, not yet to be fully felt.[137] A certain

134. See Georg Berber, 'Die Arbeitslosigkeit im deutschen Steinkohlenbergbau', in Manuel Seitzew (ed.), *Die Arbeitslosigkeit der Gegenwart*, second part: *Deutsches Reich I*, Munich and Leipzig 1932, pp. 1–31.
135. Ibid., p. 23. See also Ernst Brandi and Ernst Jüngst, 'Das Ruhrrevier', in *Die deutsche Bergwirtschaft der Gegenwart. Festgabe zum deutschen Bergmannstag 1928*, Berlin 1928, p. 50.
136. On pit closures and the consolidation of pit operations, see the *Statistisches Heft des Bergbau-Vereins*, 33rd edn, June 1933, pp. 10ff. The number of German hard coal mines as a whole dropped by roughly a third (1924: 376, 1929: 266), the production per pit nearly doubled (316,000 tons: 614,000 tons) and the manpower level per pit increased by about a third (1,487: 1,945); Berger, *Arbeitslosigkeit*, p. 22. On the effects of rationalization as a whole, see Tschirbs, *Tarifpolitik*, pp. 241–59.
137. See Ausschuß zur Untersuchung der Erzeugungs- und Absatzbedingungen der deutschen Wirtschaft (Enquete-Ausschuß). *Verhandlungen und Berichte des Unterausschusses für Arbeitsleistung (IV. Unterausschuß)*, vol. 2: *Die Arbeitsverhältnisse im*

role may also have been played by the improvement of the age and qualification structure of the labour force. The percentage of productive workers underground rose until 1927 to reach the prewar level of about 50 per cent (1922: 42.8 per cent) while the percentage of young people dropped to 1.3 per cent (1913: 3.7 per cent).[138] However, the productivity increase ran up against two factors limiting its profitability. On the one hand, it was the prerequisite for a re-entry into the costly competition, especially with British coal, in an ever tighter energy market. On the other hand, it justified the union demand for an appropriate share in the proceeds of rationalization.

However, hard coal was far from re-establishing its prewar growth rate either in the national or international energy market.[139] The stiff competition of the less labour-intensive brown coal had just as great an impact on this as had the efficiencies in fuel consumption among all important consumers. The situation was made even worse by the expansion of capacity in Germany during the war and the inflation, as well as in traditional export markets such as Holland and Russia. The production potential in Upper Silesia, separated by the peace accord, also contributed to these difficulties. Above all, the competition with British coal needed to be warded off for, with public support, it was also seeking an export market for similar reasons.[140] The RWKS reacted to this threat by increasing its control of the trading companies; beyond this, it joined the competition for state subsidies and tried to compensate for the losses incurred in the contested areas by imposing a general levy on production that also taxed the growing self-consumption of the mixed mines after 1927.[141] Both this conflict of interests within the

deutschen Steinkohlenbergbau in den Jahren 1912 bis 1926, Berlin 1928, pp. 65ff. For the effects of rationalization on the workforce see also Michael Zimmermann, *Schachtanlage und Zechenkolonie. Leben, Arbeit und Politik in einer Arbeitersiedlung 1880–1980*, Essen 1987, pp. 135ff.

138. See Brandi and Jüngst, *Ruhrrevier*, p. 49; see also Enquete-Ausschuß, *Arbeitsverhältnisse*, pp. 174ff, as well as Enquete-Ausschuß, *Verhandlungen und Berichte des Unterausschusses für Gewerbe: Industrie, Handel und Handwerk (III. Unterausschuß), Die deutsche Kohlenwirtschaft*, Berlin 1929, pp. 60ff.

139. See the following in the Enquete-Ausschuß, *Kohlenwirtschaft*, pp. 6ff.

140. See the most recent contribution: Barry Supple, 'The Political Economy of Demoralization: The State and the Coal Mining Industry in America and Britain between the Wars', *Economic History Review* sec. ser. 61, 1988, pp. 566–91, and the contribution by Roy Church in this book.

141. See Enquete-Ausschuß, *Kohlenwirtschaft*, pp. 87ff., 151ff. For the differences between pure mines and foundry mines, see Muthesius, *Ruhrkohle*, pp. 177ff.

industry and the price controlling potential of the Coal Mining Act (Kohlenwirtschaftsgesetz) increased the pressure of market conditions on industrial relations in Ruhr mining, all the more so since the unions had recovered from their defeat during the stabilization crisis and were now in a position to claim higher wages.

In the dispute on wages and state arbitration, which escalated during the stabilization phase of the Weimar Republic, the large, affiliated groups of heavy industry increasingly assumed a leading role. With their foundry mines, these *Konzerne* introduced their interests as producers and employers from the iron-producing side of their operations into Ruhr mining. Among the large mining companies with more than 5 per cent participation in the syndicate, only two still belonged to the group of 'pure' mining operations following the inflation (Harpen: 6.5 per cent, Hibernia: 9 per cent). Some foundry mines (belonging to Krupp and Thyssen) had signed plant contracts with their labour force during the stabilization crisis, while others (Rheinstahl) even temporarily left the Zechenverband or voiced their opposition to the business policy in the district organizations in which they dominated (Gutehoffnungshütte[GGH] in Oberhausen).[142] It proved impossible to fuse the employer organizations of heavy industry for practical reasons, but the direction of policy slowly shifted away from the Zechenverband, which was kept in close personal and political contact with the employer association of the Northwestern Group of the Iron and Steel Industry (Arbeitnordwest), to the heads of the large coal and steel groups, who also set the tone in the more elite executive committee of the Zechenverband. The chairman of the Northwestern Group and of the Langnamverein, Paul Reusch (of GHH), was increasingly co-ordinating the common interest policy of heavy industry despite the prominent influence of the Vereinigte Stahlwerke, which was founded in 1926 and held about 47 per cent of the shares of the Rohstahlgemeinschaft and 22 per cent of those of the RWKS. The differences in interests between the integrated groups thus began to overshadow those between coal and iron.[143]

---

142. Sée Tschirbs, *Tarifpolitik*, pp. 234ff.
143. On this, see in detail Weisbrod, *Schwerindustrie*, pp. 96ff. (Vereinigte Stahlwerke) and passi.n. Ernst Brandi, the head of the Dortmund mining division of Vereinigte Stahlwerke, assumed the chairmanship in the Zechenverband in 1927 (and in the Bergbau Verein) from Fritz Winkhaus (Hoesch), who had replaced Hugenberg in 1925 (who remained on the board as second vice-chairman). The

By way of a memorandum war that was completely tailored to the specific problems of mining, the Bergbau Verein and the Zechenverband supported the wage policy confrontations of the 'Konzerne', the profitability of which was often less dependent on miner's wages than on the advantages to be gained from vertical integration and cost advantages from greater capacity utilization.[144] They thus reacted to the legitimation pressure institutionalized by the arbitration authorities, without always being able effectively to counter the unions' social policy claims which were brought before the state arbitrators. Although it was only gradually that private industry was opened up to public regulation this procedure threatened to erode entrepreneurial autonomy step-by-step since the public check encouraged the unions' demand for economic co-determination in line with their programme of economic democracy.[145] This political danger was certainly one of the most important reasons for the assault of heavy industry on state arbitration in the later phase of the Weimar Republic.

In any case, it still remained a matter of controversy at the public balance sheet audit in April 1928 whether Ruhr mining was sufficiently profitable to withstand a further increase in wages. The unions' special vote on the Schmalenbach report correctly criticized the fact that the rates of depreciation adopted, in which the employers wanted to have future damage due to mining operations taken into consideration but not future appreciation, concealed secret reserves for modernization investments. This reduced the current leeway for distribution, although, it was upheld, the 'strictest conformity of wages to the rise in productivity . . . [was] the most

---

chairman of the board of the Vereinigte Stahlwerke, Albert Vögler, had been third vice-chairman since 1924, whereas his deputy, Ernst Poensgen, and the chairman of the supervisory board, Fritz Thyssen, directed the employer and market associations for iron industrialists.

144. See *Denkschrift zur Lage des Ruhrbergbaus, 31. Juli 1925*, reprinted together with countering memorandums and supporting statistics from the unions in Ernst Jüngst, *Richtige Zahlen beweisen! Ein Beitrag zur Lage des Ruhrbergbaus*, Essen 1925; *Die wirtschaftliche Lage des Ruhrbergbaus*, Essen 1929, as well as *Zur Lage des Ruhrbergbaus. Eine Antwort auf die Denkschrift des Essener Bergwerksvereins*, Essen 1929. See also Ernst Jüngst, *Wirtschaftsfragen des Ruhrbergbaus*, Essen 1929.

145. Union representatives participated in the official commission of inquiry (work conditions, coal industry) as well as on the recommendation of the Schmalenbach commission for the National Ministry of Economics on the occasion of a wage conflict: *Gutachten über die gegenwärtige Lage des rheinisch-westfälischen Steinkohlenbergbaus*, Berlin 1928.

effective means to prevent excesses in the competition over quotas and an over-extension of capacity'.[146] In response to this the Zechenverband calculated that, with a constant level of wages, the labour costs per each ton produced between 1925 and 1929 should have dropped by a quarter as a result of the productivity gain. Instead, the drop was insignificant, so that the entrepreneurs, who had to carry the weight of rising interest rates, social benefit costs and taxes, came out 'completely empty-handed'.[147] The actual cost problem was not due, however, to the relatively constant percentage of labour costs (about 50 per cent of the cost price), but to the fact that the development of capacity, induced by the competition over quotas and relatively independent of the market, enhanced the pressure of increased fixed costs with under-utilization.

The ballooning cost crisis in Ruhr mining that resulted from this was, however, attributed by the Zechenverband primarily to the 'political wages' of the state arbitration institutions. Yet, in fact, the arbitration authorities were able on the whole to tone down the swings in the wage fluctuations in the economy.[148] In mining, as in all heavy industry, the governmental arbitrators obviously made an effort to come to a settlement. Because they were authorized to render an award without the approval of one of the parties involved – an award which could be declared valid by the national labour minister in the public interest – they had to be certain to submit proposals that were acceptable to both sides if they wanted to be at all serious about fulfilling the task of the arbitration decree, namely to help bring about collective wage contract settlements. Actually, a complicated system of compensatory contracts for industry-wide agreements, wages and work hours developed, the individual components of which could not have counted on getting the approval from both negotiating parties. This system did, however, offer employers as a whole a certain degree of compensation for

146. See 'Sondergutachten Dr. Baade', *Gutachten*, p. 60f.
147. See Jüngst, *Wirtschaftsfragen*, pp. 166ff. During the Great Depression, the labour costs (per ton) should have dropped by nearly half according to this calculation; thanks to the productivity-increasing effect of further company cutbacks, they did at least drop to about 70 per cent of the 1924 level. See Osthold, *Zechenverband*, pp. 429ff.
148. See Hans-Hermann Hartwich, *Arbeitsmarkt, Verbände und Staat 1918–1933. Die öffentliche Bindung unternehmerischer Funktionen in der Weimarer Republik*, Berlin 1967; see also now Johannes Bähr, 'Sozialer Staat und industrieller Konflikt. Das Schlichtungswesen zwischen Stabilisierung und Weltwirtschaftskrise', in W. Abelshauser (ed.), *Die Weimarer Republik als Wohlfahrtsstaat. Zum Verhältnis von Wirtschaftsund Sozialpolitik in der Industriegesellschaft*, Stuttgart 1987, pp. 185–203.

wage increases, for example, by way of the renewal of the overtime agreement, the reduction of the wage gap (of the 'piece-work wages' great lead') and also by way of supervising export credits, rate discounts for freight or price increases.[149] What appeared outwardly to be 'state wage control' was not only the result of a need to protect wages, not the least from the assaults of heavy industry during the stabilization crisis; it was also the result of a distribution struggle that had long gone beyond the framework of industrial labour relations and now included the administrative control over prices, the availability of state subsidies, and the pliability of political institutions.

Once it was clear that the agreements linking higher wages and extra hours no longer paid off for the iron industry, heavy industry launched its decisive assault against the arbitration system and especially against the 'one-man award'. In the dispute in the Ruhr iron industry in November 1928, about 230,000 workers of the Rhenish-Westphalian iron and steel industry were locked out for four weeks.[150] And in mining the employers also switched to an offensive strategy well before the Great Depression. The reduction of work hours above ground in April 1928, which was supposed to compensate the unions for the retention of the eight-hour shift (seven plus one) underground, prompted the employers to openly demand the return of prewar work hours (eight-and-a-half hours). In addition, their answer to the state's subsidizing of wage increases by exempting the workers from part of their contributions to social insurance was a dose of firings.[151] The fact that the two sides of the wage-negotiating table commenced regularly to reject the arbitration awards instead of searching for compensatory advantages in them was, however, due not only to the employers' confrontation course; it also reflected the respective weaknesses of each side. Between 1925 and 1929, the unions were able to organize only about a quarter to at most a third of the miners. Over the long run the Alte Verband alone even lost as much as half of its postwar membership and only managed to improve on the modest level of organization (16 per cent of the workforce) during the crisis years.[152] Although it was able to exact the benefits of the collective wage contract starting in 1923, if only by relying on the state, it

149. For details on this point, see Tschirbs, *Tarifpolitik*, pp. 259ff.
150. See Weisbrod, *Schwerindustrie*, pp. 415ff.
151. See Tschirbs, *Tarifpolitik*, pp. 314ff.
152. See *Statistisches Heft*, ed. by Verein für die bergbaulichen Interessen, 33rd

could not always count on the workers. For their part, the employers ran up against the economic limits of export dumping: the levy, which increased by leaps and bounds in 1928, amounted to as much as 20 to 25 per cent of the payroll in some mines.[153] Even though upper ceilings for levies were fixed for the foundry mines in a new syndicate contract in 1931, the actual cost problem, namely that of idle capacity, would only grow worse with the coming of the Great Depression.[154] Although deflationary competition could hardly promise to provide the necessary savings of labour costs in order to cushion the cost escalation resulting from under-utilization, employers in heavy industry commenced to rid themselves not only of the state arbitration system, but of actual contractual commitments as such in an exemplary labour dispute at the turn of the year 1930–1.[155] It was their intention to use the pressure of the crisis to bring about a general revision of the social relations of the Weimar Republic or at least to prevent a return to the system of collective bargaining, so that they would be in a position to enter the expected competition for export markets following the crisis with a minimum of cost to themselves. In order to do this, the mining industry also used coal prices as a weapon. In a reverse situation from that of 1928, the Zechenverband demanded state guarantees for a wage decrease before the RWKS was willing – despite a general price decline – to concede the otherwise necessary price cuts. For the miners, such a wage decrease was doubly painful in light of the short-time work. However, it was clear that the purpose of the manoeuvre was not only to decrease wages but also deliberately to overtax state arbitration, which stood in the way of undermining compulsory collective bargaining.[156]

Because Brüning's presidential government was dependent on the toleration of the SPD after the electoral success of the National

edn, 1933, pp. 75, 104, 106; see also Weisbrod, *Schwerindustrie*, p. 125; Tschirbs, *Tarifpolitik*, pp. 312, 326.
153. See Tschirbs, *Tarifpolitik*, p. 352. The ratio for the mines of the Rheinische Stahlwerke was 1:8.5 in 1927–8 and it continued to deteriorate until it reached 1:2 in 1932–3.
154. See Muthesius, *Ruhrkohle*, pp. 186f. As early as December 1929 the plant utilization rate for the mines of the Vereinigte Stahlwerke equalled about 70 per cent and for the total RWKS members 66.4 per cent (Tschirbs, *Tarifpolitik*, p. 347).
155. On the following see Bernd Weisbrod, 'Die Befreiung von den "Tariffesseln". Deflationspolitik als Krisenstrategie der Unternehmer in der Ära Brüning', *Geschichte und Gesellschaft* 11, 1985, pp. 295–325, and Tschirbs, *Tarifpolitik*, pp. 367ff.
156. See Tschirbs, *Tarifpolitik*, pp. 381ff. On the general strategy of heavy industry during the Great Depression, see also Weisbrod, *Schwerindustrie*, pp. 479ff.

Socialists in September 1930, such demands could be stigmatized as encouraging a 'right-wing radical dictatorship' (Labour Minister Stegerwald). In any case, the mining industry was not put off by such political implications in its deliberate efforts to put an end to collective bargaining following the September elections and the first contacts with Hitler. It even threatened to leave the National Association of German Industry in order to advance its radical position. However, the iron industry dampened the moves made in this direction by the coal industry – Reusch himself threatened to pull out of the coal associations – and once again backed the National Association in a pro-government policy in view of the threatening capital drain. The Zechenverband, which had argued against a 'co-operation on a dishonest basis' as early as May 1930, [157] thus also torpedoed the top-level negotiations that were then being conducted with the unions on re-establishing the Arbeitsgemeinschaft within the framework of Brüning's deflationary policy. They opted for this tactic in spite of the threat to unions by the communist-inspired Revolutionäre Gewerkschaftsopposition (RGO), which had initiated wildcat strikes and obtained 23 per cent of the votes in 1930 and 29 per cent in 1931 in the company council elections in Ruhr mining. [158]

Under these conditions, the position of the officially recognized unions in industrial relations in Ruhr mining was only to be saved by way of state emergency decrees. In order to avoid the situation sought by the Zechenverband in which no collective bargaining would take place and mass firings would, Brüning's government ordered the practical reintroduction of the one-man arbitration award in mining in January 1931 – it had been declared to be unlawful in connection with a dispute in the Ruhr iron industry. It also reintroduced state intervention in valid collective wage contracts and subsidized the further wage decrease by taking over the contributions to unemployment insurance. [159] But the situation in mining worsened dramatically despite the more than proportional cutback in manpower by nearly half, the improvement in productivity

157. See 72. *ord. Generalversammlung des bergbaulichen Vereins, 14. Mai 1930*, Essen 1930, pp. 15f.
158. See Michael Grübler, *Die Spitzenverbände der Wirtschaft und das Erste Kabinett Brüning. Vom Ende der Großen Koalition 1929/30 bis zum Vorabend der Bankenkrise 1931*, Düsseldorf 1982, pp. 329ff.; see also Reinhard Neebe, 'Unternehmerverbände und Gewerkschaften 1929–1933', *Geschichte und Gesellschaft* 9, 1983, pp. 302–30. In 1932 the elections for company councils were suspended by emergency decree.
159. See Tschirbs, *Tarifpolitik*, pp. 399ff.

linked to this, and the decrease of roughly 21 per cent in wages cushioned by the state. The savings made in wages were in part lost again due to the raises in levies necessary to compensate for the losses in contested markets due to the devaluation of the pound sterling. In 1931–2, only 37 per cent of the syndicate's capacity could be utilized. The integrated groups of affiliated companies were hit especially hard. Only 28 per cent of the self-consumption quota of the Vereinigte Stahlwerke's mining operations in the RWKS was used in 1931–2 (1927–8: 100 per cent); the operating profit from integrated coal production could no longer balance out the gigantic losses in iron and steel.[160] In order to prevent the bankruptcy of the Vereinigte Stahlwerke, the Brüning government had to take over Flick's Gelsenberg package for more than the market value. There was no need to fear that the following presidential government, which brought down the Social Democratic-led Prussian government by means of a quasi-coup d'état, would use this majority position in the steel trust to revise the wage policy of heavy industry. On the contrary, using emergency decrees, Papen's government finally created the long-demanded possibility of waiving the principle of collective bargaining for extra employment at the company level. However, this failed to have any great concrete effect.[161]

With this, the makeshift tripartism of industrial relations in mining, the collective bargaining partnership which had been enforced and reshaped by the state, was finally turned on its head. Even though the Zechenverband warned against the setting of wages by the state – it was especially displeased that in the emergency decree of December 1931 both wages and prices were equally reduced in answer to the devaluation of the pound – it accepted the risk that its assault on collective bargaining would play into the hands of those at the political level who favoured an authoritarian solution to the class conflict. However, at this point it was not yet predictable just how a Reichskanzler such as Hitler would solve the union question – a chancellor whose appointment

160. See ibid., pp. 370f. 'Pure' mining operations, such as the Harpener Bergbau AG, were able to keep the losses within bounds.
    161. On the Gelsenberg affair, see Henning Köhler, 'Zum Verhältnis Friedrich Flicks zur Reichsregierung am Ende der Weimarer Republik', in Mommsen et al. (eds), *Industrielles System*, pp. 878–83. For the connection between the wage demands of heavy industry and Papen's emergency decree, see Weisbrod, 'Tariffesseln', pp. 316ff., and Tschirbs, *Tarifpolitik*, pp. 437ff.

was also supported by 'very prominent men of the mine district', who were expecting in return a certain degree of moderation in the party's behaviour towards the Papen government and in its quasi-socialist mass propaganda.[162] With a few exceptions, the majority of the Ruhr entrepreneurs in heavy industry hoped to the very end for a Papen option that possibly incorporated Hitler. However, the attack of heavy industry on collective bargaining not only contributed decisively to the delegitimation of the social order of the Weimar Republic, but it also tied the hands of these entrepreneurs politically. Precisely because the heavy industry magnates – incorrectly – viewed the political consensus behind collective bargaining to be the real cause of their economic problems, a restabilization of the Weimar system was considered unacceptable in their eyes under any circumstance, even if the price for this was Hitler's chancellorship.[163]

The Price of Freedom: From National Socialist Werksgemeinschaft to Co-determination

The 'healthy economy in a strong state', in Carl Schmitt's words,[164] demanded by the entrepreneurs of heavy industry, meant not only being freed from the constraints of collective bargaining, but also overcoming the political system of Weimar. This political view of entrepreneurial freedom coloured their understanding of industrial relations, making it into a political defensive against the unions. Independent of the still controversial issue as to whether the Weimar economy as a whole threatened to fall apart under the pressure of the supposedly unjustifiable development (measured

162. These were the words of the mining lobbyist August Heinrichsbauer in a letter to Gregor Strasser dated 20 September 1932, reprinted in Günter Plum, *Gesellschaftsstruktur und politisches Bewußtsein in einer katholischen Region 1928–1933*, Stuttgart 1972, pp. 301ff. On the behaviour of the Ruhr mining industry towards the NSDAP, see Reinhard Neebe, *Großindustrie, Staat und NSDAP 1930–1933. Paul Silverberg und der Reichsverband der Deutschen Industrie in der Krise der Weimarer Republik*, Göttingen 1981, pp. 117ff., 142ff.
163. The political implication of this corporate interest thus cannot be overruled by the evidence that 'big business' is not to be made personally and directly responsible for Hitler's seizure of power as Henry A. Turner argues in *German Big Business and the Rise of Hitler*, Oxford 1985.
164. See the speech given by Carl Schmitt at the convention of the Langnamverein on 23 November 1932 in *Mitteilungen des Langnamvereins* 1, 1932, pp. 13–32.

against the success of rationalization) in real wages,[165] they did not
attempt to come to terms with organized labour even in the periods
of economic recovery. During the period of prosperity before the
First World War, mining employers saw little reason for such a
settlement due to their entrenched position in the syndicate; during
the stabilization period of the Weimar Republic, only the state
protection of collective bargaining deterred them from fundamentally rearranging the postwar social order. In the end, it was their
organizational strength, protected by the syndicate and backed up
by the tie-in with iron and steel, that made them considerably
resistant to union demands. This was so even despite the essentially
self-induced cost pressure that resulted from having not made the
necessary adjustments to the market.

Characteristically, their own vision of an alternative order of
industrial relations evolved to counteract and match the growing
union pressure during periods of economic prosperity or recovery.
Thus they pitted the non-strike company union movement against
the prewar idea of collective wage contracts, and the idea of the
so-called plant community (Werksgemeinschaft) against the Weimar tripartism. Ultimately, both were aimed at maintaining or
re-establishing the union-free company and the 'free' labour contract, which would indeed become a reality in the Third Reich,
although not at all in the form of autonomous self-determination
for employers. The mining employers would soon painfully realize
that the long-sought entrepreneurial freedom had its political price.

The Werksgemeinschaft idea was first conceived as an intensification of the company welfare policy to counter the work of the
company councils, and was developed during the stabilization
period of the Weimar Republic into a global concept of 'social
rationalization' that concealed neither its anti-union thrust nor its
mission of national education.[166] Basically, the purpose of the

165. On the controversy over this thesis, see Knut Borchardt, 'Zwangslagen und
Handlungsspielräume in der großen Wirtschaftskrise der frühen zwanziger Jahre'
(first published in 1979), in K. Borchardt, Wachstum, Krisen und Handlungsspielräume
in der Wirtschaftspolitik, Göttingen 1982, pp. 165–82; Carl-Ludwig Holtfrerich, 'Zu
höhe Löhne in der Weimarer Republik?', Geschichte und Gesellschaft 10, 1984, pp.
326–76; and Charles S. Maier, 'Die Nicht-Determiniertheit ökonomischer Modelle.
Überlegungen zu Knut Borchardts These von der "kranken Wirtschaft" der Weimarer Republik', Geschichte und Gesellschaft 11, 1985, pp. 275–94.
166. See Michael Schneider, Unternehmer und Demokratie. Die freien Gewerkschaften in der unternehmerischen Ideologie der Jahre 1918 bis 1933, Bonn–Bad Godesberg
1975, pp. 68ff., 175ff.

Deutsches Institut für Technische Arbeitsschulung (DINTA, German Institute for Technical Training) of the Gelsenberg engineer Karl Arnhold, supported by heavy industry since 1925, was to strengthen the loyalties of the labour force to the company through increased training of both an occupational and nationalist nature. Simultaneously, it aimed at de-escalating the conflict over workplace control with the help of scientific labour studies along the lines of the rationalization movement.[167] In this way, the nostalgic wishful thinking of the employers of 'fusing' the labour force to the purpose of the national task of production, to the exclusion of the troublesome unions,[168] received a modern, technocratic jolt which anticipated the linkage of archaic and modern elements in the National Socialist ideology of the company community (Betriebsgemeinschaft).[169]

Following the brutal break-up of the unions, which began with an assault on union headquarters in March 1933 and was symbolized by the National Socialist version of a 'National Labour Day' on 1 May, the way was clear for the realization of the ideals of the Werksgemeinschaft in union-free industrial relations.[170] However, the demanded 'free' labour contract seemed not so desirable afterall. The Zechenverband preferred instead to renew the collective contracts it had routinely cancelled by mail. This was done not in order to protect the threatened unions, but instead to avoid the protest of the National Socialist Factory Cell Organization (NS-Betriebszellenorganisation or NSBO) on the one hand, and on the other, to be able to enjoy the benefits of a collective wage contract

---

167. See Schwenger, *Sozialpolitik*, pp. 68ff.; see also Peter Hinrichs and Lothar Peter, *Industrieller Friede? Arbeitswissenschaft und Rationalisierung in der Weimarer Republik*, Cologne 1976, pp. 70ff.
168. See the comments by von Löwenstein at the nineteenth regular general assembly of the Zechenverband, 23 March 1926, p. 13: the struggle for 'the emotional attitude of the working class to work', in Arnhold's view, would have to lead to 'the most intimate co-operation' without the insertion of political personalities from outside the company, 'in order that the fusion process of person to person, of entrepreneur to worker can take its course undisturbed and can lead us to more constructive heights of economic, social and human solutions and to true freedom'.
169. See Tim Mason, 'Zur Entstehung des Gesetzes zur Ordnung der nationalen Arbeit vom 20. January 1934: Ein Versuch über das Verhältnis "archäischer" und "moderner" Momente in der deutschen Geschichte', in Mommsen et al., *Industrielles System*, pp. 322–51.
170. See Tim Mason, *Arbeiterklasse und Volksgemeinschaft. Dokumente und Materialien zur deutschen Arbeiterpolitik 1936–1939*, Opladen 1975, introduction, pp. 17ff.

without collective bargaining.[171] This ideal was to materialize in the official wage-setting policy of the National Socialist labour trustees (*Treuhänder der Arbeit*) who issued wage ordinances following a trial period in which collective bargaining was officially forbidden. The trustee administration was set up in the former arbitration districts and the majority of the posts were filled to begin with by personnel friendly to the employer. However, in 1934 a personnel reshuffle took place which brought in more people close to the party. Even the trustee for Westphalia, Thyssencollaborator Klein, whose plans for a corporative state were no longer opportune, had to bow to a company lawyer and party-man from Saxony. In the consultation committees attached to the trustees the employers could exert their influence on the workers' representatives, who were, however, more committed to the party than to the labour force. The shift wages were fixed at about the crisis level (the reintroduction of the unemployment insurance contributions waived by the Brüning government compensated for the small wage increase) and the granting of production bonuses was left to the discretion of the company management.[172] The 'de-unionization' of collective industrial relations was virtually complete above the company level.

However, a prerequisite was that the danger of a united fascist union be warded off. This danger first arose from the union agitation of the NSBO, and once union property had been taken over by the German Labour Front (Deutsche Arbeitsfront, DAF), it emerged from the expansionist social dynamics of a mass organization with quasi-compulsory membership. In Ruhr mining the NSBO, under the cover of alliances with other lists, had acquired a minority position in the company council elections of 1931 (4 per cent) that was quite considerable on the national scale but modest compared to its own goals. With the violent elimination of communist competition, it was able to expand its position in the spring of 1933 to as much as a third of the vote in some districts.[173] The

171. See Tschirbs, *Tarifpolitik*, pp. 436ff. See also Klaus Wisotzky, *Der Ruhrbergbau im Dritten Reich. Studien zur Sozialpolitik im Ruhrbergbau und zum sozialen Verhalten der Bergleute in den Jahren 1933 bis 1939*, Wuppertal 1983, pp. 75ff.
172. On wage policy in general, see Tilla Siegel, 'Lohnpolitik im nationalsozialistischen Deutschland', in Carola Sachse et al. (eds), *Angst, Belohnung, Zucht und Ordnung. Herrschaftsmechanismen im Nationalsozialismus*, Opladen 1982, pp. 54–139; for the same in mining, see Wisotzky, *Ruhrbergbau*, pp. 42ff.
173. See Wisotzky, *Ruhrbergbau*, pp. 31ff. See also Gunther Mai, 'Die Nationalsozialistische Betriebszellenorganisation. Zum Verhältnis von Arbeiterschaft und

Germany 181

new regime acted quickly to check the social revolutionary impulse by denying the NSBO union status and to prevent the threatening infiltration of the NSBO by the leaderless union rank and file. The Charter of Labour (Arbeitsordnungsgesetz) of January 1934, prepared by the former legal adviser of the Zechenverband, Dr Mansfeld, regulated the scope of responsibility of the trustees and replaced the company councils with so-called councils of trust (Vertrauensrat) under the direction of the 'company leader'. This law satisfied the expectations for advancement of the NSBO functionaries and at the same time established the legal form for the National Socialist relationship between 'leaders' and 'followers' in the context of entrepreneurial Werksgemeinschaft.[174]

The 'disciplinary authority of the private sector of the economy' (Mason) that was granted the employer in the Charter of Labour, was not in any way effectively limited either by the propagandized demand of the German Labour Front to establish model company ordinances or by the mild assertiveness of the councils of trust. Already after two elections (1934/35), the true results of which were not published due to the impact of the election boycott, the politically reliable representatives simply remained in office. In the Ruhr region the labour force did show signs of dogged resistance in the council elections despite propaganda claiming the opposite: unpopular candidates were crossed off the lists that had been worked out by the NSBO factory cells and company management, ballots were invalidated or elections boycotted altogether.[175] The results were, however, subsequently corrected by trustee appointments, and members of the few councils of trust who were continuing to exercise company council functions and thus placing demands on the company were quickly redeployed from their posts or fired.[176] One could no longer speak of industrial 'relations' on

---

Nationalsozialismus', *VfZ* 31, 1984, pp. 573–613, as well as Volker Kratzenberg, *Arbeiter auf dem Weg zu Hitler? Die nationalsozialistische Betriebszellenorganisation, ihre Entstehung, ihre Programmatik, ihr Scheitern 1927–1934*, Frankfurt 1987.

174. See Wolfgang Spohn, 'Betriebsgemeinschaft und innerbetriebliche Herrschaft', in Sachse, *Angst*, pp. 140–208; see also Andreas Kranig, *Lockung und Zwang. Zur Arbeitsverfassung im Dritten Reich*, Stuttgart 1983.

175. See Wisotzky, *Ruhrbergbau*, pp. 104ff. and individual results in the appendix, pp. 273ff. On the other hand, the Bergbau Verein interpreted the relatively high voter trim-out and the confirmation of most of the NSBO spokesmen as a success for the new Betriebsgemeinschaft: *Jahresbericht 1934*, Essen 1935, pp. 73ff.

176. See Michael Zimmermann, '"Betriebsgemeinschaft" – der unterdrückte Konflikt. Aus den Protokollen des Vertrauenrates der Zeche "Friedrich der

the company level in the terrorized, shrunken form that the
National Socialist organization of labour had taken.

The conservative coalition partners and functional elite, however, still warned against a 'second revolution'. When Hitler, for
his own reasons, grasped the opportunity to sacrifice his storm-
troops in the so-called 'Röhm putsch' of June 1934 in the course of a
brown 'massacre of St Bartholomew', these fears were permanently put to rest. The German Labour Front was also finally
forced to give up union functions, and its responsibility was reduced to the administration of social welfare and efficiency
campaigns.[177] The DINTA became the labour research institute of
the Labour Front in which '*Menschenökonomie*' was further conducted now under a National Socialist outlook. The well-tried
'Harpen method' of training to acquire a hewer's licence (*Hauer-
schein*), which had been required since 1925 by mining authorities,
now gave mining work the semblance of a career. Such company
training schemes, through which the employers intervened for the
first time in the occupational domain of recruiting the piece-work
crews underground in addition to the nationalistic and athletic
schooling for character and leadership, thus competed increasingly
with the efforts of the Labour Front in the area of occupational
education.[178] The demand of the Labour Front to assume responsibility for apprenticeships and advanced occupational training from
the companies was successfully countered by the employers, who
could count on the support of the mining authorities, for their
expertise as well as their Herr-im-Haus position. It was not only
the technical requirements of conveyor operations that made better
training necessary; the training periods also enabled a better supervision, especially of the young miners, who were actually committed by the labour offices to training contracts for several years,

Große"', in Lutz Niethammer (ed.), *Die Menschen machen ihre Geschichte nicht aus
freien Stücken, aber sie machen sie selbst . . .*, Bonn 1985, pp. 171f. See also
M. Zimmermann, '"Ein schwer zu bearbeitendes Pflaster": der Bergarbeiterort
Hochlarmark unter dem Nationalsozialismus', in Detlev Peukert and Jürgen
Reulecke (eds), *Die Reihen fast geschlossen: Beiträge zur Geschichte des Alltags unter dem
Nationalsozialismus*, Wuppertal 1981, pp. 65–84.
  177. See Tilla Siegel, 'Rationalisierung statt Klassenkampf. Zur Rolle der Deut-
schen Arbeitsfront in der nationalsozialistischen Ordnung der Arbeit', in Hans
Mommsen and Susanne Willems (eds), *Herrschaftsalltag im Dritten Reich*, Düsseldorf
1988, pp. 97–224.
  178. See Schwenger, *Betriebliche Sozialpolitik*, pp. 84ff.; as well as Wisotzky,
Ruhrbergbau, pp. 193ff.

starting from 1938. The Labour Front did not succeed either in
intruding on the competence of employers in matters of training
with their favourite propaganda project, the annually held occupational competitions.
  Finally, the increased efforts of company welfare policy won the
self-serving support of the Labour Front. The office 'Schönheit der
Arbeit' (The Beauty of Labour) looked after the shower rooms, and
the Frauenamt (Women's Office) attempted unsuccessfully to secure a greater influence over both the deployment and reporting of
plant social workers.[179] Even in the continual competition for
responsibility in company welfare policy, the general management
of the mines maintained the upper hand. The Labour Front was
able to pull off some spectacular individual actions, as in 1934
when, by unilaterally declaring its readiness to foot the bill it
extracted from the reluctant mining industry the payment of 100
per cent vacation remuneration that had been suspended since the
crisis. Just like the multifarious leisure opportunities offered by the
Labour Front office 'Kraft durch Freude' (Strength through Joy),
such psychological successes were the only effective social policy
achievements of the specific National Socialist form of industrial
relations for the miners in the early years of the Third Reich and
affected the mining employers only minimally.[180] They were able
to keep the Labour Front out of their enterprises for the most part
and made use of it for the purpose of their own production and
company ideology. GHH director Kellermann, chairman of the
RWKS, even refused all co-operation or direct support whatsoever,
whether it was in the form of collecting Labour Front dues or
merely providing a telephone connection for the councils of
trust.[181]
  Apart from their political disfranchisement, the material benefits
for the miners were rather modest. Unemployment in mining
decreased only gradually because the shift production yield of the
miners underground increased between 1930 and 1936 by nearly a
third (from 1,687 kg to 2,199 kg) due to further rationalization
measures, so that the same total yield could be produced by a

  179. See Wisotzky, *Ruhrbergbau*, pp. 201ff.
  180. Ibid. pp. 92ff. Typical for the symbolic upgrading of miners was, for
example, the preference given them for cruises or the readiness to provide special
trains to the soccer championship games of Schalke 04 in Berlin in 1934.
  181. See John Gillingham, *Industry and Politics in the Third Reich. Ruhr coal, Hitler
and Europe*, London 1985, pp. 44ff., and J. Gillingham, 'Die Ruhrbergleute und
Hitlers Krieg', in Mommsen and Borsdorf, *Glück auf*, pp. 325–43, especially pp. 334f.

184   *Bernd Weisbrod*

labour force that was smaller by a third. An improvement in
income of about 20 per cent was caused chiefly by the reduction of
short-time work, but was roughly compensated for by inflation
and shortages of food, as even the Dortmund Gestapo admitted.[182]
From 1936 on, however, a manpower shortage was created by the
full employment triggered by the rearmament boom within the
framework of the four-year plan. It was this shortage which caused
the revival of spontaneous and individual protest behaviour and
rudimentary negotiation forms of industrial relations in Ruhr mining. The mining employers tried to avoid wage increases by
resorting to special company benefits such as potato money or rent
reductions, but were in the end forced to renegotiate wage rates,
although without being able effectively to counter the malingering
and short-time work that was seriously on the rise in 1938.[183]

Since the Zechenverband had become superfluous as a party in
collective bargaining, it was left to the social policy department of
the Bergbau Verein and the RWKS-dominated Ruhr district group
of the mining branch of industry, which pooled their personal and
organizational resources, to co-ordinate the defence strategy. The
miners, who were no longer so easily intimidated politically, made
barely camouflaged use of the advantageous market conditions.
They clearly benefited from the surge in demand, which was
accelerated at home by the changing production targets of the
four-year plan, and abroad by the diversion of British coal to the
French market, where the Popular Front government had introduced the forty-hour week. Yet mining did not succeed in making
a breakthrough either into the expanding energy industry or into
the profitable by-product business, where the chemical industry
got the better of it especially in the production of synthetic gaso-
line. The European cartel agreements also never got off the ground,
so that little could be done for strategic reorientation away from the
production of raw materials, in which there were few opportunities
left for more rationalization.[184]

182. See Wisotzky, *Ruhrbergbau*, pp. 74f., 86f. On the general development, see
also Rüdiger Hachtmann, 'Lebenshaltungskosten und Reallöhne während des
"Dritten Reiches"', *Vierteljahresschrift für Sozial- und Wirtschaftsgeschichte (VSWG)*
75, 1988, pp. 32–73.
183. See Wisotzky, *Ruhrbergbau*, pp. 215ff. See also Gillingham, *Ruhrbergleute*,
pp. 336f. for examples of punitive action taken at the beginning of the war.
184. On this see extensively Gillingham, *Industry*, pp. 68ff. On the rise of the
chemical industry, see Peter Hayes, *Industry and Ideology. IG Farben in the Nazi Era*,
Cambridge 1987.

As the shortage situation continued after 1936, the room for manoeuvre open to mining employers narrowed politically and economically. Both the excessive number of shifts and the worsening of the conditions of Knappschaft pensions ran up against resistance from the labour force and from Ruhr-Gauleiter Terboven. Considering the sparse labour market, the Ruhr mining industry had to make do, despite protest, with politically unpopular Saar miners who had been recruited forcefully, while National Socialist prestige projects, such as the construction of the Hermann-Göring-Works to smelt the low-grade Salzgitter ores, or of the Siegfried Line, enjoyed priority status.[185] In mining even the typical, pre-union means of a worker improving his wages by changing jobs emerged again. However, this time there was a clear trend towards switching to the nearby armaments industry, which was in a position to pass on the higher wage costs, as opposed to mining which stood practically under a price tutelage. In such a 'fight of one against all' (Wisotzky), so distant from any well-planned deployment of labour, the mining industry was left out in the cold because it could not muster a political counterweight to the Hermann-Göring-Works and because it had lost its leading position in wages to the metal industry. Press reports in NS publications on the 'slave-driving' and the 'tone underground' increased the pressure on mining all the more, yet they were more a reflection of the fears of the functionaries than of any actual danger.[186]

What finally forced the mining industry to compromise was not so much escalating labour resistance of crisis proportions as it was the simple economic interest in keeping one's ground and taking advantage of the politically prescribed task of increasing production. In this situation, the Labour Front assumed the function of a surrogate union and demanded that piece-rates as set in the company regulations be negotiated – outside the general wage scales – in the competent trustee committee. The negotiations on piece-work rates from 1936 to 1938 revealed the mining industry to be on the defensive, but the simulated tripartism did not prevent it from asserting its traditional Herr-im-Haus position. In some cases, better pay had been offered since 1936, but a support of the local

185. On the conflict of the Ruhr with the Hermann-Göring-Works, see Matthias Riedel, *Eisen und Kohle für das Dritte Reich. Paul Pleigers Stellung in der NS-Wirtschaft*, Göttingen 1973, pp. 155f., and Gerhard Mollin, *Montankonzerne und 'Drittes Reich'. Der Gegensatz zwischen Monopolindustrie und Befehlswirtschaft in der deutschen Rüstung und Expansion 1936–1944*, Göttingen 1988, pp. 52ff.
186. See Wisotzky, *Ruhrbergbau*, pp. 122ff.

aldermen in wage negotiations by elections, as the Labour Front had proposed, was rejected by mining management as being a 'return to parliamentary-Marxist customs'.[187] Mining management was also successful in retaining control of decision-making on the type of contract and especially in establishing the individual pay rate, which was so disliked yet which was facilitated by technical development. It was able to do this despite the trustee decision to grant a formal veto right to the work crews, as had been recommended by the Labour Front. The employers only compromised on general, long-term contracts. Above all, they clung to shift wages, regardless of the consequences. They attempted to relieve the wage issue by providing special private benefits – after 1936 with considerations ranked according to family and seniority status, after 1937 with holiday pay – or, like the Weimar model, by taking state subsidies for insurance contributions, which took about 6 per cent of the burden off miners and only increased that of the employers by about 2 per cent.[188]

However, none of this could prevent the shift production yield from falling after 1936, due to the deterioration of the age and qualification structure of the enlarged labour force, on the one hand, but also due to the 'normal' effect of a full employment economy without significant material production incentives. Therefore, certain developments such as absenteeism, a rate of job mobility equalling that of 1928, hidden wage conflicts and increased unrest within the Labour Front, the agency actually responsible for labour morale, should not necessarily be interpreted as signs of a widespread 'worker opposition', even if the struggle for the 'the soul of the worker' had not been clearly won either by the employer or by the regime.[189] In any case, the mining employers left no doubt that they would resort to calling on the Gestapo

---

187. Ibid., pp. 167ff., quote on p. 171.

188. On the pension insurance law of December 1937, see ibid., pp. 161ff. See also Martin H. Geyer, *Die Reichsknappschaft. Versicherungsreform und Sozialpolitik im Bergbau 1900–1945*, Munich 1987, pp. 319ff. It is typical that in order to gain their approval, the mining employers were promised that no price decreases would be made.

189. See Wisotzky, *Ruhrbergbau*, pp. 238ff., in disagreement with Timothy W. Mason, 'The Workers' Opposition in Nazi Germany', *History Workshop Journal* 11, 1981, pp. 120–37; see also the sceptical evaluation by Wolfgang Franz Werner, *'Bleib übrig'! Deutsche Arbeiter in der nationalsozialistischen Kriegswirtschaft*, Düsseldorf 1983, pp. 29f. On the dwindling potential for resistance in the miners' milieu see Detlev J. Peukert and Frank Bajohr, *Spuren des Widerstands. Die Bergarbeiterbewegung im Dritten Reich und im Exil*, Munich 1987.

should real insubordination occur. It was the mining employers, not the Labour Front, who first called for police and labour camps in the case of collective disobedience, once it became obvious that the terrorist elimination of the workers' organizations and their resistance groups had not been enough in the early years of the Third Reich to unlock the gate to eternal 'labour peace'.[190]

The 'November syndrome' (Mason), which evoked a panic fear, especially in Hitler, of a repetition of the workers' revolt of 1918–19, prompted the regime to strike out hard against the supposed hotbeds of political opposition, while reacting to the growing unrest in the labour force with well-chosen concessions. Yet the Labour Front had quickly to drop the union-like demands it made, such as a 15 per cent 'home front work bonus'.[191] No real danger threatened the mining employers until the shortage caused by the four-year plan occurred. Without additional manpower, the gigantic demand for coal materializing from the Hermann-Göring-Works for the smelting of inferior German ore against the impediments put up by the Ruhr enterprises, could only be met with an extension of work hours. Therefore in March 1939, Göring himself ordered work hours to be extended even beyond the prewar level to eight-and-three-quarter hours against the advice of the mining management. Although the proposal of Hugo Stinnes, which was adopted, to reward the extra output with a 200 per cent bonus could be watered down by the dismayed employers, the increase of shift pay by as much as 20 per cent placed a very heavy burden on total production costs without drastically improving production or productivity.[192]

These ad hoc measures were symptomatic of chaotic planning, and the consequences of such measures were disastrous. The unrest among the labour force due to the privileges awarded certain individual performances had to be countered with special actions (for instance, the distribution of bacon and pig-raising). Yet

190. In the only strike held in Ruhr mining during the Third Reich, the Gestapo was called in immediately (Wisotzky, *Ruhrbergbau*, p. 230).

191. Such a Labour Front memorandum from November 1938 led to the resignation of the responsible director, who was taken over by the mining industry, ironically enough.

192. See Klaus Wisotzky, 'Der Ruhrbergbau am Vorabend des Zweiten Weltkriegs. Vorgeschichte, Entstehung und Auswirkung der "Verordnung zur Erhöhung der Förderleistung und des Leistungslohns im Bergbau vom 2. März 1939"', in *VfZ* 30, 1982, pp. 418–61. See also the documents in Mason, *Arbeiterklasse*, pp. 563ff.

contrary to the past, mines such as Hibernia that were again entering into the red could not count on governmental compensation.[193] The politically blocked price of coal continued to stay fixed at the crisis level, and Göring's man in the new party combine (Reichswerke Hermann-Göring), Paul Pleiger, successfully banked on the disunity and weakening of heavy industry in the hopes of obtaining an independent coal base, be it through an exchange of aryanized brown coal pits (Petschek for Flick's share of Harpen) or through the takeover of the state-owned mines. When the war broke out, Thyssen emigrated and thus resigned himself to the effective expropriation of rights and shares of the Vereinigte Stahlwerke by the Salzgitter works and the Austrian Alpine Montangesellschaft respectively. Following the confiscation of Thyssen's mining property, Pleiger rose to become the largest coal producer of the Reich. Beginning in March 1941, he made himself, as director of the private general cartel Reichsvereinigung Kohle, the mining industry's intercessor against a plan for compulsory regulation modelled after the national food producing industry, the so-called *Reichsnährstand*, which the national coal commissioner threatened to implement.[194]

During the war, the paramount interest in production prevented any decision in this fundamental conflict between private 'monopolistic economy' and 'command economy' (Mollin). The catastrophical manpower shortage increased the dependence of the employer on state labour assignment offices and on their own labour force. Without a special wage incentive, workers could not be motivated to work the mandatory shifts on Sundays and holidays or to work productively in co-operation with workers inexperienced in mining for group wages.[195] The mining management harboured mixed feelings about the unavoidable militarization and brutalization of industrial relations that resulted from the deployment of civilian foreign workers, forced labourers from the east and prisoners of war. Yet the playing on the racist feelings of superiority and the self-interest of the German workers paid off to a degree: in December 1943 43 per cent of the (foreign) labour force, three-quarters of which were prisoners of war, were subordinate to them.[196] Following initial losses of up to 20 per cent, a dramatic

193. On the profit situation, see Gillingham, *Industry*, pp. 85ff.
194. See ibid., pp. 60ff.; see also Riedel, *Eisen*, pp. 271ff., as well as Mollin, *Montankonzerne*, pp. 102ff.
195. See Werner, '*Bleib übrig*', pp. 235ff.
196. See Herbert, *Fremdarbeiter*, pp. 220ff., 281ff., as well as Gilligham, *Industry*,

slump in production was prevented less by the 'occupational ethic' linked to the introduction of the mining apprenticeship in the autumn of 1940 as it was by the instinct for survival of the miners and the exploitation of Russian prisoners of war.

No calculated stock-taking can illuminate the entire extent of the destructive exploitation of manpower and natural resources during the war, let alone the human sacrifices and loss of life due to slave labour. The hewers suffered from physical exhaustion, deteriorating supplies and bomb damage, although their income per shift increased by 17 per cent (7 per cent in real wages) during the war (mining pay on the whole was up plus 6 per cent, that is minus 4 per cent in real wages). At the same time the asset erosion of the mines greatly exceeded the accountable losses (from 1.25 RM/ton to 8.49 RM/ton).[197] Above all employers in heavy industry lost the 'economic freedom' for which they had supported the union-free system of industrial relations in the Third Reich, and were even forced finally once again to give up their hard-fought control over the labour force. Hard driving and the threat of a transfer to the front could not change the fact that the output of the miner in the pits depended primarily on conditions that were out of the reach of the influence of the mining managements and even the intervention of the Gestapo. The remaining German labour force may have been disunited and apolitical, but from the contradictory experiences of re-establishing autonomy in mining practices and the absolute delegitimization of the politically compromised mine-owners, these miners were able to derive the demand to redefine 'free' industrial relations on their own terms after the German defeat – this time without the mining entrepreneurs.

The socialization of mining, a goal that the Christian union leader Imbusch had last made part of the programme during the Great Depression, appeared to become reality all on its own. In the mines and district communes, worker committees were formed spontaneously to keep operations going during the political power vacuum of the transition period, to rid management and administrations of Nazi activists, and to maintain the meagre supply.[198]

pp. 118ff. See also Christian Streit, *Keine Kameraden. Die Wehrmacht und die russischen Kriegsgefangenen 1941–1945*, Stuttgart 1978, pp. 268–85.

197. See Werner, '*Bleib übrig*', p. 238; Gillingham, *Industry*, p. 135.

198. See Lutz Niethammer, Ulrich Borsdorf and Peter Brandt (eds), *Arbeiterinitiative 1945. Antifaschistische Ausschüsse und Reorganisation der Arbeiterbewegung in Deutschland*, Wuppertal 1976, pp. 281ff. On the political power vacuum from the

This self-organization of the districts in a 'society falling apart', a pragmatic variation of the socialization movement that followed the First World War, established the grass-roots model of politics that would be typical of the miners' milieu for the entire postwar period.[199] At the same time, it cleared the way for the rebuilding of the political labour movement which the British occupational authorities at first limited to the lower levels of organization. Because of the deliberate delay in developing the unions and the parties, the company became the centre of political activity. The communist company councillors were able to win out on this level, clearly so in October 1945 and afterwards barely so and only in the northern districts of the Ruhr region: in March 1946 they won 42.1 per cent of the mandates in forty-four large-scale plants (SPD 25.8 per cent, CDU 17.9 per cent) and 38.8 per cent in the entire Ruhr region (SPD 36.8 per cent, CDU 14 per cent).[200]

The founding in December 1946 of the highly organized Industrial Trade Union for Mining (Industrieverband Bergbau) did result in a certain shift in power. This industry-wide union with its 320,000 members deliberately brought together all the political sections of the German labour movement after its defeat at the hands of the Nazis. With a small majority, the Social Democrats were able to assert their opinions over those of the Communists in the issues of both the chairmanship and organization (the principle of locally-based associations instead of factory-based groups).[201] The outbreak of the Cold War (beginning with the London Foreign Ministers' Conference in December 1947) and the decision to

point of view of the company councils, see also Alexander von Plato, *'Der Verlierer geht nicht leer aus'. Betriebsräte geben zu Protokoll*, Berlin 1984, pp. 93ff.

199. From the various results of the oral history project 'Lebensgeschichte und Sozialkultur im Ruhrgebiet 1930 bis 1960' (LUSIR) led by Lutz Niethammer, see Michael Zimmermann, '"Geh zu Hermann, der macht das schon." Bergarbeiterinteressenvertretung im nördlichen Ruhrgebiet', in Lutz Niethammer (ed.), *"Hinterher merkt man, daß es richtig war, daß es schiefgegangen ist.' Nachkriegserfahrungen im Ruhrgebiet*, LUSIR vol. 2, Berlin 1983, pp. 277–310. See also Zimmermann, *Schachtanlage*, pp. 212ff.

200. See Christoph Kleßmann and Peter Friedemann, *Streiks und Hungermärsche im Ruhrgebiet 1946–1948*, Frankfurt 1977, pp. 69f., and Christoph Kleßmann, 'Betriebsräte und Gewerkschaften in Deutschland 1945–1952', in Heinrich-August Winkler (ed.), *Politische Weichenstellung im Nachkriegsdeutschland 1945–1953*, Göttingen 1979, pp. 44–73.

201. On the founding history, see *Die Gewerkschaftsbewegung in der Britischen Besatzungszone. Geschäftsbericht des Deutschen Gewerkschafts-Bundes (Britische Zone) 1947–1949*, Cologne 1949. The degree of unionization in mining can be estimated to have been already about 90 per cent in 1947, see Martin Martiny, 'Die Durchsetzung

establish the Marshall Plan (April 1948) also finally undermined the strong position of the KPD among miners and prepared for the 'Social Democratization' of the district in the long run.[202] However, the communist danger played an important role in the early British occupation policies in so far as the socialization of the primary resource industries appeared in the view of the new Labour government to be the only progressive alternative both to politically discredited private capitalism and to communism, even if national security interests were to be decisive for the course of events in the end.

In any case, the resolve of the British to expropriate the heavy industry conglomerates could not be questioned.[203] Forty-four RWKS representatives and 116 – in part politically uninvolved – leaders of heavy industry were arrested (September/November 1945); the mines as well as the plants of the iron and steel industry were confiscated (December 1945 and August 1946); and British Foreign Minister Bevin pledged in October 1946 to the House of Commons that the basic resource industries had to be 'run as the property of the people by the people, notwithstanding the international controls which will guarantee that they will never again be able to endanger Germany's neighbours'. Thus, both the French plan to separate the Ruhr region from Germany as well as the original British plan to internationalize the key industries were given up in favour of the idea of advancing socialization on the level of the new 'Länder' in order to counter the 'Russian danger' on the central level of decision–making for Germany as a whole.

---

der Mitbestimmung im deutschen Bergbau', in Mommsen and Borsdorf, *Glück auf*, pp. 389–414 (on p. 392).
202. On the general constitutional conditions, see Lutz Niethammer, 'Rekonstruktion und Desintegration: Zum Verständnis der deutschen Arbeiterbewegung zwischen Krieg und Kaltem Krieg', in Winkler, *Weichenstellungen*, pp. 26–43, and on the meaning of the private economy option of the Marshall Plan for the labour movement, see Othmar Nikola Haberl and Lutz Niethammer (eds), *Der Marshall-Plan und die europäische Linke*, Frankfurt 1986, especially chapter 3.
203. For the following see Horst Lademacher, 'Die britische Sozialisierungspolitik im Rhein-Ruhr-Raum 1945–1948', in C. Scharf and H.-J. Schröder (eds), *Die Deutschlandpolitik Großbritanniens und die britische Zone*, Wiesbaden 1979, pp. 51–92. See also Rolf Steininger, 'Reform und Realität: Ruhrfrage und Sozialisierung in der anglo-amerikanischen Deutschlandpolitik 1947/48', *VfZ* 27, 1979, pp. 167–240; R. Steininger, 'Großbritannien und die Ruhr', in Walter Först (ed.), *Zwischen Ruhrkontrolle und Mitbestimmung*, Cologne 1982, pp. 11–63; as well as R. Steininger, 'Die Sozialisierung fand nicht statt', in Josef Foschepoth and Rolf Steininger (eds), *Die britische Deutschlandpolitik 1945–1949*, Paderborn 1985, pp. 135–50.

For practical reasons alone this concept was hardly convincing,
since the British relied on the principle of managerial continuity in
the business organizations set up in the Third Reich as well as – for
'reasons of efficiency' – in the practical running of the mines. With
the increasing dependence of the British on the financial involvement of the USA in the Bizone, this vision finally faded in the face
of American reservations that such a far-reaching decision as
socialization should be reserved for a freely-elected central government to provide for legislatively and should anyhow take a back
seat to the extremely necessary increase in production.[204] With this,
the occupation authorities found themselves in growing contradiction to the broad political consensus in favour of an immediate
socialization of the coal industry. The first elected state parliament
of North Rhine-Westphalia and the industry-wide Industrieverband
band Bergbau considered creating a sort of 'gemeinwirtschaftliche
Ordnung', or co-operative order, in which the mines were conceived of as becoming 'social unions' – with equal representation of
the miners and the holding company that would hold the confiscated mines in trust in a 'shareholder assembly' and a 'spokesman
for the miners' on the 'executive board' of the mines – and in which
concrete ideas were to be developed on the democratic management of the coal industry by a 'coal council', oriented on the
Weimar model.[205]

The law on the socialization of the coal industry that was passed
by the North Rhine-Westphalian state parliament in August 1948
with the votes of a 'union axis' (Rudzio) against those of the Free
Democratic Party (FDP) – with part of the CDU abstaining – failed
to go into effect due to the objections raised by the occupation
force, just as had the Hessian efforts at socialization.[206] In connection with the renegotiation of the Bizone treaty in November 1947,
the British and the Americans had already agreed on a substantial
preliminary decision for the coal industry while postponing the

---

204. See Wolfgang Rudzio, 'Das Ringen um die Sozialisierung der Kohlewirt-
schaft nach dem Zweiten Weltkrieg', in Mommsen and Borsdorf, *Glück auf!*, pp.
367–88; see also Werner Abelshauser, *Der Ruhrkohlnbergbau seit 1945. Wiederaufbau,
Krise, Anpassung*, Munich 1984, pp. 20ff.
205. See the documentation in Martin Martiny and Hans-Jürgen Schneider (eds),
*Deutsche Energiepolitik seit 1945. Vorrang für die Kohle. Dokumente und Materialien zur
Energiepolitik der Industriegewerkschaft Bergbau und Energie*, Cologne 1981, doc. 1–3.
206. The American position was, however, more differentiated than that of
Military Governor Clay. See Dörte Winkler, 'Die amerikanische Sozialisierungs-
politik in Deutschland 1945–1948', in Winkler, *Weichenstellungen*, pp. 88–110.

issue of ownership. Managerial control of the British North German Coal Control was turned over to the Deutsche Kohlenbergbauleitung (DKBL). In turn, the sole responsibility for the DKBL was transferred to the Haniel director Heinrich Kost, who acted as general director. The IV Bergbau could appoint two of the eight managing directors and enjoyed parity representation on the advisory board. It did criticize the close personnel connections of the DKBL to the conglomerates, in which even politically incriminated business managers were again being employed. But despite considerable opposition from its own members, it recognized a chance for the future organization of an integrated, industry-wide mining corporation with central direction of production and marketing, a chance which took priority over questions of ownership and union co-determination at first.[207]

The policy of increasing production through productivity premiums, which the unions had to accept in the interests of the labour force, also acted as a precedent against an immediate socialization.[208] As a result of the destructive exploitation of resources and manpower as well as the over-aging of the labour force, production had dropped to less than half of the prewar level. The food supply became even worse than it had been in the last winter of the war. The physically exhausted miners 'played hooky' in order to organize the necessities of life – half of the miners' housing was destroyed or heavily damaged. They refused to work extra shifts even for an improved supply of domestic fuel without a good, comparable value in foodstuffs as long as coal deliveries abroad were being requisitioned at far below the international market price. Under such circumstances, the introduction of a points system for special rations in January 1947 acted as a 'miniature Marshall Plan' (Borsdorf).

The execution of the plan rested with the 'supply centre of German mining', the successor organization of the Bergbau Verein, or more specifically with the district groups that mediated between the mines and the union. Although all appearances of 'slave driving' were avoided and the production plans were set at the company level in joint committees, the plan was received with a general dissatisfaction that was expressed in hunger marches and in

207. See Martiny, *Mitbestimmung*, pp. 396f.
208. See Ulrich Borsdorf, 'Speck oder Sozialisierung? Produktionssteigerungskampagnen im Ruhrbergbau 1945–1947', in Mommsen and Borsdorf, *Glück auf!*, pp. 345–66. See also Abelshauser, *Ruhrkohlenbergbau*, p. 30ff.

the increase of KDP votes to over 20 per cent in the cities of the Ruhr in the first state parliamentary elections in April 1947.[209] Not until the premium for good work attendance was reworked into one for productivity, and CARE packages and imported goods were distributed in special allocations, a practice viewed sceptically by the union, did the labour force once again reach the prewar level of production, and the miners were once again kings in the calorie hierarchy. This provided the prerequisite for overcoming the coal shortages, but it also weakened the argument that an increase in production was first of all dependent on the socialization of mining.

For the time being the danger of socialization of the mining companies had thus been averted, but not the danger that stemmed from the stipulations of the Potsdam agreements on disarmament and demilitarization and on the decentralization of German economic life. The iron and steel industry had been subordinated after October 1946 to a German trustee administration under the Vereinigte Stahlwerke director Dinkelbach, and a form of company-based parity co-determination had been worked out in the course of the break-up of the integrated Ruhr Konzerne after 1947 which guaranteed the employees in the now independent smelting works half the seats on the advisory council and one seat for a labour director with equal rights on the executive management board. This was the essential form of co-determination in the coal and steel industries which was finally established by legislation in May 1951, following intense debates, and was to place labour relations in mining on a completely new basis.[210]

The 'steel model' of company co-determination had emerged from the conflict over the effects of the restrictive Allied Company Council Act no. 22 from April 1946.[211] The employers were pressed to make concessions which were meant at least to re-establish the competence stipulated in the 1920 law on company

209. See Kleßmann and Friedemann, *Streiks*, pp. 40ff. Statewide, the KDP obtained 14 per cent of the vote (SPD 31 per cent, CDU 37.5 per cent); p. 30.
210. On this point, see extensively Horst Thum, *Mitbestimmung in der Montanindustrie. Der Mythos vom Sieg der Gewerkschaften*, Stuttgart 1982; see also the introduction to the source edition: *Montanmitbestimmung. Das Gesetz über die Mitbestimmung der Arbeitnehmer in den Aufsichtsräten und Vorständen der Unternehmen des Bergbaus und der Eisen und Stahl erzeugenden Industrie*, prepared by Gabriele Müller-List, Düsseldorf 1984, as well as G. Müller-List, 'Die Entstehung der Montanmitbestimmung', in Först, *Ruhrkontrolle*, pp. 121–42.
211. See Gloria Müller, *Mitbestimmung in der Nachkriegszeit. Britische Besatzungsmacht – Unternehmer – Gewerkschaften*, Düsseldorf 1987, pp. 86ff. See also Thum, *Mitbestimmung*, pp. 31ff.

councils, including the participation of workers' representatives on the advisory board. In order to keep the company councils in check, the British steel authorities, however, considered it necessary to strengthen union influence by additionally ordering *union* representatives in the advisory council and by granting it the right to appoint a member of the executive management board who would be responsible for issues concerning the labour force. Several conglomerates such as Klöckner, Otto Wolff and GHH were prompted by the Economic Association of Iron and Steel (Wirtschaftsvereinigung Eisen und Stahl) to attempt even in January 1947 to undercut any such definite commitment to parity co-determination on the side of the steel trustee by themselves offering the union 'practical equality, that is equal rights but also equal responsibility on the part of "capital and labour"', in order to win its co-operation in their effort to keep the break-up of Ruhr Konzerne to a minimum. This tactical offer itself says little about the 'sympathy for co-determination' of heavy industry employers. But in any case, it cannot be ruled out that because of credit or structural considerations they perhaps even considered making a partial sacrifice to socialization in order to save their reprocessing operations, especially since even the limit on steel production set well below capacity in the first industrial plan could not be fully met at the time.[212]

The issue of co-determination in mining did not become acute until the task of industrial restructuring was transferred to the DKBL as stipulated in Law no. 75 of November 1948 'on the organization of German coal mining and the German iron and steel industry', which officially postponed socialization and linked for the first time the deconcentration of both branches with one another.[213] The coal side of the conglomerates had increasingly lost its character as an autonomous industry as early as the 1920s within the framework of vertical integration and, compared to the iron side, had continued to be disadvantaged during the rearmament boom of the Third Reich. The idea of completely separating the

212. See ibid., pp. 141ff. Also Volker Berghahn, *Unternehmer und Politik in der Bundesrepublik*, Frankfurt 1985, p. 212, evaluates the entrepreneurs' offer as a 'tactical move'. These are reprinted in *Montanmitbestimmung. Dokumente ihrer Entstehung*, Cologne 1979, doc. 15–17. On steel production, see Werner Abelshauser, *Wirtschaft in Westdeutschland 1945–1948. Rekonstruktion und Wachstumsbedingungen in der amerikanischen und britischen Zone*, Stuttgart 1975, pp. 147ff.
213. On the following, see Abelshauser, *Ruhrkohlenbergbau*, pp. 50ff.

mining companies and regrouping them into ten mining groups according to geological considerations, as proposed by the Allies, appeared to be problematic from the start due to technical reasons of production (fuel economy). The Industriegewerkschaft Bergbau (IG Bergbau) agreed in principle with the organizational reservations of the mining entrepreneurs, but it developed its own ideas in which mining companies of a sufficient size and with a central marketing organization were to seek out a Konzern-free co-operation with the iron industry. But just as before in iron and steel, union demands had to be met if their collaboration was to be secured for keeping the restructuring of industry within acceptable bounds.

The IG Bergbau however, did not use this opportunity primarily to improve company co-determination, but instead was content with parity staffing of the DKBL management and planning apparatus that had been reorganized against Kost's will. The union thus continued to hope to be able to control the supposedly socialization-ready mining industry as if from above.[214] The proposal made by representatives from both the entrepreneurs and the union in the DKBL to link about 30 per cent of the production of twenty-three new coal companies directly to iron and steel quickly became obsolete. The Allies did not want to permit a rate of more than 15 per cent and exempted the non-affiliated companies and the companies in foreign possession, representing roughly 40 per cent of the entire production, from the mandatory industrial reorganization. The Allied Law no. 27 of May 1950, that replaced Law no. 75, finally separated mining operations from the large iron and steel combines without any consideration of the regional economic context and even promised the old shareholders suitable compensation. Thus, the speculation of the IG Bergbau on a transformation of property relations and a permanent reorganization of the coal industry under its direct influence seemed to be thwarted forever.[215]

Even the 'steel model' of company co-determination came under pressure again. The British had intended this model only as a means to upgrade the unions and to 'deflect' (Müller) the debate over the

---

214. See Martiny, *Mitbestimmung*, pp. 398ff., and Norbert Ranft, *Vom Objekt zum Subjekt. Montanmitbestimmung, Sozialklima und Strukturwandel im Bergbau seit 1945*, Cologne 1988, p. 26.
215. The union ideas are documented in Martiny and Schneider, *Energiepolitik*, doc. 13 and 14.

rights of the company councils whereas the unions had hoped it
would open the door to socialization. By signing the Petersburg
protocols in November 1949, the new Adenauer government –
supported by the Klöckner owner Henle – had accepted international control of Ruhr coal distribution by the International Ruhr
Authority but had taken over the responsibility for decartelization
and monopoly control.[216] In return, the government enhanced its
sovereignty and increased the chances to put an end to the industrial
dismantling, especially of the steel works in the Ruhr region. A
year later, the Federal Ministry of Economics attempted on the
basis of its new responsibilities, while implementing Law no. 27, to
reinstate the property rights of the old owners in the new
companies by way of a stock trade-off and at the same time to
revise the occupation law on co-determination in independent steel
works according to the German laws regulating stock
companies.[217] This was all the more threatening since the unions
should have recognized in the continual top-level talks held since
January 1950 with the employers that a general extension of co-
determination could not be achieved by way of negotiation.
Neither did the union's expectation of an accord with 'progressive'
entrepreneurs to counter the 'reactionary' line of the North Rhine-
Westphalian employer associations materialize, nor could help be
expected from the government, whose legislative bill from
October 1950 did not even incorporate all of the concessions which
the employers were prepared to make.[218]

The governmental negotiations on the Schuman Plan that were
being held simultaneously were to bring a common control of the
market for Western European heavy industry by way of supra-
national control. These negotiations were like aces in the unions'
hand since the government, which had basically adopted the line of
the Konzerne on the issue of vertical integration, was highly
dependent on the unions in order to be able to present a unified
German position.[219] As a counter-deal the unions expected that

216. See Horst Lademacher, 'Das Petersberger Abkommen', in Först, *Ruhrkontrolle*, pp. 67–87.
217. See Thum, *Mitbestimmung*, pp. 62ff., and Abelshauser, *Ruhrkohlenbergbau*,
pp. 59ff.
218. On the negotiations being held between the social parties in Hattenheim,
Bonn and Maria Laach during the course of the year 1950, see Thum, *Mitbestimmung*, pp. 38ff., as well as *Montanmitbestimmung*, pp. XLIVff.
219. See John Gillingham, 'Die französische Ruhrpolitik und die Ursprünge des
Schumanplans', in *VfZ* 35, 1987, pp. 1–24. For the general background, see Alan S.
Milward, *The Reconstruction of Western Europe, 1945–1951*, Berkeley 1984, pp. 362ff.

parity co-determination be guaranteed by legislation. Although this meant that the IG Bergbau idea of a uniform reorganization of an effectively unaffiliated coal mining industry was finally given up, mining could now be incorporated into the 'steel model' of co-determination as if in return for giving up this concept.[220] The union action introduced by IG Metall to support this strategy of defence was fully supported, after some hesitation, by the IG Bergbau, 92.8 per cent of whose membership voted to strike in January 1951.

In the light of the 'politicization of this partial class conflict' (Thum), a labour dispute was no longer necessary to realize the unions' immediate goal.[221] The opposition of the leading industrial associations was dispelled by Adenauer with the assurance that the 'special regulation' for the coal and steel industry would not pre-judge any further legal regulation of workers' rights, which was finally to predetermine the defeat of the unions in the confrontation over the Company Constitution Act of 1952.[222] During the crucial phase of negotiations, the entrepreneurs' attempt to water down the parity representation on the advisory board and to establish in mining the subordination of the labour director on the executive management board also failed. On the issue concerning labour directors, it was hardly possible to retreat from the practice in the iron and steel industry. So, in order to prevent the failure of the entire negotiated package, the mining industry was forced to give up its opposition to the labour directors as 'intruders in the realm of the Bergassessoren'.[223] In addition, due to the stronghold of the politically united labour union in the left-wing of the CDU, even the stability of the government coalition was at stake in the final analysis.

The Montanmitbestimmungsgesetz (the law on co-determination in the coal and steel industry) of May 1951 cleared the way for signing the treaty for the European Coal and Steel Community and thus immediately for the re-establishment of private property relations in heavy industry, which the government did not hesitate

220. Thum, *Mitbestimmung*, p. 66.

221. According to Abelshauser (*Ruhrkohlenbergbau*, p. 61) the willingness to strike was indeed 'necessary', but it was not a 'sufficient stipulation' for success.

222. On the government's commitment, see doc. no. 75 and 78f. in *Montanmitbestimmung*, pp. 227f., 236f.

223. See Ranft, *Objekt*, pp. 30ff., 103ff. (quote on p. 107), and Thum, *Mitbestimmung*, pp. 79ff.; the documentation of the negotiated positions in *Montanmitbestimmung*, doc. no. 80ff.

to initiate by way of a trade-off of stocks. Moreover, the federal government gained a greater amount of leeway on the issue of rearmament, which, however, did not play a decisive role in the conflict itself. Last of all – and this reveals in perhaps the clearest way the consolidation of entrepreneurial power and the way they received as much support from the government as it could afford – it was now possible to start a private investment aid programme to supply the necessary means to the neglected coal and steel area and to other shortage industries from a compulsory industrial levy, so that they would once again be able to compete internationally. Except for the last line of defence in the coal and steel co-determination, what remained from the high-flying postwar plans of the unions of the actual democratic reorganization of the economy was only a very subordinate participation in the areas of investment policies and capital allocation, both of which were actually led by industry. The unions had nothing more with which to counter this 'concerted action' between the federal government and industrial associations.[224]

Nevertheless, industrial relations in mining, which together with the steel industry received the bulk of the investment aid programme (over 50 per cent of the total transfers), were founded anew and based over the long run on the co-determination in the iron and steel industry. However, in practice, the mining companies still put up barriers for the union colleagues in management, and would have preferred to dump on them the department 'Mensch im Betrieb' (the individual in the company).[225] The ideas that were still predominant here were those which the DKBL director Kost had developed in the context of a 'social order for mining' in agreement with the entrepreneurial management in anticipation of the dispute over co-determination. As he maintained in retrospect on the occasion of the liquidation of the DKBL in 1953, the union had unfortunately given preference to a 'strictly institutional idea', although the real issue was actually that of the 'moral mission' of finding a new relationship not only between individuals, but also between the individual and the company. Kost, who would assume the chairmanship of the Wirtschaftsvereinigung Bergbau, openly gave co-determination a familiar meaning when he saw it as being

224. See Heiner R. Adamsen, *Investitionshilfe für die Ruhr. Wiederaufbau, Verbände und Soziale Marktwirtschaft 1948–1952*, Wuppertal 1981.
225. See Ranft, *Objekt*, pp. 122ff.

primarily 'an intellectual concern that should help the individual in the company to become more responsible at his workplace, to contribute both in thought and action to the things that immediately involve him, and to advance in the company'.[226] Therefore, it was in no way due to a new entrepreneurial 'feeling of responsibility' (Müller-List), certainly not in the mining industry, that the 'steel model' finally became established throughout the entire coal and steel industry. The 'socially responsible ideology of partnership', as it was propagated in employer circles to deliberately distance it from the earlier 'gobbling up of socialists' should also not be interpreted as being the confirmation of a new phase of free corporatism. The base line of entrepreneurial thinking in the conflict over co-determination was 'anti-parity' and 'anti-union', and even the tripartite structure of negotiations should not blind one to the fact that at the same time a new *bilateralism* between the state and industry excluding the unions was being established.[227] In any case, the defence of co-determination in the coal and steel industry in a politically favourable constellation does not mark the 'beginning of a new economic balance of relations', as Böckler proclaimed, but rather 'the beginning of the end' of the postwar compromise which the occupation power had enforced.[228] Especially in mining, it became as if the 'last hope' (Ranft) for the democratization of the economy once the union plans for a new order had finally failed.

The development of co-determination did not, however, proceed without problems. In the wake of the emerging reconcentration, co-determination could usually also be transferred through collective contracts to the new holding companies. Yet it meant having to overcome continual opposition, using strikes where necessary or legislation, in order to prevent a return to the below-parity regulation on co-determination of the Company Constitu-

226. See the final report (28 December 1953) by Heinrich Kost, 'Die Tätigkeit der Deutschen Kohlenbergbauleitung', reprinted in *Glückauf* 90, 1954, p. 16.
227. See Berghahn, *Unternehmer*, pp. 221, 230, who distances himself from Abelshauser's interpretation of a 'corporative permeation of the West German market economy' (see now Abelshauser, 'Ansätze "korporativer Marktwirtschaft" in der Koreakrise der frühen fünfziger Jahre', *VfZ* 30, 1982, pp. 715–56). Instead, Berghahn considers the decisive change to be in the 'Americanization' of the German industrial culture which evolved later.
228. See Ulrich Borsdorf, 'Der Anfang vom Ende? – Die Montanmitbestimmung im politischen Kräftefeld der frühen Bundesrepublik (1951–1956)', *WSI-Mitteilungen* 3, 1984, pp. 181–95.

tion Act. Co-determination in the coal and steel industry was finally safeguarded even in controversial cases by a protest and warning strike in January 1955, by the Amendment Co-determination Act of 1956, and a series of private co-determination agreements.[229] The actual trial period for the new legal form of industrial relations in Ruhr mining had to be endured by the labour directors themselves, whose entrenchment in the board of directors the mining assessors had resisted the most. The more important management functions in economic and social policy in mining reverted to the newly founded entrepreneurial association, the Unternehmensverband Ruhrbergbau (1952) or directly to the mines following the disbandment of the DKBL. But the pressing problems of the labour force fell within the competence of the labour director, problems such as the recruitment of young miners – who then quickly turned their backs on mining – during the phase of forced growth in the 1950s and the safeguarding of social policy during the wave of pit closures in the major mining crisis after 1958.[230].

This period of conflict came to a temporary end only with the founding of the Ruhrkohle AG in November 1968. The 'integrated company' that the IG Bergbau demanded, having since given up its reservations on the issue of ownership, was basically a financial rescue plan under which the steel groups were released from their responsibilities for unprofitable coal at the expense of the public sector and at the same time created a sufficient liquidity cushion for their own investment plans. On the other hand, the union managed decisively to extend co-determination against the opposition of the Unternehmensverband Ruhrbergbau; that is, the labour directors of the managing companies were to be backed up by subordinate managers at pit level in the form of personnel and social directors who were subordinate to the plant director but equal to the production director.

229. On this safeguarding strategy, see Erich Potthoff, 'Zur Geschichte der Mitbestimmung', in E. Potthoff et al. (eds), *Zwischenbilanz der Mitbestimmung*, Tübingen 1962, pp. 45ff., and Ranft, *Objekt*, pp. 66ff. and appendix C. See also Ulrich Borsdorf, '"Die Belegschaft scheint geschlossen in den Betten zu liegen." Ein Streik für die Montanmitbestimmung', in Lutz Niethammer (ed.), *'Die Menschen machen ihre Geschichte nicht aus freien Stücken, aber sie machen sie selbst'*, Berlin 1984, pp. 196ff.
230. See Ranft, *Objekt*, pp. 149ff. On the further development, see Abelshauser, *Ruhrkohlenbergbau*, pp. 87f., and Karl Lauschke, *Schwarze Fahnen an der Ruhr. Die Politik der IG Bergbau und Energie während der Kohlenkrise 1958–1968*, Marburg 1984.

Bernd Weisbrod

This further development of co-determination in mining was also due in the final analysis to a political change, just like the repeated changes in the state of industrial relations in the history of mining. The Social Democratic minister of economics of the Great Coalition had brought the 'concerted coal action' into being in 1967 and expressly integrated the union into the private economic system of demand management.[231] The political change of climate also finally made it possible for the compensated companies, which had continued to cultivate 'adversarial' rather than 'co-operative' relations with the IG Bergbau despite co-determination, to be prepared to come to a tripartite solution to the crisis.[232]

For the mining union the price of freedom proved in the end to be rather high despite the favourable conditions of the postwar period. Its original goal of establishing a fundamental, where possible socialist reorganization of mining was to take the form instead of an integrated private industrial sector of the economy under the imprint of the Ruhrkohle AG.[233] By international comparison, a favourable prerequisite was thus created for a socially acceptable crisis management that was also safeguarded politically by the broad-based establishment of the 'co-determination culture'. However, 'bilateral management' burdened the union with the task that under favourable conditions the mining assessors had claimed for themselves, namely to assume the political responsibility for mining interests in general. Yet mining had now become a shrinking, crisis-ridden branch of industry.

231. See Peter Schaaf, *Ruhrbergbau und Sozialdemokratie. Die Energiepolitik der Großen Koalition 1966–1969*, Marburg 1978.
232. See Walter Müller-Jentsch, Helmut Plass and Hans-Eckbert Treu, *Industrielle Beziehungen im Kohlenbergbau der Bundesrepublik Deutschland. Ein Forschungsbericht erstellt im Auftrag des International Institute for Labour Studies Geneva*, Paderborn, February 1986: manuscript, pp. 28, 86f.
233. See Martiny, *Mitbestimmung*, pp. 402f.

5

Industrial Relations in the Belgian Coal Industry since the End of the Nineteenth Century

Ginette Kurgan-van Hentenryk and Jean Puissant

Introduction

Between 1830 and 1914 Belgium presented a contrast. On the one hand, it had undergone a precocious and vigorous industrial revolution which turned the country into the first modern industrial state on the Continent at the beginning of the nineteenth century. On the other, its social relations were archaic by comparison with those of its neighbours and competitors. The coal industry, which constituted one of the principal bases of the country's development, more or less conformed to this pattern in being characterized by a mixture of modernity and social conservatism.

From the Westphalian saddle, including the Aix-la-Chapelle and Dutch coalfields, the Belgian coal industry was based in five coalfields. To the north, there was the Campine coalfield, exploited only after 1917 (the Northern coalfield). Moving towards the south-west, there were the coalfields of Liège (principally medium and low grade coal); the Lower Sambre and Charleroi (coal of the same quality as Liège) and the Centre (high and medium grade coal). Finally, there was the Borinage coalfield, the modern development of which dated from the eighteenth century but which was the least favoured from the point of view of natural conditions (firedamp and high grade coal), which marked the beginning of the French coalfields of the Nord–Pas-de-Calais (see figure 1).

If the Belgian coalfields are characterized by the long period of their exploitation, going back certainly to the thirteenth century, for the southern coalfields the extraction of coal came up against great problems resulting from their very nature. On average the quality of the seams was inferior to that of neighbouring and

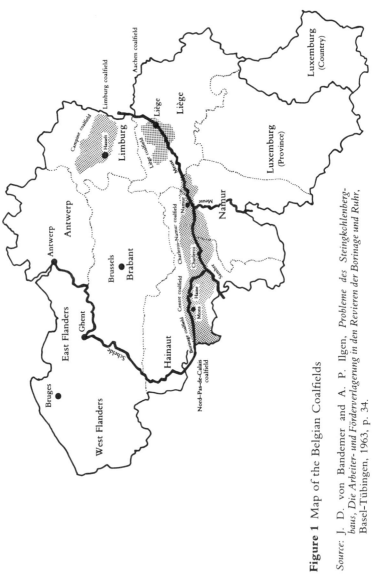

Figure 1 Map of the Belgian Coalfields

Source: J. D. von Bandemer and A. P. Ilgen, *Probleme des Steingkohlenberg-baus, Die Arbeiter- und Förderverlagerung in den Revieren der Borinage und Ruhr*, Basel–Tübingen, 1963, p. 34.

competing coal regions and there was a significant number and variety of geological faults. These characteristics explain the low average productivity of Belgian workers, despite the fact that their competence was recognized throughout Europe and America, where appeals were made to recruit them to open up numerous workings.

Between 1820 and 1870, coal extraction was rapidly modernized through the implementation of new techniques. Production grew by a factor of seven and went from 2 to 15 million tons, of which a third was exported. This strong growth was made possible by two economic and social factors. First, the size of the investment required explains the rapidity with which finance capital took control. The Société Générale, the principal holding company in the country, controlled one-third of production in the Borinage coalfield, the most important in Western Europe from 1834 to 1840, and more than 50 per cent from the time of the great depression at the end of the century. Other banks such as the Banque de Belgique and the Rothschild bank also participated in this process of taking over control. These new companies took the form of limited companies. The majority of coal companies followed suit. Others delayed changing their status. However the role of banks, and of the Société Générale in particular, was more developed in Hainaut (Borinage, Charleroi) than in Liège where vertical concentration remained dominated by metal-working. Secondly, the size of the labour force mobilized in this activity was quite remarkable. From this point of view it took first place in the country. It numbered 116,000 workers in 1896. One in every five industrial workers was a miner. In addition, the labour force was the most concentrated both in terms of the size of the enterprises and of geographical density. It was often mixed – as in Liège, Charleroi and the Centre – with other categories of workers in basic industries situated in the coalfield (133,000 workers).

The main players in nineteenth-century industrial relations

Industrial relations constituted a 'game' with two principal partners: the employers, who justified their tactics and strategy by reference to the cost-price of coal and the hazards of competition, and the workers, who pursued the aims of maintaining and reproducing their labour power and enforcing respect for satisfactory working conditions.

But a third partner also existed, playing an intermittent but important role – the state. Its police function ensured it could always intervene when necessary to maintain or re-establish order. But where the two principal partners (the employers and their organizations, the workers and their organizations) both pursued a constant strategy, marked only by purely tactical variations, the attitude of the state depended on the balance between those who controlled it, and it took on, from the nineteenth century, the roles of referee, partner or spectator in the social 'game'. Its role turned out to be decisive in giving advantage to one of the two protagonists (the employers in the nineteenth century) or in seeking a consensus between them (the organization of industrial relations in the twentieth century).

The employers Though the coal-owners' associations dated back a long way into the past, they had not, initially, had a permanent character. Under the influence of the Société Générale and the state these associations stabilized themselves shortly after Belgium became independent. The Société Générale, which controlled one-third of production in the middle of the century and more than half of the production of the most important coalfield, the Borinage, created an internal department for 'Industry', then 'Coal', responsible for controlling, stimulating and informing the companies within the group. It also brought into existence, in each coalfield, regular meetings of managing directors of the companies which it owned (1839 in the Borinage) and later of chief engineers (around 1844 in the same coalfield). The former occupied themselves primarily with the market and general policy, the latter with techniques of exploitation. Both of them, as a secondary but important function, dealt with the management of labour and wages. From 1838 the managing directors of the majority of Borinage coalfields had meetings of this kind. It was committees such as these which responded to the request of the state to create insurance funds for mine-workers. In return, they obtained from the state the reaffirmation and enforcement of a legal obligation which had fallen into disuse, namely the requirement that workers should have a labour book (1813). This ensured effective control over the labour force.

The colliery associations, organized separately in each coalfield, put in an official appearance later – 1840 in Liège, 1841 in the Centre, 1865 in the Borinage. Charleroi, where one had existed from 1831, was the exception. Alongside them there appeared associations of chief engineers (1874 in the Borinage). There is no doubt in the minds of observers that these associations acted not only on a social level but also on a political level. In the Borinage in particular, the Société Générale group and then the colliery association within which it was dominant, imposed and extended uniform regulation of labour for the whole of the coalfield (1836 to 1861). In 1861 a liberal professor from the Mons School of Mining, a convinced theoretician of economic liberalism, vigorously denounced the concealed coalition of owners, intended to impose working conditions on the workers (such coalitions being forbidden by article 414 of the Penal Code). He even went so far as to threaten the owners that he would advise miners to form unions, like those which already existed in the large towns, to defend their interests. In 1884, 1886 and 1889, the colliery associations appointed lawyers and engaged in lobbying against government proposals to regulate labour.

At this time, which marks the apogee of the urban middle class, MPs elected in mining constituencies were, with few exceptions, linked directly or indirectly with the coal-owners and this 'allegiance' was particularly noticeable in Hainaut. In Liège, the main influence was that of the owners of steel and metal-working plants, who also controlled a good number of mines. Thus from the 1860s (despite the constitutional fiction which attributed a national character to all who were elected to Parliament) the 'industrial deputies' took their seats only to defend the interests of their own economic sectors. Their particular interventions were accepted to the extent that the sectors they represented constituted the backbone of the economy. For this reason Parliament as a whole, and the government in particular, listened to them and took notice of them.

The commercial associations – sales agencies – were often derived from the original coalfield associations. They took commercial decisions on prices and quantities, thereby making provisions for real quotas on coal output. At the end of the century, specific agencies and unions appeared in order to respond, from the time of the economic upturn of the 1890s, to the challenge of growing competition in the country and in Europe. These unions were always formed on a coalfield basis: the Sales Agency for Fine Medium Grade Coal of the Centre Coalfield in 1891 (which, in 1896, became the Sales Agency Ltd), the Union of Liège Collieries (1897), the Association of Producers of Low Grade Coal of the Charleroi and Lower Sambre Coalfield (1891). In the Borinage, where the Société Générale was dominant, this type of association did not exist. This great universal bank played an equally central role within the organization of bidders for state railway contracts. This organization was revamped in 1904 by the director of the 'Industry' department of the Société Générale.

An understanding also existed between the various regional associations over certain external markets such as the French one. The country was sufficiently small and centralized, and coal-mining circles sufficiently narrow (from the middle of the century the principal technical staff and directors were increasingly graduates of the same mining schools) for those in charge of the coal industry to meet regularly, exchange information and take decisions, sometimes weekly.[1] In Liège, particularly close relations existed between the various industrial sectors, the extent of vertical integration leading to the formation of an association of collieries, mines and metal factories (before 1878).

On the social level, the owners generally considered the labour force simply as one factor among others. But, on top of this, they had to bear in mind that they were being supervised by the administration. This forced them to take measures on health and safety which had no equivalent in other industries. They were thus persuaded, more than anyone else, to fulfil their obligations to the labour force. None the less, managers of limited companies and owners of family companies differed. The former invested more in machines, equipment and the organization of production, and kept the cost of labour to a minimum, while explicitly recognizing the need for the reproduction of the labour force. The latter saw management of labour, at around 60 per cent of cost, as the decisive element in profitability. But paternalism and management of the social factor are not interchangeable terms. Family collieries seemed more innovative in the provision of housing for workers and in social paternalism.

Catholic circles, at the time of the congresses of Malines (1864, 1865, 1867, the time when the first International Working Men's Association appeared in Belgium) denounced the absence of paternalism (in the sense used by Charles Perin) among contemporary industrialists, particularly in the coal industry. These comments led

1. G. De Leener, *L'organisation syndicale des chefs d'industrie*, vol. 1, *Les faits*, Brussels 1909, pp. 92ff. The centres of the coalfields are all situated within 100 km of Brussels. The furthest one. Liège, is one hour from Brussels by train. The book by De Leener is the only secondary work on employers' organizations.

to the creation of the School of Mines and Civil Engineering of the Catholic University of Louvain (1867), which was intended to produce 'directors of industrial establishments and mining engineers who understood their duties'. An Association of Employers Who Favoured Workers was founded in 1877. The Catholic congresses at the end of the century (1886, 1887, 1890) took up these attacks on the indifference of employers. The French follower of Le Play, Urbain Guérin, violently criticized the role of limited companies from this point of view (1887).

This voluntary policy had some effect in certain coalfields considered to be 'catholic' (Bois-de-Luc, Centre, Charbonnages Unis de l'Ouest de Mons, the Borinage) but for the most part the atmosphere was liberal, not only politically but, above all, economically. In the second half of the century, managers, executives and also owners (the managers often being associated with ownership through being paid, in part, with shares) were educated in the same Schools of Mines (the state school at Liège which dated from 1825 and the provincial school at Mons founded in 1838) or, sometimes, the same law faculties of universities. In 1873 in Liège, seventeen out of thirty posts of director and managing director were occupied by engineers with degrees from the University of Liège. Between 1880 and 1890 there was hardly any coalfield which did not have engineers from Liège.[2] Engineers employed by public authorities and engineers in private concerns, little by little, came to share the same convictions and the same view of the working class. In 1884, at the height of the depression, Eudore Pirmez, an important industrial liberal from the Charleroi region, a colliery administrator, deputy and former minister, wrote, 'It is the situation of the owners and the capitalists which is worse. They are the ones who suffer. There are no complaints on the part of labour.' This reflection roughly characterizes the attitude of employers, as did the speech, made a little earlier in the Chamber, by Balisaux, a banker, industrial liberal and member of the Colliery Association of Charleroi, in opposition to all regulation of the labour of women and children in the mines. Alluding to the numerous enquiries conducted on this subject, notably that of the Royal Academy of Medicine, he declared:

2. Nicole Caulier-Mathy, 'Le patronat et le progrès technique dans les charbonnages liégeois (1800-1914)', in G. Kurgan-van Hentenryk and J. Stengers (eds), *L'innovation technologique, facteur de changement (XIXe–XXe siècles)*, Brussels 1986, pp. 50–1, 86.

There are witnesses whom the Academy was not able to consult on the healthiness of our mines for the simple reason that, while they might be able to think, they are not able to speak. These are the horses we employ deep in our workings. They come to us thin and puny from the Pyrenees. We put them down into the mines and, when we take them out, we find them happy, fat and healthy. One can read in their eyes a bitter regret at having to leave these delightful spots . . .[3]

In 1886, after the dramatic disturbances in Liège and Charleroi, the Colliery Association of the Couchant of Mons (Borinage) replied to the question on how to improve relations between capital and labour, posed by the Commission of Enquiry on Labour, by saying that, apart from the very liberal proposition that the education of the working class should be improved (while completely refusing compulsory schooling), it was essential to 'prevent disorderly meetings from the very beginning, repress attacks on the freedom to work as rapidly as possible, and considerably increase the number of gendarmes in industrial centres where the local police is everywhere completely inadequate'.[4] This was a reaffirmation without alteration of the doctrine of the police state and the defence of freedom of contract between employer and worker.

The workers From 1830, particularly after the formation of the new limited companies, numerous large strikes took place in Liège (1830) and the Borinage (1830, 1841 in protest at the labour books which were burnt collectively, 1836 and 1861 against labour regulations imposed in the coalfield by mines owned by the Société Générale).[5] This spontaneous and disorganized social agitation began to spread. Strikes over wages and working conditions multiplied in the 1860s, in Charleroi (1867–8), the Centre (1865) and Liège (1869). They became endemic during the great recession of the 1870s and 1880s but remained local, occurring in individual mines or neighbouring mines, by region or by coalfield. No more

3. *Annales parlementaires. Chambre.* Meeting of 14 May 1878.
4. *Bulletin de l'Association Houillère* (of the Couchant of Mons), 1886, pp. 74-127. Reply to question 41 of the enquiry of 1886 quoted in Jean Puissant, *L'évolution du mouvement ouvrier socialiste dans le Borinage*, Brussels 1982, p. 154. The Association 'is opposed to the state intervening constantly in relations between employers and workers . . .' It formally condemned the tendency 'to demand from positive laws the regulation of economic facts of a superior order and the application to industry of rigorous measures which limit liberty and swallow up common rights'.
5. As far as the Borinage coalfield is concerned, see Puissant, ibid.

generalized strike movement occurred, except as a result of simple coincidence (in 1869 in Liège and the Borinage).

It was only following the spontaneous movements in the Charleroi coalfield in 1867–8 that the strike movement and union practices became linked with the socialist political thought of the capital. A large number of workers' associations were created. They were not necessarily organized by trade but, because of the structure of employment in the various localities, many of them were, in fact, based on a single trade. This was certainly the case in the Borinage. In Charleroi and in the Centre the majority were essentially formed by miners. This was less marked in Liège.

With several thousand members, the internationalist organizations were able to set up strike funds. However, it was not so much action which was the order of the day but the strengthening of the organizations themselves. The strikes of 1869 in Liège and the Borinage were ultimately attributable to agitation led by the First International in the industrial areas, but they did not result from a direct appeal to strike. The setting up of trade associations, of strong and organized unions, had the aim of obtaining various benefits from employers through collective bargaining. The sole important strike led by workers' organizations, directed towards this goal, occurred in the Centre in December 1875 to January 1876. It was opposed to wage reductions in a major coal mine. This was the final and ill-fated manifestation of the First International in Belgium. In 1869 the miners' organizations similarly claimed (from the state, in this instance) a role in the management of mine-workers' insurance funds, which were due to be renewed in 1870. This demand reappeared in the strike of January–February 1885 in the Borinage.

The weak and precarious state of the 'trade' associations of the First International in 1869–74, and of the socialist and anarchist 'political' associations of 1879–84, plus the absence of clear tactical perspectives, prevented the labour movement from engaging in, or imposing a relationship on, the employers. For their part, the latter used this weakness and precariousness as a pretext for ignoring the agitation, appealed to the repressive intervention of the public authorities and devised the tactical means necessary to pre-empt social movements. In periods when the market was active and prices were rising, the engineers perfected various procedures to exert close control over rising wages without putting the workers on their guard. Conversely, in periods of recession, they decided to avoid general wage cuts but to feel their way forward in the more passive mines, even to the extent of deliberately accepting conflicts which allowed stocks to be reduced. The powerlessness of the workers was total. Only the evolution of the markets for labour and products allowed a temporary improvement in wages, as in 1871–2 for example. Neither the internationalists nor the socialists and anarchists in Brussels were able to control or guide the angry, disorganized, spontaneous social movements of the miners.

The state between two stools   Roughly speaking, from 1847 to 1857, control of the state fell into the hands of the liberal bourgeoisie. Alongside this, economic liberalism was dominant, not only in the Liberal Party but also in the Catholic Party. On both sides of this chronological divide the state played the same range of roles – intervention, indifference and policing employer/employee relations.

Belgium had inherited the mining legislation of the French Revolution and of the Empire, namely, the granting of concessions to individuals and private companies and the establishment of an administration responsible for receiving the fees, supervising the concessions and keeping an eye on safety (1810–13). In fact, the state took on more general powers of control and supervision over labour such as labour books and the forbidding of strikes. In certain respects, it was capital that was placed under supervision within the mines. Such at least was the feeling of the employers towards the administration and its engineers. In the early days the mining administration, which often had to deal with concessionaires who had learned their business as they went along, put forward proposals for modernizing and rationalizing the workings, circulated information on technological innovations and also analysed the labour factor as such. The engineers underlined the importance of improving health and safety from the point of view of profitability. This attitude led some of them to defend proposals for regulating the labour market and labour itself (1843).

The director of the mining administration, Auguste Visschers, was the originator of the mine-workers' insurance funds in 1840–1. In 1830, in 1841, and again in 1861, ministers and provincial governors intervened as arbitrators in social conflicts. They were called upon by strikers, who appealed to the King (1841, 1861). In 1861 during the general strike in the Borinage against the regulation of labour unique to that coalfield, the Liberal minister, Charles

Rogier, went to Mons. He wrote to the Minister of Public Works:

> If it is true to say that as a rule it is appropriate to leave to the free
> initiative of the parties involved the business of debating the interests
> which will bring them into relationship with one another, one cannot
> disagree that the absence of the state is only completely justified in so far
> as the law has assured to everyone the means of allowing their legitimate
> rights to prevail and that in no case should these decrees result in dangers
> for public order.[6]

The task of maintaining public order would thus draw the state into intervening in social conflicts. This 'modernist' affirmation, coming from a liberal who had been a follower of Fourier in his youth, was already an isolated one within his party. Also, in 1869, Rogier drafted an unsuccessful bill to regulate the labour of women and children.

The influence of economic liberalism, the appearance of the First International and the return to explaining strikes through the role of 'outside agitators' distracted the state from the business of industrial relations, though not from its role of surveillance, of formal control of public order and of repression, which it exerted by means of banning strikes (until 1867), imposing the labour book (1831 to 1841) and, above all, by utilizing the army to put down social movements. With the exception of the textile industry in Ghent, the coal industry was the only one affected by repeated strikes before the end of the century. The most important were accompanied by deadly shootings which culminated, in March 1886, in the death of twenty-eight workers.

The changing attitude of the state did not prevent it from regularly observing the social situation. Vast enquiries were conducted. In 1843 there was an enquiry into The Condition of the Working Classes and Child Labour. In 1869, the mining administration conducted an enquiry on The Situation of Workers in the Mines and Metal-Working Industries. The Labour Commission held enquiries into Industrial Labour (1886) and The Length of the Working Day in Mines (1907). In 1843 and in 1869 the conclusions of these enquiries remained a dead letter, but in 1886 and 1907 some of them were put into effect.

6. Letter dated 27 July 1861, mentioned in Puissant, ibid., p. 102.

Economic Conditions

Structures of Ownership

The Walloon coalfields   An analysis of the structures of ownership
in the coal industry in the early twentieth century shows great
diversity, linked to the history of the exploitation of the various
Walloon coalfields, situated in the south of the country.[7] Earlier,
the decline of the Belgian mines located in the great colliery axis
stretching from Westphalia to northern France had given way to a
concentration of enterprises during the Great Depression of 1875 to
1895. None the less, in 1900 there were 119 mines occupying an
area of 95,000 hectares, which produced altogether something in
the order of 23 million tons.

Even though, at this time, the majority of concessionaires
organized their enterprises as limited companies, this legal form
concealed important differences with respect to both the ownership
and management of capital. Despite the presence of a score of
enterprises, the coal industry of the Borinage was dominated, in
reality, by the Société Générale, the most important financial
institution in the country, which owned five of the main coal
companies and, thereby, controlled half of the production of this
coalfield. Within the Société Générale, the director for industry
centralized the auctions of state railway contracts, auctions which
served as a basis for fixing the prices of the various types of coal on
the Belgian market. Because of this, the Société Générale exercised
considerable influence over prices. The management of each
company in the Société Générale group was put in the hands of a
managing director who kept in touch with his colleagues in the
coalfield in order to come to an arrangement about transport,
distribution and labour conflicts. He reported to the board of
directors which was chaired by a member of the Société Générale
board. At a lower level, the chief engineers of the coal-workings of
the Société Générale met once a month to co-ordinate production

7. Most of the information about the structure of ownership of the Walloon coal
companies was obtained from the *Recueil Financier* and from the annual reports of
the Société Générale, the Banque de Bruxelles and Brufina. The following works
should also be mentioned: Fernand Baudhuin, *L'industrie wallonne avant et après la
guerre*, Charleroi 1924; J. M. Wautelet, *Dynamique de l'accumulation dans les charbonnages belges (1886–1914): une approche par les bilans*, Crehides no. H7605, Louvain-
la-Neuve c.1980; Puissant, *L'évolution*; Caulier-Mathy, 'Le patronat'; Léon Dubois,
Lafarge Coppée 150 ans d'industrie, Paris 1988.

and to come to an agreement on workers' wages, in order to avoid competition. In this way, the Société Générale set the tone for industrial relations in the whole coalfield.

Of all the Walloon coalfields, that of the Centre was the one where the concessions were the most extensive and coal production linked to the development of local industry. Several family enterprises were involved. Two of them, the Mariemont-Bascoup mines, run by the Warocqué family, and the coal enterprises of Ressaix, Leval, Péronnes and St Aldegonde, a group of concessions purchased by the well-known makers of coke ovens, Coppée, were the main Belgian producers. Others belonged to steel firms which wished to improve their supply of combustible materials.

The Charleroi and Liège coalfields, both of which were highly industrialized but were also producers of domestic coal, were characterized by being broken down into many small concessions and by the small size of the coal-workings. Even though the Société Générale was involved in the Charleroi coalfield, where it owned five coal companies and accounted for more than a third of output, its position was less dominant because of the widespread dispersal of production between forty or so enterprises, a good number of which belonged to local industrialists and had customers on the spot.

As for the Liège coalfield, family concerns were dominant. They systematically recruited managers from among the engineering graduates of Liège University. This strengthened the regional character of ownership and management, which resulted in a pronounced local particularism which came out when attempts were made to organize the coal industry on a national scale. On the other hand, large metal-working concerns in the area, such as Cockerill, Ougrée-Marihaye and la Nouvelle-Montagne, had acquired concessions which they exploited themselves.

In 1914, a new stage in the concentration of the Walloon coal industry was reached when the two largest concerns in the country, Raoul Warocqué and Evence Coppée, entered into an association with the Banque de Bruxelles. Although the Société Générale's involvement in industry had begun in the coal sector, the new group of coal interests resulted from a different strategy. As producers of coke ovens with a worldwide reputation, the Coppée family had, from the beginning of this industry, developed a subsidiary activity of producing coke and coals destined to supply their coking plant. In order to finance their investments in Belgium

216 *Ginette Kurgan-van Hentenryk and Jean Puissant*

and abroad, they sought special links with bankers which the Société Générale could not enter into because of competition between their coal interests. For its part, the Banque de Bruxelles, which had become the second most important universal bank in the country, sought stable industrial customers and wanted to extend its interests in the coal industry which was considered to be essential at that time. The association took off after the First World War. With the assistance of Coppée, the Banque de Bruxelles acquired a share in the control of various coal companies, assuring it of a dominant position in the Centre coalfield. Management of the companies it owned was put in the hands of the Coppée group, which charged a fee related to output. At the time of the banking reform of 1934–5, the industrial interests of the Banque de Bruxelles were transferred to a new holding company, Brufina. In 1937, the arrival as chairman of Brufina of Baron Paul de Launoit, the architect of the transfer of control of the largest steel company in Belgium to the Cofinindus holding company, provoked a loosening of the association. In 1946 the management contract with the Coppée group came to an end. Brufina returned the management of the coal companies it owned to its own Coal Division.

The Campine coalfield   From the 1920s onwards the exploitation of the Campine coalfield was organized on a totally different model from that of the Walloon coalfields. By their nature the mines of Campine consisted exclusively of *flénus* and high-grade coals for industrial use. The configuration of the terrain and the size of the seams allowed extraction from a limited number of centres. But the depth at which the seams lay, plus the relative softness of the rock compared to the south, meant that large investments were necessary. Consequently, from the outset, the principle of the concentration of resources was established and the state granted concessions of several thousand hectares to groups of companies which had sound financial backing. The number of companies active in Campine was limited to seven, for an output which reached at least 10 million tons at the beginning of the 1960s.

These seven companies were made up of the principal coal companies of the Walloon coalfields, plus Belgian and French steel companies looking for a better energy supply. In this way the Société Générale was associated with the Pont-à-Mousson Blast Furnaces and Foundries in the Beeringen mine, while Coppée was associated with Le Creusot and the exploitation of Winterslag. At the end of the 1930s, as a result of the rapid changeover of participants, the Société Générale controlled three companies, Brufina two, while Coppée and the Cockerill steel company were the dominant influences in each of the other two. This division of influence was accompanied by an interlinking of the principal investors in the Campine coalfield companies. It was not unusual for a director of Société Générale to sit on the board of a Brufina coal company next to representatives of the major industrial customers of the Campine coal companies.

The interpenetration of the interests of the financial groups and the steel-makers became more marked after the Second World War and was accelerated after the creation of the European Coal and Steel Community. Thus, after the merger in 1955 of the Cockerill and Ougrée-Marihaye steel companies, the Société Générale and de Launoit group found themselves side by side in the Les Liégeois mine exploited by Cockerill in Campine.[8]

The speeding-up of concentration after the 1920s As early as the interwar period, the degree of concentration resulting from the initiative of the financial groups, plus the crisis of the 1930s, had as a consequence a reduction in the number of coal companies in all the Walloon coalfields except for the Centre. From 98 in 1913, the number fell to 55 in 1939 and to 46 in 1958 at the time of the coal crisis. This last brought about the merging of Société Générale and de Launoit companies in the Borinage in the same way as Coppée and Brufina merged their interests in the Centre. With 76 per cent of Belgian mines considered to be dependent on financial groups, there were none the less sixteen independent companies, mainly situated in the Liège and Charleroi coalfields.

However, it would be quite wrong to imagine that the control of the coal companies by the financial groups was exercised according to a single model and in a perfectly integrated and systematic fashion. In reality the nature of the group and its history influenced the management of its coal interests. A comparison between the Société Générale and the Coppée group demonstrates this. The diversification of the banking and industrial interests of the Société Générale had created 'fiefdoms' within the group. From 1950, each

8. For the coal industry in Campine, see K. Pinxten, *Het Kempisch steenkolenbekken, een economische studie*, Bibliotheek van de Vereeniging voor economische wetenschappen, Brussels, Antwerp, Louvain, Ghent, 1937, and J. Moons, *De economischestructuur van de Kempische steenkolenmijnnijverheid*, Hasselt, Louvain, 1957.

Table 1 Mining concessions and coal production in Belgium

	1900	1929	1939	1950	1960
Number of concessions	219	158	130	121	115
Concessions in operation	119	100	77	74	50
Surface area of workings (hectares)	95,188	134,096	109,902	141,485	121,970
Number of workings	277	228	162	156	75
Output (thousands of tons)	23,463	26,940	29,844	27,321	22,469
Output per coalfield (as % of both)					
South	100	88	76	70	58
North	–	12	24	30	42
Average number of workers	132,749	151,253	130,549	135,775	71,460
Workforce per coalfield (as % of both)					
South	100	88	84	76	64
North	–	12	16	24	36

member of the board was, as a matter of course, put in charge of a particular sector of the company's activity. The violent crisis which hit the Société Générale in the spring of 1988 revealed the absence, before the 1980s, of any integrated management system in Belgium's number one financial conglomerate. In this way, conflicting interests within the group regularly caused conflicts between the 'coal men' and the 'electricians', the steel-makers and other consumers such as the cement manufacturers. As a result, there were strong tensions in times of crisis and compromises were arrived at according to the balance of power within management.[9]

In the interwar period, an important role was played by Alexandre Galopin, who was responsible for both coal mining and metalworking. In 1935, he became governor of the Société Générale and continued to intervene actively in the coal business during the crisis and the Second World War. After the war, its coal interests became less important, mainly because of the expansion since the 1920s of its colonial enterprises and of its interests in non-ferrous metals and the electricity industry.

9. 'L'organisation du patronat charbonnier belge et son influence politique', in *Courrier hebdomadaire* of the Centre de Recherche et d'Information Socio-Politiques (CRISP), no. 7, 20 February 1959, P. Joye, *Les trusts en Belgique: la concentration capitaliste*, 2nd edn, Brussels 1960; *Morphologie des groupes financiers*, 2nd edn, CRISP, Brussels 1966; *La Société Générale 1822–1972*, Brussels 1972, p. 197.

Table 2 Coal companies owned by the *Société Générale* and the *Banque de Bruxelles**

Year	Walloon coalfields				Campine				Total			
	Société Générale		B.B.–Coppée		Société Générale		B.B.–Coppée		Société Générale		B.B.–Coppée	
	Companies	Output**	Companies	Output**	Companies	Output**	Companies	Output**	Companies	Output**	Companies	Output**
1921	11	5,307	4	2,833	2	–	2	–	13	5,307	6	2,833
1929	11	6,046	6	3,971	3	1,526	3	1,380	14	7,572	9	5,351
			Brufina-Coppée				Brufina-Coppée				Brufina-Coppée	
1935	10	5,101	8	3,947	3	2,155	3	2,661	13	7,256	11	6,608
			Brufina				Brufina				Brufina	
1939	8	5,136	7	3,531	3	2,561	2	2,242	11	7,697	9	5,773
1950	8	5,601	7	3,333	3	3,148	2	2,375	11	8,749	9	5,708
1960	6	3,813	2	1,675	4	5,041	2	2,958	10	8,854	4	4,633

* From 1921 to 1935, the coal companies owned by the *Banque de Bruxelles* were managed jointly with those of the Coppée Group. Following the bank reform, they came under the control of Brufina, a holding company created as a result of the dividing up of the activities of the *Banque de Bruxelles* in 1935. In 1937 when de Launoit became president of Brufina, the coal mines of the Coppée Group were no longer taken into account and the complete separation of the interests of the two groups was completed in 1946.

** Figures for output are given in thousands of tons.

By contrast, the Coppée group, which had originally been a family industrial enterprise, had experienced extremely dynamic centralized management up until the time of the death of Evence Coppée III in 1945. Its coal companies were part of an integrated industrial strategy which left a lasting mark on the company until its merger with Lafarge in 1980. And yet, in the course of time, rivalry with the Société Générale stimulated an expansion of its coal interests, where the pure preoccupation with integration was eclipsed by the desire to control the largest possible tonnage with a view to improving the balance of power within the coal industry.

Production

As figure 2 shows, from 1900 to 1913 coal production remained relatively stable at 22 to 23 million tons.

After a rapid fall during the First World War, it was only in 1920 that the pre-war level was once again reached. In the interwar period, the size of variations in output grew; combined with the greater instability, this meant that production oscillated between 21 and 29 million tons with two peaks, the first from 1927 to 1931 and the second from 1937 to 1939. The economic and social crisis of 1932 caused a serious fall of 25 per cent, which was gradually recovered in 1933. Contrary to the experience of the First World War, German occupation from 1940 to 1944 did not result in a spectacular drop in production because the German authorities wished to keep the mines going to satisfy the needs of their war economy. The privations and constraints imposed on the labour force during the war provoked a massive drift from the mines after Liberation and an alarming shortage of coal. It was only in the early 1950s that 1939 production levels were once again attained. The continuous decline of the coal industry set in from the time of the 1958 crisis, so that production in 1980 was 6 million tons, that is 27 per cent of output at the beginning of the century.

During the interwar period, the Société Générale had controlled a good quarter of Belgian coal production. The rise of the Coppée–Banque de Bruxelles group reached its apogee around 1935 when it controlled a quarter of the production of industrial coal. After the Second World War, the Société Générale's share of the coal industry rose, to around 40 per cent in 1960, whereas the ending of collaboration between the Coppée group and Brufina kept the share of the latter down to about a fifth of total output.

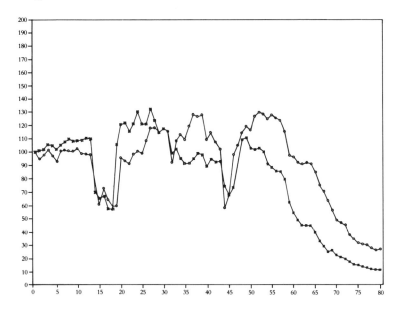

Figure 2 Output and employment in the mining industry in Belgium,
1900–1980

Index (1900 = 100)
○ output
□ employment

Labour

A strong consumer of labour, the coal industry employed, on
average, 132,000 to 145,000 workers before 1914. During the First
World War the number of employees fell to 57 per cent of the
pre-war level and the interwar period was characterized by instability in employment, with the additional feature that, from
1919 to 1928, employment consistently grew more rapidly than
production. Employment reached its highest level of 174,533
workers in 1927, a year when the Belgian industry benefited from
the British general strike, which lasted until December 1926 in the
pits. Starting from the crisis of 1932, with the sole exception of the
year 1944, employment grew more slowly than production, the
difference between the two being accentuated from the 1950s, a
phenomenon linked to the policy of rationalization imposed by the
European Coal and Steel Community.

Productivity

As a great consumer of labour, the remuneration of which constituted almost the whole of the cost of production, the Belgian coal industry constantly pointed to the low productivity of workers to justify a policy of low wages intended to make up for the poor quality of the coal deposits. In the interwar period, the difficulty of finding customers had made the employers very conscious of the productivity of labour. Through a policy of concentrating the mine-workings, of mechanization and of analysis and division of tasks, they had attempted to increase productivity with varying results according to coalfields.

On the eve of the Second World War, production in the Campine coalfield was almost entirely mechanized. By contrast, in the Southern coalfields, the cutting of coal and the digging of tunnels was mechanized but transport within the tunnels and the pits was only about 50 per cent mechanized.[10] As a result, it is not surprising to notice a more rapid growth in the productivity of face-workers compared to the labour force as a whole, with 72 per cent as opposed to 40 per cent in 1938.[11]

The superior productivity of the Campine coalfield did not compensate for the feeble productivity of the Walloon coalfield. As a result, Belgium found itself in an inferior position compared to other European producers, and in particular vis-à-vis Germany, its principal competitor for industrial customers. In 1938, the average daily output of the Belgian worker was 752 kg compared to 1,547 kg in Germany, 1,170 kg in Great Britain and 833 kg in France, the country's principal commercial partners.[12]

Working conditions in the mines during the Second World War caused a drop in productivity. It did not reach the pre-war level until 1951, the time of the formation of the European Coal and Steel Community (ECSC). The Belgian mining industry found

10. Moons, *De economischestruktuur*, p. 139.
11. In 1913, the daily output of a face-worker was 3,160 kg. It went up to 5,443 kg in 1938. On the other hand, the daily output of the workforce as a whole (underground and pithead workers) increased from 538 kg in 1913 to 753 kg in 1938 (according to the *Annuaire statistique de la Belgique*). See also H. Dollard, *L'avenir de la production charbonnière en Belgique*, Brussels 1950, pp. 1–4. In Campine, the daily output of a seam-worker increased from 6,520 kg in 1929 to 9,302 kg in 1936 while that of underground workers doubled and went from 826 kg to 1,779 kg. Pinxten, *Het Kempish steenkolenbekken*, p. 81.
12. Société de Bruxelles pour la Finance et l'Industrie (Brufina), *Rapport à l'assemblée générale des actionnaires du 26 octobre 1939, exercice 1938–39*, p. 13.

itself in a clear position of inferiority vis-à-vis its partners, with a daily output per face worker of 1,054 kg against 1,290 kg in France, 1,457 kg in Germany, 1,617 kg in the Saar and 1,725 kg in Holland.[13] This position of inferiority was the reason for the special dispositions made by the European Coal and Steel Community to allow Belgium to adapt to the new market conditions.

In the following decade, productivity grew more rapidly in Campine than in the Southern coalfield, to such an extent that in the middle of the fifties productivity in the Campine coal enterprises had considerably reduced the gap compared to the mines of the Ruhr.[14]

The market

Because of the characteristics of the Belgian economy in the first half of the twentieth century – namely an open economy based on heavy industry and largely oriented towards exports because of the small size of the internal market – the coal market showed itself to be sensitive to fluctuations in the economy while at the same time having itself no little influence on the industries which consumed coal, especially steel-making. Even then, it is necessary to distinguish the market for domestic coal from that for industrial coal.

The demand for domestic coal represented around 20 per cent of total production and was very inelastic in relation to price until the Second World War. By contrast, producer goods industries made up at least 40 per cent of total demand. Since such a large part of these industries was geared to exports and since the proportion represented by coal in their costs of production was a large one, they were very sensitive to changes in price so that demand for industrial coal was relatively elastic.

Until the First World War, the coal companies adapted their costs to market conditions by aligning wages with sales prices, which meant that the labour force had to bear the brunt of any slowdown in the economy. Because of this, variation in stocks was also fairly weak. On the other hand, inelasticity of coal supplies in periods of strong demand led to a massive turn towards imports.

13. ECSC, *Rapport sur la production de charbon. Situation en 1952*, ECSC archives, Brussels, CEAB 32, 1952–4.
14. In June 1955, the daily output per underground worker was 1,020 kg in the Southern coalfields, compared to 1,515 kg in Campine, 1,435 kg in the Nord–Pas-de-Calais and 1,570 kg in the Ruhr. *Brufina. Rapport pour l'exercice 1954*, 1955, p. 16.

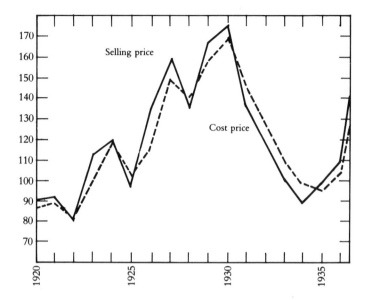

Figure 3 Selling price and cost price of Belgian coal 1920–1937 (in Belgian Francs)

Source: A. Coppé, *Problèmes d'économie charbonnière*, Bruges 1940, p. 98.

This was only weakly compensated for by exports, principally to France.

After the war, the coal market was modified by the enacting of collective agreements in the coal industry which profoundly altered the relationship between sales price and cost price (figure 3). From that time on, there was a certain rigidity in cost prices in so far as the employers were no longer completely free to fix wages. The price of labour was the principal element in the cost price, accounting for 65 per cent. So it is no surprise that the principal argument put forward by the industry to explain its difficulties was that the charges imposed to pay for the post-war social reforms weakened it vis-à-vis its principal competitors, Germany, Great Britain and Holland. In reality, Belgian coal imports, other than coking coal, underwent a strong tendency to fall after the First World War even before the producers and the state had intervened to organize the market.[15]

15. A. Coppé, *Problèmes d'économie charbonnière*, Bruges 1940. As far as statistics

Throughout the interwar period, exports varied from 3 million tons to a maximum of 5.5 million tons in 1931, only to be reduced to a negligible quantity after the Second World War. The complete mobilization of the coal industry in the service of reconstruction, and the weakening of its competitiveness with respect to its principal competitors, considerably reduced its external outlets.

From the 1920s, coal companies, in particular those of the Société Générale group and Evence Coppée, attempted to make up for the uncertainties of the market. The Coppée group set up a central sales bureau for all its coal output. The links between the coal industry and the steel industry, its principal industrial customer, were strengthened. One has only to recall the participation of French companies in the exploitation of the Campine coalfield. Similarly, by building up an interest in the Espérance-Longdoz steel company, Coppée assured an outlet for the production of the Winterslag coalfield by means of a long-term contract.

However, it was through the production of synthetic coal and coke, through the carbon-based chemical industry and by utilizing the by-products of the coking plants that the owners attempted to improve their market position. In 1930, 28 per cent of Belgian coal was consumed in the coal processing industries.[16]

At the end of the twenties, on the initiative of Coppée and the Société Générale, the cartelization of industrial coal producers was undertaken. We will come back to this issue later. The efforts made to stabilize the market were only a partial remedy for fluctuations in sales prices and stocks. Even though the enterprises had agreed to finance stockpiling in order to stabilize production conditions, which was a novelty after the war, the size of variations in stocks went beyond their capacity completely to master the policy, as the figures in table 3 show.

---

are concerned, the study of Belgian foreign trade between the wars is complicated by the fact that, from 1922, statistics were drawn up for the Belgium–Luxemburg Economic Union. Figures for coal exports are not affected because there was no coal mining in the Grand Duchy of Luxemburg. By contrast, Luxemburg industry was a large consumer of coal and it is difficult to work out what proportion of imported coal went to each of the countries of the Union (Coppé, pp. 229ff.).
16. Coppé, ibid., p. 185.

Table 3 Belgian coal stocks at end of month (in thousands of tons), 1920–1960

Period	Minimum	Maximum
January 1920–August 1930	116	1,860
September 1930–July 1936	1,980	4,063
August 1936–May 1941	496	2,696
June 1941–June 1948	210	808
July 1948–May 1953	837	2,682
June 1953–December 1954	2,815	4,095
January 1955–February 1958	179	2,546
March 1958–December 1960	3,230	7,763

1886–1919: Between Two Centuries

The year 1886 is widely considered to be a turning point in the behaviour of the state in social matters; in actual fact 1919 was the time when present forms of industrial relations were established.

Change and Continuity

Change In March 1886, strikes, disturbances and looting broke out, first in Liège then in Charleroi. Thousands of young miners were the ones mainly involved, the victims of the worst of the Great Depression at the end of the century (high unemployment, low wages). These events caused the adoption of legislative measures affecting labour and workers.

The appearance of the Belgian Workers Party (BWP) in 1885, the rallying of urban trade unions to political action, the demand for universal suffrage and the violence of the social explosion in spring 1886 constituted the necessary, but not sufficient, causes for change. Faced with a labour movement which was beginning to organize itself, the political authorities had to choose between greater repression or controlled evolution. The Catholic government opted for the latter. Unlike the Liberal Party, the Catholic Party was not dominated by industrial interests. It enjoyed its greatest influence in Flanders, which was still mainly agricultural and rural. Christian workers' charities developed in various circles. Ideas of industrial paternalism such as those of Charles Périn or La Tour du Pin were widespread. Part of the Catholic university world, eminent people and political figures, were influenced by the doctrines of the counter-revolutionary sociologist Frédéric Le Play. 'Social reform' appeared to be indispensable to stop the rising revolutionary tide in Europe (1848–1871–1886). Belgian Le Playsian circles clearly influenced the policies followed by the government after the disturbances.[17]

Already in the minority in Parliament, the Liberal Party was torn between its two wings, the progressives who defended the idea of legislative reforms, and the doctrinaires, dominated by industrialists, who remained opposed. The direct political influence of the coal-owners on the government and on Parliament was marginalized. After 1884 no more coal directors held ministerial portfolios, and after 1894 no more coal-owners sat in Parliament. The relative democratization of the political system, obtained through pressure from the BWP and the massive strikes in its support (1891, 1893) – strikes which the miners had triggered off and been in the forefront of – resulted in 1893 in the introduction of universal suffrage tempered by plural voting. After the 1894 elections, a homogeneous socialist group represented the coalfields. Proportional representation, implemented from 1900 on, once again permitted the election of liberal deputies for some of these constituencies, but they were not, in general, industrialists. Thus the coal-owners became a pressure group outside the political institutions.

In 1895, the representatives of large sections of the basic industries – mining, metallurgy, glass-making – formed what was to become the Central Industrial Committee. Pride of place in its programme went to 'relations with ministerial departments so as to be kept informed of preliminary studies, proposed parliamentary bills and legislation'. This was a genuine adaptation to the more active policies of the state. It also envisaged circulating information and documentation and entering into relations with similar groups in Belgium and abroad. But there was nothing about possible relations with the labour force. The creation, in 1909, of a National Coal Federation arose from the same preoccupations. By contrast, the Hainaut Union of Coal Producers (1891) arose from the broadening of the social movement and had as its aim the mutual assurance of its members against the risk of local strikes. The union was not imitated. It disappeared because of the founding of trade

17. See J. Puissant, '1886. La contre-réforme sociale?', in *100 ans de droit social belge*, Brussels 1986, pp. 67–100.

unions and also, paradoxically, because they were weak, which rendered the employers' group unnecessary.

If the BWP can be considered as the political expression of the urban trade unions, it was, in fact, the originator of unions in the large Walloon industries, notably coal. The Trade Union Commission (1898) was seen less as a means of regrouping and co-ordinating the action of existing unions than as a means to multiply them, to support new ones and to encourage them to federate. In the coal areas, where the socialist party recruited the bulk of its forces during the general strikes in favour of universal suffrage (1891, 1893, 1902, 1913), unionism appeared as a means to control the working class and its spontaneous movements. The basic framework of the socialist movement was made up of numerous co-operative bakeries, which assured it of representation in the localities, and the unions had to allow it specific control over workers' professional interests even at the workplace. This logic of starting from the top and working down explains the founding of the National Miners Federation even before the setting up of most of the local unions. 'The Miners Federation . . . was founded rather like those towns in the Far West where monuments are built before houses, where schools, churches and railway stations are opened before the inhabitants arrive, in order to attract them', wrote Emile Vandervelde.[18]

A first inter-coalfield congress was held in 1887 but the BWP did not recognize it because the assembly, under the influence of the radical elements in the party, pronounced itself in favour of an immediate general strike for universal suffrage. The radicals thought that, in this period of social agitation resulting from the Great Depression, a general strike of miners would deprive industry of its 'bread', bring about a total paralysis of the economy and force the government to accept political reform. The reduction of these tensions among the socialists made it possible to set up the National Miners Federation in Charleroi in 1890, a few weeks before the International Miners Federation at Jolimont (Centre). Regional federations followed. That of the Centre was re-established in December 1890 (it had unquestionably existed in 1874), and that of the Borinage in 1892, in imitation of the French union of the Nord–Pas-de-Calais.

18. E. Vandervelde, *Enquête sur les associations professionnelles d'artisans et d'ouvriers*, vol. 1, Brussels 1891, p. 180.

The socialist miners' unions thus enjoyed certain advantages, in particular the support of a rapidly expanding party which had more than thirty deputies in Parliament and which was to seek, in a voluntarist way, to reinforce the unions in 1905 and 1912. But they also suffered from some disadvantages. Their links with a political party made them suspect in the eyes of the employers and also in the eyes of various sections of the working class. One part of it, in the Charleroi area, continued to follow the corporatist unionism of the Knights of Labour, which had been set up in 1882 in the glass industry and which spread, in 1885, to the coal mines. As a result, the miners of Charleroi were organized in two competing federations which did not come together until 1905.

A small, but growing, proportion of Catholic workers were put off by the militant anti clericalism of the socialist groups and began to organize, under the aegis of philanthropists and ecclesiastics. At first they had an anti-socialist perspective. Later they became actors in the social arena. The first regional federation was set up in the Centre in 1895. It was at La Louvière (Centre) that the National Federation of Free Miners was founded (1906). It existed mainly in Hainaut. In 1909, it took part in the founding of the Walloon General Confederation of Christian and Free Unions and, in 1912, in the organizing of a National Christian Trade Union, a part of the National General Confederation.

The goal of the Free Miners was 'the improvement of the material and moral conditions of the workers through the defence of their rights, the organization of their interests and the restoration of Christian brotherhood in relations between labour and capital'. Under the auspices of the basic trinity of social order, 'religion, family, private property', they proposed that all conflicts should be solved 'through conciliation or arbitration, through good relations between employers and workers based on mutual respect for rights and on the carrying out of reciprocal duties'. The strike fund which was envisaged had as its aim 'to make peace reign between capital and labour'.[19]

The principles of Christian unionism were very clear. They

19. E. Van Den Driessche, *La Centrale des Francs-Mineurs. Son histoire, sa vie*, Brussels 1984, pp. 91, 96. On the BWP Centrale des Mineurs, see N. Dethier, *Centrale syndicale des travailleurs des mines de Belgique. 60 années d'action 1890–1950*, Brussels 1950. On the Knights of Labour, see J. Michel, 'La Chevalerie du Travail (1890–1906). Force ou faiblesse du mouvement ouvrier belge', *Revue belge d'histoire contemporaine*, IX (1978) 1–2, pp. 117–64.

Table 4 Belgian trade union membership according to unconfirmed
internal sources (and as a % of the total number of workers)

	Socialist		Christian		Total number of workers
1890	16,539		–		118,983 (in 1891)
1902	–		1,346	(1%)	134,092
1908	34,528	(23%)	3,350	(2.3%)	145,277
1913	39,417	(27%)	5,864	(4%)	145,337

aimed at pacifying relations between capital and labour. None the
less, the existence of a strike fund meant that the use of the strike
weapon was not excluded 'in order to bring about the triumph of
the just claims of the workers'.

Affirmed as a principle by Christian miners, practised as a fact by
Socialist miners, the search for dialogue, for negotiations and for
conciliation with the employers, was a characteristic feature of
newborn unionism.

The rise in the number of trade unionists before 1914 is uneven
rather than uniform and depended on developments in the economy,
the evolution of the social movements and the political situation.

The rise of unions linked to the BWP was undoubtedly a new
feature. And yet it did not bring about any change in the balance of
power.

Continuity The workers' movement had gone on to the offensive.
The employers, remaining on the defensive, were no less victorious
than before. In fact it was only the Knights of Labour of Charleroi
who obtained, by means of a strike in 1890, following a meeting
with the representatives of the Coal Association and the mediation
of the mining inspectorate, the reduction by an hour of the working day of the coal-cutters. The strike, which was a response to
improving economic conditions, took the employers by surprise.
As a consequence, they elected to their leadership the toughest
among them. From then on, the order of the day was to avoid all
contact with unions, to refuse them any legitimacy. The numerous
repeated approaches of the unions never received any response
(1898–1905). Clearly this attitude reinforced the profound belief of
the socialist unions in opposition and in class struggle. At their
congress in 1910, Lombard, one of the principal union leaders in
Charleroi, declared

The Federation is twenty-three years old and the employers wish to continue treating us as children . . . We demand discussions, and the employers do not wish to listen to us. We must denounce this authoritarianism to public opinion and continue our campaign by all conciliatory means. To my mind, if a conflict breaks out there is no other solution. They will bear all the responsibility.[20]

The weakness of the National Miners Federation, the localism of the grass-roots unions and of the regional federations in particular, reinforced even more by the structural characteristics of the coalfields, and the spontaneous nature of many of the strikes explain the ineffectiveness of union pressure. The 1899 national strike in favour of a rise in wages was a failure, as was that of 1905. Following the strike in the Ruhr, through solidarity with it and profiting from the cessation of German imports, the Federation tried to get the International Federation to issue an appeal for a European strike. In Belgium, 113 managers of coal mines were contacted. Only three responded, negatively, to the demand for a rise in wages. The chaotic strike did not spread through the whole country. Once again it ended up as a serious failure.

Local and regional strikes had more force even though they did not achieve any better results. Joël Michel has described the Belgian Federation as the weak link in the chain of European mining unionism. The situation of the unions followed the fortunes of industry and of the European unions, but at a qualitatively lower level. From 1890, the limitation of the working day to eight hours and wage increases were claimed repeatedly in the course of numerous strikes. Eighty per cent of them remained fruitless, more than for industry as a whole (60 per cent). The strikes which did not fail ended in compromise.[21]

Certain organizers were prepared for anything to get around this problem – the exploitation of a mine by the union (in the Borinage), political pressure on local authorities controlled by socialists (towns and then provinces) and support for the Lewy proposal for international control of coal output, a proposition defended several times in the international miners' congress. Also, it is not surprising that socialist unionists should respond favourably to two proposals

20. Quoted by J. Michel, 'Un maillon plus faible du syndicalisme minier, la Fédération Nationale des Mineurs Belges avant 1914', *Revue belge de philologie et d'histoire*, LV (1977) 2, pp. 458–9.
21. Ibid. See also the thesis (Thèse d'Etat) submitted by J. Michel in Lyons, 1987.

by the employers for conciliation and arbitration committees, at Mariemont-Bascoup in 1888–99 (Centre) and at Pâturages and Wasmes in 1892–3 (Borinage).[22]

These two committees were set up in mines independent of the Société Générale, one a family firm, the other a small limited company in decline. The committee at Mariemont-Bascoup, set up on the English model, was organized by Julien Weiler, a rather progressive liberal engineer, son-in-law of the liberal economist who, in 1861, had threatened the employers that he would encourage the workers to set up unions. In 1880, after a strike in the Borinage, Weiler wrote, 'If you had heard all those lying accusations, those unjust claims, all those cries of hatred and envy, you would have trembled at the idea of seeing those ignorant masses, hungry for some sort of enjoyment, start to move. And you would have demanded, not without fear, what light could stop them? what force could contain them?'[23]

After being approached by the socialist organizations in 1886, Weiler developed the concept that the force capable of stemming the tide was the Conciliation Council. It functioned in Bascoup from 1888 to 1915 and in Mariemont from 1888 to 1899. The council of Pâturages and Wasmes, led at that time by E. Lewy, was also organized with the co-operation of the Federation of the BWP. But Lewy's colleagues in the region condemned him for participating. Certainly this strange administrator had unwanted attitudes, declaring that the employers ought to be favourable to the formation of unions and should 'push their workers into taking part'!

These two restricted examples are the exception that prove the general rule of the employers' refusal to engage in relations with the workers, other than as individuals, or to recognize the principles of delegation and representation. In addition, the companies suspected that the unions were incapable of controlling the workers.[24] This

22. Concerning the first, see P. Cochez, 'Le conseil de conciliation et d'arbitrage des charbonnages de Bascoup (1888–1915)', in *Le Centre (1830–1914)*. *Mémoires d'une région*, La Louvière 1984. Concerning the second, see Puissant, *L'évolution*, pp. 315ff.

23. Scloneux (J. Weiler), 'La grève de décembre 1879 dans le Borinage', 15 February 1880, p. 193 (quoted by Puissant, *L'évolution*, p. 162).

24. 'The (union) federation, which did not succeed in preventing the strike in principle, is still, at the present moment, unable to win the support of the majority of workers. It lacks the necessary authority to have its decisions respected. The truth is that organizations of this type do not represent the working class and the

attitude was not specific to coal employers, one can find it among all employers. Everywhere contact between employers and unions remained the exception. Among Belgium's major industries the sole regional collective agreement before 1914, which included the setting up of a joint conciliation organization, came into existence in the textile industry at Verviers in 1906. The disappointing experience of the Councils for Industry and Labour in 1887 confirmed this position once more. These joint councils, which brought together elected industrialists and workers, were convened by the provincial executive, the Permanent Delegation, at the demand of one of the two parties in order 'to seek the means of conciliation which can bring an end' to a conflict. They could also be convened by the government, in a consultative capacity, to give their advice 'on proposals of general interest relating to industry or labour'.[25] If this second purpose, that of consultation, was exercised effectively, the first, conciliation, was practically non-existent. The unions tried to use these councils to defend their claims, but the employers refused to sit on them for that purpose.

One could also question what the reasons were that allowed these Councils for Industry and Labour to be set up. The coal employers, who had put up lively, determined opposition to all measures relating to the regulation of labour,[26] did not voice their opinion during the voting for this law, which was potentially full of implications of the most profound changes in industrial relations. The unanimity with which it was voted was revealing. It implied, in the atmosphere of permanent, brutal confrontation of the end of the century, a kind of calming of the spirit. But the

---

employers are completely correct to refuse to have contact with them and intend to deal directly with their workers'. *Procès-verbal de l'Association Charbonnière de Liège*, 20 January 1911 (response to repeated demands for a bilateral meeting made by the Union Federation since July 1910), quoted by Dethier, *Centrale syndicale*, p. 130.

25. Set up on the initiative of the Commission du Travail. See the report of V. Brants, professor at the Catholic University of Louvain (a Leplaysian). The law proposed in the Chamber by Frère-Orban, the chief 'doctrinaire' of the liberal opposition, and approved by H. Denis, professor at the Free University of Brussels, was passed unanimously in 1887. See J. Neuville, *L'évolution des relations industrielles*, vol. 1, Brussels 1975, pp. 385ff.

26. Notably in 1878, at the time of the first great parliamentary debate on regulation of female and child labour; in 1884, when certain measures were implemented by royal decree; in December 1889, during the vote on the law of 13 December on female and child labour and, even later, in 1909, in connection with the law on the nine-hour day.

contents were sufficiently weighted to prevent any radical change. It was very much the kind of social legislation intended to legalize a form of social paternalism and to act as a sort of social protection rather than to modify social relations. It is in the forces controlling the state, the political authorities, that one must seek out the origin of these changes (and, as a consequence, their range and significance), not in a victorious advance of unionism. In struggling for universal suffrage, the BWP had given priority to political struggle. The weakness of the unions stemmed, in part, from the same root. They, too, turned towards political struggle in 1902 and 1913 and gave their main attention to the role of socialist deputies in the Chamber and to the miners' deputies who dominated the National Federation until the second decade of the twentieth century.

The Breakdown

Of all categories of workers, miners – because they were militant, because they took part in strikes and demonstrations, because they were used as an archetype of exploitation and of struggle by the BWP – appeared, from the social point of view, to be the group which best illustrated the need for reform. The nature of their production (main source of energy, the 'bread' of industry), the type of work (underground, surrounded with mystery), and its dangers (the symbolic role of mining disasters) linked up with political efforts to ensure special legislative status for miners (1897, representatives on the mining inspectorate; 1909, nine-hour day for face-workers; 1911, pension scheme for all at the age of fifty-five for face-workers and sixty for pithead workers who had completed thirty years of service).

In the course of discussions on modification of mining legislation, necessitated by the imminent opening up of the Nord coalfield in 1905, BWP MPs proposed direct state control (without success, of course) and the regulation of labour in the new coal mines, especially the eight-hour day. A parliamentary majority favourable to limiting the length of the working day took shape in 1907 and, as a delaying tactic, set up a series of enquiries.[27] After a

27. Neuville, *L'évolution des relations industrielles*, vol. 2, *La lutte ouvrière pour la maîtrise de temps*, I, Brussels 1981, pp. 185ff. Also: Enquête du Conseil Supérieur du Travail (unpublished); Enquête du Conseil Supérieur de l'Industrie et du Commerce (unpublished); Commission d'Enquête sur la durée du travail dans les mines de houille, published in 13 volumes between 1907 and 1909.

long parliamentary struggle, a nine-hour day for face-workers (excluding travel time) was voted on 29 December 1909 with effect from 1 January 1912. It was passed by a majority consisting of social Catholics – the Young Right (the centre-left of the Catholic Party), Christian Democrats (CD), the left of the Liberal Party plus the BWP – against the traditional right of the Catholic Party and the right wing of the Liberal Party. The debate brought about the resignation of the government in 1907 and the entry into the new Catholic cabinet of representatives of the Young Right and of the CD.

In the course of the enquiries, the coal employers rehearsed their usual arguments: the Belgian coal industry had to be supported against international competition; the evidence of 'honorary workers' in favour of regulation was illegitimate; the workers themselves were not demanding changes; the principle of liberty had to be upheld. They also denounced the 'sociologists' who wrote the official reports, who were said to be 'manifestly influenced by certain consumers, dominated by unjust prejudices': 'The question that has to be decided is whether the heavy hand of the state should fall on the whole of industry or whether liberty should continue to reign in relations between employers and adult workers.'[28] This argument was relayed to Parliament by numerous orators including P. Boël, a liberal industrialist from the Centre (coal and steel), and the Minister of State, J. Devolder, who acknowledged having coal interests (as well he might since he was no less than vice-governor of the Société Générale).

This development in the area of social legislation was clearly the precursor of changes in industrial relations. Following a strike in the Liège coalfield in 1911 (against a new set of workplace regulations adopted in response to the law on working hours), and then a strike in the Borinage in 1912 (against a change in the customary weekly method of payment, which had been modified because of the law on pensions) the Minister of Labour, at the request of the workers' unions, entered into relations with the Coal Association and set up a system of trilateral negotiations. Because of the strike in the Borinage, Parliament even modified the law to bring the conflict to an end.[29] The employers had always refused contact

28. De Smet de Naeyer, the former Catholic Prime Minister (he resigned in 1907), an industrialist in Ghent, *Annales parlementaires, Sénat*, 22 July 1909, quoted by Neuville, ibid., p. 263.
29. Concerning the strike in the Liège coalfield, see Dethier, *Centrale syndicale*,

with the workers but they did accept ministerial mediation. The balance of power was in the process of changing. The absence of this type of relationship before 1911 was a clear sign of the absence of political will on the part of the government to put pressure on the social actors to persuade them to accept negotiations. This also showed the ineffectiveness of the Industry and Labour Councils.

The war interrupted this kind of gentle evolution of social relations but it speeded up the changes which had made this evolution possible.

Industrial Relations between the Wars

Employers' organizations

The difficulties of organizing a pressure group The experience of war stimulated the desire of the coal employers to organize themselves. On 11 December 1918, a co-operative society, the Coal Group, was created. Eighty coal companies became members.[30]

Cohesion was far from reigning within these institutions, which were constantly torn between the organizational model of the Rhine–Westphalia Union and the idiosyncracies of companies and coalfields. However, faced with the intervention of the state and post-war social agitation, the employers felt themselves to be isolated. They followed the suggestion of Evence Coppée III that they should systematically organize the defence of coal interests. When, at the beginning of 1920, the shortage of fuel forced the government to make plans for the distribution of coal, the board of directors of the Coal Group decided to collaborate and take on the distribution of coal according to quotas established by the Ministry of Economic Affairs.

As a result of this initiative, a statistical bureau was set up in July 1920. It was followed, on 8 January 1921, by the launching of a weekly publication, *Bulletin des Charbonnages* (Coal Mines Bulletin),

---

pp. 122ff. Concerning the strike in the Borinage, see Puissant, *L'évolution*, pp. 433–4.

30. This account of the employers' organizations is based, for the most part, on the archives of the Coal Federation (FEDECHAR), in particular the annual report of the committee of the general assembly and part of the minutes of the meetings of the committee. A note of 20 October 1953 entitled *Historique de la Fédération* summarizes the development of the Federation and its member associations.

intended to inform enterprises on the economic situation in the major coal-producing countries and to supply statistical information on the coal industry as well as 'a social chronicle' of Belgium and abroad. A new stage was reached on 1 June 1921 when the Federation of Coal Associations merged with the Coal Group in a co-operative society called the Federation of Belgian Coal Mines.

If the leaders of the Federation put at the head of the agenda the defence of employers' interests in the area of wages and coal prices, the aim of their policy was to struggle constantly against encroachment by the state and workers' control in enterprises. They relentlessly attacked the eight-hour day, the collective agreement which linked wages to the retail price index and the burdensome financial policy which was conducted with regard to coal prices. All parliamentary bills on social matters were systematically submitted to ad hoc commissions, examined by the coalfield associations and then analysed in a general report aimed, according to circumstances, at obtaining the support of the government or of MPs for the employers' point of view.

The Federation did not hesitate to point out the 'harmful tendencies':

> The result, of some of these at least, is not just to further develop state involvement in our country but also to prepare the right of supervision, that is to say to open up the way to workers' control in industrial enterprises with respect to management, production and finance and, as a consequence, to prepare for workers' participation in the management of enterprises, in a word to establish, little by little, the dictatorship of the proletariat in industry and to paralyse the employers' authority in order to stifle it later.[31]

It was in this way that its initiatives with respect to the Ministry of Industry brought about the temporary collapse of the socialist attempt to introduce worker inspection in the mines by means of representatives elected directly by the workers.

In early 1925, anxious as a result of difficulties in finding outlets for their output, the employers blamed the poor sales of coal on the high wages of Belgian miners. When their delegates tried to get a wage reduction in the Mixed Commission for Mines, they were energetically taken to task by the workers' representatives. The

31. FEDECHAR archives. *Rapport de l'exercice 1922.*

latter reproached them with not having brought their equipment to a sufficiently high level, with not having an adequate sales organization and with having accumulated excessive stocks which had reached record levels of 1.6 million tons. Faced with criticisms over its lack of strategy, its deficiencies in production and in its commercial methods, the employers' organization decided to fend off all enquiries on the subject because, as its leaders said, 'in order to gain time in the struggle against the reduction of wages, we believe that such an enquiry, which cannot come to anything, will soon be demanded by the workers' representatives'.[32]

Although there was a respite because of the British strike, the result of which was a spectacular reduction in stocks between May 1926 and January 1927, foreign competition became a major preoccupation after 1924, with the prospect of Germany having its commercial freedom restored. Indeed, before the conclusion of the Belgian–German commercial treaty of 4 April 1925, German coal imports had been in the nature of reparations. A German Coal Distribution Agency, on which the Coal Federation was represented, fixed prices and organized sales. From the day when the Belgian–German commercial treaty came into effect, restoring free exchange, German coal, especially from the Ruhr, which was produced much more cheaply than Belgian coal, flooded into Belgium at lower prices than reparations coal.

The re-entry of British coal on to the Belgian market in 1927, the growth of imports from Holland, the vigour of foreign competition with respect to Belgian railways, a major coal customer, made it necessary to think about a powerful organization to face up to international competition. This competition was all the more formidable in that the employers had difficult relations with the railways over tariffs applied to its products.

While the Coal Federation recognized that a better sales organization, a good transport policy and the lightening of taxes would contribute to the competitive position of Belgian coal, it none the less believed that any durable solution would have to be based on lengthening the working day and reducing wages.

Even though, in the joint commission, they ensured that 25 per cent of the 'index number' from the retail price index should be taken up by the price of coal, the coal employers multiplied their

---

32. FEDECHAR archives. *Projet de rapport à l'assemblée générale du 18 mars 1925.*
Procès-verbaux de la Commission Nationale Mixte des Mines, février–mars 1925.

initiatives with the government in order to propose the adoption of measures intended to remedy the crisis. This campaign ended in the creation by the government of a Commission for the Study of the Problems of the Coal Industry, consisting of three MPs, three civil servants from the Ministry of Industry, seven employers' representatives and seven workers' representatives, under the chairmanship of the financier and former Prime Minister Georges Theunis.

The hopes the employers' organizations had for the modification of the balance of power between employers, unions and the state were disappointed. Nevertheless, by its work, the new commission contributed to the promotion of a common commercial organization for the coal industry, and at the same time, the King, and the Prime Minister of the day, Henri Jaspar, were eagerly encouraging the coal employers to come to an agreement between themselves.[33] But, and this shows the relative powerlessness of the Federation and the centrifugal forces within it, the initiative began to take shape outside it. Indeed, because of the way it was organized, the employers' association did not have the means to conduct an overall policy for its sector.

When its statutes expired in 1926, the request of the employers from the Liège coalfield to put an end to the co-operative society and to substitute an organization which, preferably, did not have any legal standing, was agreed to and, on 8 December 1926, the Federation of Belgian Coal Associations was created, without any profit-making goals but with the social aim of examining all questions of general interest to the Belgian coal industry. But, and this was one of the provisions of the statutes which turned it, as one of the major employers of the time said, into 'an amorphous entity which could not act', its decisions could not bind the associations it was made up of beyond the degree to which they wished to give their support.[34] In addition, all direct links between the Federation and the enterprises had disappeared.

33. Speech of King Albert I on the occasion of the seventy-fifth anniversary of the Mons Association of Engineers, 21 October 1928, qouted by A. Abrassart in *La politique charbonnière en Belgique. Réponse au mémoire publié sous le couvert de la Chambre de Commerce d'Anvers en juillet 1934*, August 1934, FEDECHAR archives.
34. These incisive words were spoken by Lucien Guinotte, an influential member of the Banque de Bruxelles–Coppée group during a meeting of a committee of the Coal Federation. He was replying to complaints from representatives of the Liège coalfield that they did not benefit from the special transport tariff for the coal industry obtained by the Belgian Industrial Coal Agency from Belgian railways.

This new form of organization did not reduce the activity of the Federation as pressure group and spokesman for the coal employers in numerous national and international bodies. Its representatives took part in the international economic conferences in Geneva and in the mining section of the International Industrial Employers Association, where part of the International Labour Organization (ILO) enquiry into wages and working conditions in the mines was conducted. This international activity reinforced its insistence on the inferiority of the Belgian coal industry 'which held a sad record of having the highest labour cost per ton extracted'.[35]

It was in its defence of employers' interests over wages and international activities that it was entrusted with the mission of recruiting foreign labour to alleviate the shortage of Belgian miners caused by work for the exhibitions in Antwerp and Liège. In order to keep competition over paying wages to a minimum, it set up its own Immigration Commission in 1929 to co-ordinate the recruiting efforts of the different coalfields and to act as 'an employers' organization of the highest authority' in the eyes of foreign governments. During that year, the Commission organized the immigration of 5,824 foreign workers from Poland and Czechoslovakia.[36]

Towards cartelization of the coal industry   The first efforts at a general organization of the coal industry go back to 1929. They resulted, in part, from the agreement between the two principal coal groups, the Société Générale and the Banque de Bruxelles–Coppée Group, achieved on the initiative of Alexandre Galopin and Evence Coppée III. Two limited companies were set up on 5 January 1929, the Belgian Industrial Coal Agency and the Belgian Coke and Coking Coal Union.[37]

The Belgian Industrial Coal Agency brought together all the coal companies of the promoting group plus four independent owners from the Charleroi coalfield, to which was later added the John Cockerill steel company, which had an important coal division, as well as several other independent coal mines. The agency centralized the sale and purchase of coal by its shareholders.

(FEDECHAR archives, *Procès-verbal de la réunion du comité du 5 novembre 1930.*) Statut de la Fédération des Associations Charbonnières de Belgique, *Annexes au Moniteur Belge*, 24 December 1926, no. 781.

35. FEDECHAR archives. *Rapport du comité pour l'exercice 1927.*
36. FEDECHAR archives. *Rapport du comité pour l'exercice 1929.*
37. Statutes in *Annexes au Moniteur Belge*, 28–9 January 1929, nos. 1249, 1252.

The Belgian Coke and Coking Coal Union brought together producers and consumers of coke and coking coal belonging to the two promoting groups. In addition to commercial operations connected with coke and coking coal, it had the authority to participate in all enterprises aimed at creating outlets or developing any of its areas of activity. At the same time as the principal producers of industrial coal and coke in the Hainaut and Campine coalfields were involving themselves in a common commercial policy under the aegis of the principal financial groups, the owners in the Liège coalfield, always anxious to preserve their independence, started their own commercial organization with the creation of a co-operative company, Liège Collieries.

The crisis of 1930 showed that these efforts were insufficient. The coal employers, through the Federation and with the vigorous support of Charles Demeure, an engineer and professor at Louvain University, made energetic demands that, in the absence of an international coal agreement, the Belgian market should be protected.[38] Even though Belgium's fundamental attachment to free trade would suffer, only import controls would enable them to struggle against foreign competition.

The principal obstacle to the establishment of quotas was article 6 of the Belgian–German commercial treaty of 4 April 1925 which forbade all import restrictions on trade without the prior agreement of Germany. In addition, the claims of the coal-owners clashed with the coal consumers' interests; the latter, through the Antwerp Chamber of Commerce, the Union of Blast Furnaces and Steel Foundries and the Inland Navigators' Union, protested against all measures limiting imports. But the accumulation of stocks at the pitheads – they reached 2 million tons in October 1930 and continued to rise – induced the government to set up a Commission on the Coal Crisis in March 1931 and to start negotiations with Germany. At the beginning of October, a system of licences limiting imports to 74 per cent of 1930 tonnage came into effect. But it was shown to be insufficient and negotiations with Germany were continued. In 1934, Germany agreed to renounce article 6 of the 1925 treaty but concluded an agreement which would limit exports of Belgian products into Germany if German coal exports were greatly reduced. The controls put into effect from 1933 were impaired as a result. Only a ten-franc tax was maintained.

38. A bibliography of Demeure can be found in Coppé, *Problèmes*, p. 301.

The worsening of the economic crisis after 1932 and the extent of the depression in the coal sector encouraged the Catholic Minister for Economic Affairs, Frans Van Cauwelaert, to promote the setting up of a compulsory cartel, through the use of full powers voted to the government in July 1934. Given the government's desire to set up several organizations of this type, the coal producers engaged in negotiations. Thanks to the tenacity of Van Cauwelaert and of his successor Van Isacker, who endeavoured to overcome the reticence of employers who had not joined the Belgian Industrial Coal Agency, the first compulsory cartel in Belgium was created on 7 January 1935 in the form of a cooperative society, the National Coal Office.[39] The weight of the Société Générale and of the Coppée–Banque de Bruxelles Group in the coal sector, and the fact that their interests were deployed both in the coalfields of the south and those of the north, had greatly facilitated this concentration. The result, from 1935 to 1944, was that Belgian coal policy was conducted by Alexandre Galopin and Evence Coppée III and that the cartelization brought about by the state left an enduring mark. The linchpin of the agreements and sales agencies, which succeeded one another under different forms until the beginning of the 1950s, was Herman Capiau, former managing director of one of the Société Générale coal companies and then director general of the Coal Federation.[40]

Set up at a time when the market was saturated with stocks of more than 3.5 million tons and with a production level approaching that of 1929, the new organ could not simply limit itself to the organization of sales, but had to try to adapt output to market conditions. One of its first tasks was to overcome the obstacle represented by the great variation in products through the adoption of a classification system inspired by that of the Rhine–Westphalia Union. In order to preserve the profit margin of enterprises, it held prices at a level prior to that of the 1932 crisis, while at the same time rapidly unifying prices on the market, differentiating sales prices according to the destination of the product and ensuring price stabilization. The reorganization of distribution, begun by the Belgian Industrial Coal Agency to reinforce the position of producers vis-à-vis wholesalers, was pursued. In order to reduce

39. FEDECHAR archives. *Rapport du comité pour l'exercice 1935*. Also *Annexes au Moniteur Belge*, 26 January 1935, no. 857.
40. FEDECHAR archives. Note entitled 'Vingt ans après', 9 August 1955. Note on Herman Capiau, 9 June 1952.

excessive stocks, a group of three experts, including Charles Demeure, was given the responsibility of producing an acceptable
plan to allocate production quotas.[41]

Whereas the gravity of the crisis had accelerated an awareness
among the coal-owners of the advantages of concentration, there
was sustained and vigorous criticism, in particular on the part of
the Antwerp Chamber of Commerce. It put its finger on a fundamental problem, fraught with long-term consequences, namely
that the monopoly reserved to the National Coal Office kept the
least profitable enterprises afloat, in particular those of the Borinage
defended so vigorously by Demeure. Their cost prices served as the
reference point for fixing sales prices and this was detrimental to
consumers. In response to the attacks launched against the banks,
considered responsible for this policy, the Société Générale argued
that the output of mines owned by the banks was 8.7 per cent
higher than the output in the mines of the old Walloon coalfield as a
whole and 16.6 per cent higher than that of mines which they did
not own.[42] On the other hand, a comparison of the output of
Société Générale mines in the Borinage and that of the coalfield as a
whole from 1929 to 1939 showed a disengagement, visible from
1935 on. In 1939, overall output in the Borinage was 79 per cent of
the 1929 level. Output at mines owned by the Société Générale in
this coalfield was only 65 per cent of the 1929 level.

Institutionalization of Industrial Relations – 1919 to 1947

During the war, coal production fell by more than a third, its
labour force by half (comparison between 1913 and 1917). The
collapse of the labour force was a result of wartime conditions but
also of the more deep-rooted development of disaffection with
regard to mining. During the period of growth which had preceded
the war, the coal industry had run out of steam; output had not
grown and neither had productivity. Profit per ton extracted had
fallen in comparison with wages. According to J. M. Wautelet, this
was the result of the organization of workers; according to

41. Coppé, *Problèmes*, pp. 198ff.
42. *La politique charbonnière de la Belgique et la politique des pouvoirs spéciaux. Bulletin
de la Chambre de Commerce d'Anvers*, July 1934, pp. 271–4. A. Abrassart, the author
of the reply to the criticisms of the Chamber of Commerce, was managing director
of the Hornu-Wasmes coal company, one of the first to have been purchased by the
Société Générale, and administrator of the Coal Federation.

J. Michel of difficulties in recruiting labour.[43] The employers did in fact complain that they could not find enough workers, especially qualified ones. In Campine (1917), although there existed an important labour reserve (the region being completely rural), increases in output in the 1920s were only possible through the massive recruitment of foreign labour (30 per cent in 1931).[44] Whereas the nineteenth century had been an age of relatively abundant mine labour (because of the rural exodus), the twentieth century by contrast saw recurrent recruiting difficulties because of general industrialization. In 1910, the English sociologist B. Seebohm-Rowntree described Belgium as a country of low wages, long working days, weak education of the labour force, low productivity and little mechanization, especially in the mines.[45] Legislation (1889 and 1909) obliged the coal-owners to mechanize. The use of the pneumatic drill was extended during the war. Legislative modifications and restrictions of the labour market and, above all, changes in the political balance explain the watershed of 1919.

The invasion of Belgium allowed the rapid integration of the BWP into the political system, based on a government of national unity during and after the conflict. It also facilitated its social integration, which was desired by the sovereign, by influential personalities like the industrialist E. Solvay (chemicals), the banker E. Francqui (Société Générale), and by the leadership of the Socialist Party itself. The tripartite Catholic–Liberal–Socialist government of 1918 announced the immediate implementation of universal suffrage, the eight-hour day and the abrogation of article 310 of the Penal Code. The coal-owners had little room for manoeuvre. Their traditional political supporters had accepted this programme. Economic recovery, with coal output in pride of place, was a patriotic goal which required a workforce. In addition, the working class, which had suffered for four years from unemployment and food shortages, had been encouraged by victories achieved without a fight but they were desirous of seeing them

---

43. M. Wautelet, 'Accumulation et rentabilité du capital dans les charbonnages belges', in *Recherches économiques de Louvain*, 1975, pp. 265–83, and Michel, 'Un maillon', pp. 465ff.
44. L. Minten, *De stakingen in de Limburgsche steenkoolmijnen tijdens het interbellum*, KUL, Louvain 1984 (summarized in *AMSAB Tijdingen*, 1984–5, 4, pp. 31–47).
45. *Land and Labour. Lessons from Belgium*, London, 1910, pp. 81ff. In 1913, 15 per cent of coal-cutting was mechanized compared to 62 per cent in 1926.

Table 5 Belgian strikes and union membership

	1912	1913	1919	1920	1921
Strikes	202	162	336	506	252
Strikers	60,570	15,939	178,778	330,141	148,828
Socialist union members	129,126	126,775	626,736	718,410	698,340
Miners	–	39,417	123,468	120,000	–
Christian union members	22,761	102,177	–	156,631	200,201
Miners	–	5,864	8,664	–	–

consolidated. They exerted aggressive pressure on the economic and political authorities through intensifying strikes and joining unions *en masse*.

Industrial joint commissions One day in January 1920, A. Delattre, secretary of the Miners' Union of the Borinage, had 'a surprise to which . . . a certain amount of satisfaction was added': the arrival of a managing director in his union office in the Maison du Peuple, the first meeting of this type between a 'boss' and a 'trade-unionist' in the region. Two worlds 'ignorant' of each other met, confronted one another and came to an agreement in 1919–20.

The explosive growth of unions is explained by the participation of socialists in government, which reduced the significance of affiliating to a workers' organization, by the general atmosphere of workers' struggle in Europe and also by the policy followed by the socialist Minister of Industry and Labour, J. Wauters, which systematically favoured the unions. Compensation for those made redundant came into effect from December 1918 by means of Primary Insurance Funds of communes and, above all, of unions.[46] From 1918, the Trade Union Commission of the BWP formulated principles for action in favour of an immediate reduction in the working day, wage increases, equal negotiations between unions and employers and, thus, the recognition of unions. At the grassroots, strikes multiplied.

With the support of the Catholic head of government, Delacroix,

46. This method had already been operating during the war following the decision of E. Francqui. It was reaffirmed by J. Wauters in 1920, that is, after the increase in the number of union members. None the less, the insurance–unemployment–union link explains the high level of unionization in Belgium over the long term. See G. Van Themsche, *De Werkloosheid in België tijdens de crisis van de jaren 1930*, 2 vols, Thesis, VUB, Brussels 1987, pp. 24ff.

J. Wauters, who had created mixed commissions by sector to examine ways of introducing the eight-hour day, extended their remit to include studying solutions to the demands pressed by social movements. The coal strikes resulted in the setting up of the National Mixed Commission for Mines (NMCM) on 18 April 1919, the second most important after steel-making. After rapid agreement on the eight-hour day, the unions demanded that other claims should be discussed. The employers refused. The Minister of Labour invited them, after a new round of strikes, to enter into negotiations. The union demands for negotiations and pressure from strikes, followed by the political intervention of the Minister of Labour and then of the head of government, set the pace for the establishment of the new organs: 'The goal of the Commission must be to find, through collaboration between the representatives of employers' and workers' associations, peaceful solutions to all questions concerning labour in the mines. This is an important factor of conciliation in the work of national recovery.'[47] Sixteen sectoral commissions were created from 1919 to 1921. The activities of each one complemented those of the others and thus enlarged their area of competence.

The Mining Commission played a pioneering role. Its regulations served as a model, confirmed by royal decree. It was the only one to envisage joint supervision of decisions taken and embodied in collective agreements. In 1920, provision was made for free coal, sickness benefit and, above all, the linking of wages to the retail price index. The Commission consisted of ten employers' representatives and ten union delegates (eight Socialists and two Christians) who alone had a deliberative vote, and was presided over by a high-ranking civil servant, the Secretary General of the Mining Administration or even the Minister of Labour himself. Secretarial duties were performed by a civil servant. The Centrale des Mineurs obtained, despite the opposition of coal-owners in Liège and Charleroi, the setting up of regional and local joint commissions as well as of elected pit delegates empowered to meet engineers and managing directors of the mines (1920). Convened on the initiative of one of the parties (frequently the union) but also by the minister, the joint commissions allowed discussion, nego-

47. Minister of Industry and Labour, 27 November 1919, in H. D. Antonopoulo, 'Les commissions paritaires d'industrie en Belgique', in *Revue de l'Institut de Sociologie ULB*, Brussels 1924, p. 472, an important article for this part of the account.

tiation leading to an eventual contractual agreement, conciliation and even arbitration. Conciliation and arbitration could be carried out by mining engineers, public personalities such as provincial governors and also by members of various commissions. G. Dujardin, a BWP deputy, former miner and former union activist, was delegated on several occasions to be an arbitrator, indicating thereby the role of the socialist union in the joint procedures.

Pushed into a corner in this way by the combined pressure of the social movement and of the political authorities, the employers were forced, against their will, to accept negotiation, the signing of agreements and, above all, the recognition of the existence and representativeness of the unions. In addition, this development tended to reinforce existing institutions on both sides in the direction of greater centralization and more responsibility over the decisions taken.

The Coal Federation The Coal Federation was organized in this direction in 1921. Fundamentally, the attitude of the employers to workers' unions was scarcely different from what it had been before the war but the overall balance of power had changed, all the more so as the government was seeking, to the grave disquiet of the employers, to control production and prices. 'The best way of preventing this new state interference is to establish our own [coal agency].'[48]

Opposition to legislative measures such as the abrogation of article 310 of the Penal Code and the eight-hour day was exerted directly on the political level. Opposition to the joint commissions, which was more difficult, was more low-key, involving 'putting an end as quickly as possible to their meetings' and 'temporizing'.[49] The risk that the unions would be reinforced by measures taken with respect to unemployment was clearly perceived and vigorously denounced. But it made no difference. The employers were on the defensive. In 1920 they seemed to prefer to come to agreements with the unions rather than submit further to direct state intervention. The linking of wages to the retail price index (28 July

48. The quotations which follow from the General Meeting of the Coal Federation come from the Coal Federation archives, Brussels, minutes analysed by O. Levêque, S. Rottiers and H. Van Praag, graduate students in history. The minutes of the National Mixed Commission for Mines (NMCM) have been analysed by M. J. Stallaert. I would like to acknowledge my thanks to them all. FEDECHAR archives, *Procès-verbal*, 17 December 1919.
49. FEDECHAR archives, *Procès-verbal*, 1 December 1919.

1920) was certainly the most significant and the most important element of the agreements signed in 1920. From the time of its adoption, the employers sought to modify it and threatened to denounce it. In the sphere of legislation, the disappearance of the BWP from the government allowed the coal-owners to oppose proposed legislative measures without too much difficulty but it was not as easy in the joint commissions where each union claim was countered by demand for the modification of the agreement on wages. In 1926, at the height of the British general strike, the agreement was modified. The price of coal was weighted as 25 per cent, the retail price index as 75 per cent in the setting of wages. The employers' organizations unceasingly demanded a new modification establishing a 50:50 ratio. 'We have encountered such resistance among the workers' representatives that we have preferred to hold back our claim rather than risk conflict at an inopportune moment.'[50]

Thus the tactic adopted was to present counter-proposals to any claims by workers, while the strategy was to control wages more closely by modifying the agreements. The defensive elements of the employers' policy will be considered together with the analysis of the unions' attitude.[51]

In fact, it was elsewhere that the owners were to focus their efforts to stabilize the situation, principally in the area of labour. Shortage of labour caused difficulties for production but above all for productivity and thus for profitability. It also implied increasing the benefits agreed to in order to maintain the level of labour in the mine. The effect of these measures was often contradictory in that reductions in working time and the lowering of pensionable age, intended to attract labour, reduced it in the short term. This is the reason why the coal-owners actively revived housing policies which had been pursued at the beginning of the nineteenth century. The new coal mines of Hainaut, and above all those of Campine, were given new workers' towns where systematic paternalism and strict social control, with the support of the church in Campine, allowed the social contagion of the old coalfields to be avoided. Certain towns in Campine were strictly private and access to them

---

50. FEDECHAR archives, *Rapport 1926*, 21 January 1927, p. 7. On the evolution of wage agreements, see G. Logelain, 'L'évolution des salaires dans les mines belges depuis la convention de 1920', in *Annales des mines*, January 1949, pp. 108–20.

51. Union sources and studies on unions are more numerous than on the employers. They can be used to observe employers' attitudes indirectly.

was forbidden to third parties. In addition, the immigration service for foreign labour set up by the Coal Federation allowed a large number of foreign workers to be taken on, Poles, Hungarians, workers from the Baltic states, Czechs and Italians. Most of them were Catholics and they were sometimes accompanied by priests or by government agents of extreme right-wing states.[52] In Campine, because of the sociology of this rural Flemish region and the policy of the mining companies, the Christian union was the only one to have any sort of effective presence.

The unions In 1920, according to union sources, 80 per cent of the labour force in the mines was unionized.[53] This large number, which corresponded to the unarguable explosion of union membership in Europe and in Belgium, was also the result of attempts by the socialist Centrale des Mineurs to introduce a kind of forced unionization, supervised by militants at the pitheads. This attempt was vigorously denounced by the Free Miners, who saw in it a measure designed to exclude their union. The employers also denounced this attempt as 'dictatorship of the proletariat in the enterprise' and 'workers control'. The socialist union quickly became aware of the harmful character of this large but unreliable influx. The falling cadence of social conquests after 1921, the failure of the unions to strike deep roots among pithead workers and restrictions introduced in the unemployment insurance system by the Liberal–Catholic majority from 1929 to 1935 brought about a large decline in the number of union members.

In 1920, the Centrale des Mineurs included 76 per cent of the workforce, the Free Miners 5 per cent. In 1938, the figures were 36 per cent and 13 per cent respectively. The Free Miners were most developed in Campine, where they were in the majority and played a significant role in the regional mining commission after the intervention of their leader during a localized conflict over wages in 1922. The Christian union was not really seen as a union by the employers in the Northern coalfield. Union members were sacked

52. Minten, *De stakingen*, and F. Caestecker, *Het vreemdeling beleid in de tussenoorlogse periode*, RUG, Ghent 1983 (summarized in *AMSAB Tijdingen*, 1983–4, 3–4, pp. 31–47). In 1932, there were 25,638 foreign workers in the mines (one-fifth of the total) including 5,500 Italians, 3,200 Czechs, 2,880 Poles, 2,032 Serbs and 1,284 North Africans.
53. For union membership, see Georges S. Bain and R. J. Price, *Profiles of Union Growth. A Comparative Statistical Survey of Eight Countries*, Oxford 1987.

by some coal companies but, as an element in the Catholic structure, the union was accepted, even utilized, to limit socialist influence, which took fifty years to rival that of its competitor in the region. In the Southern coalfields it was the other way round. Van Buggenhout, securely ensconced in Campine, none the less played an active part in the National Joint Commission.

But it was the Centrale des Mineurs, founded in 1919, which occupied the central position in industrial relations. Despite its centralization, it remained the sum of its regional branches. In the Borinage, local organizations even maintained a high degree of social autonomy despite the strengthening of regional and national funds, especially because of their role in the handing out of unemployment compensation. In 1920 to 1921, the Centrale was not able to maintain its political control over all its members. The Knights of Labour withdrew in a resurgence of autonomism, which had fallen at the beginning of the century. Communists, excluded from responsibility within the socialist unions, became active within it. Controlled by the revolutionary opposition, the Knights, of importance mainly in Charleroi, remained in the late twenties within the Trotskyist movement whereas the Communists created the Centrale Révolutionnaire des Mineurs. While the two tendencies were marginal in terms of their size, they helped to explain the collapse of socialist union membership because of the violent confrontations between them at strike meetings. 'The policy of the Centrale [des Mineurs] was always to accept the normal and regular implementation of agreements which it had signed even where these were unfavourable to it,' wrote A. Delattre, one of its principal leaders.[54]

The Centrale des Mineurs, throughout the whole interwar period, limited itself to the politics of negotiation within the mixed commissions. This fidelity to the system obtained in 1919 explains, to a large extent, the disaffection of some miners, accustomed to settling their problems with the owners by means of strikes. The Centrale thus sought to define new rules for implementing strikes and for subsidizing strikers, which would give it greater control over such exercises. The evolution of social movements showed that it had not entirely succeeded, to the great irritation of the

---

54. A. Delattre, *Souvenirs*, Cuesmes 1957, p. 113. Delattre was a former miner, journalist, secretary of the Centrale des Mineurs of the Borinage (1919), national secretary (1920), Minister of Labour (1935–8) and Minister for Energy (1947–8).

employers, who suspected it of practising double-speak and the tactic of having two irons in the fire. In the Borinage coalfield, the contradiction between this policy and the vigour of the social movement brought the union, on several occasions, to the verge of a split (1924, 1932) and the employers almost to the point of withdrawing from the mixed commissions. In 1930, the coal employers of the Borinage effectively did withdraw in protest against the continued waves of strikes (26 since 1919), despite the exhortations of other unionists who excused the 'Borinage hotheads' and demanded from the employers a little 'generosity' for their workers.[55]

The social movements Three important strikes showed the risks and limitations of collective negotiations (1924, 1932, 1936). The first two were strikes of miners, the last of several trades. In 1924, the miners of a coal company in the Borinage, who had suffered a wage reduction which was outside the agreement (from 1922 to 1924, wages had risen more quickly than allowed for in the agreement), drew the regional union organization into a four-month strike. Carried away by its grass-roots, it none the less found a cause – 'that which has been granted by a mixed commission can only be taken back by a mixed commission.' The National Mixed Commission for Mines, chaired by the Minister of Labour, came up with an agreement which allowed the union leadership to bring the strike to a halt. Extra-conventional alterations in wages had to be discussed in mixed commissions. The union also obtained, at a late stage, as in 1932, responsibility for maintaining equipment, undertaken until then by engineers and supervisors. For a long time, the employers had resisted the idea that the hiring of maintenance workers should be done through the authority of the union.[56]

In June 1932 the strike of Borinage miners, once again against wage reductions after two years of the government's deflationary policy, spread to all the Belgian coal mines including Campine. But it also broke out among other workers. Faced with the danger of a strike which had insurrectional aspects, contending with the public

55. FEDECHAR archives, *Procès-verbal CNMM*, 9 July 1930.
56. Dethier, *Centrale syndicale*, pp. 136ff.; A. Delattre, *Une grande bataille sociale. La grève des mineurs du Borinage (août–octobre 1924)*, Brussels 1925. Delattre, who was in no way responsible for the strike, repeated at each general meeting: 'Before the war, we had nothing to lose, today there is the agreement [to safeguard] in order to obtain a majority favourable to conciliation proposals.'

authorities and also with union power and socialist politics, the BWP and its union commission decreed a general strike. Having regained control of the movement, and following government concessions, the socialist leadership called off the strike, leaving the miners to continue it alone until September. A 1 per cent wage rise and restrictions on the import of German coal were obtained.[57] During each unofficial strike, the employers brought before the labour courts workers who had 'illegally' broken their labour contracts, obtaining heavy penalties, which were assumed by the union organization. In order to fulfil its commitments with regard to its members, the latter had to be supported by major national and international currents of solidarity. Each time, new measures were decided upon to impede the unleashing of strikes and their recognition by the union outside the regular procedures. The strikes of 1936, started through contact with events in France, were not begun by miners but affected workers from different industries. They came to a rapid end with the intervention of the political authorities, in the form of the tripartite government, which convened the National Labour Conference, a sort of national interprofessional mixed commission. This body decided on measures such as wage increases, paid holidays, the recognition of unions within enterprises and the gradual implementation of the forty-hour week. These positions were passed on to the mixed sectoral commissions which established the new conventions, while, for its part, Parliament voted the necessary laws. The 'miners' did no more than apply general decisions for the first time in half a century. The crisis had hit the coal industry hard. From the social point of view, it was no longer the most dynamic element in the Belgian economy.

The principal characteristics of interwar developments The return of the socialists to the government in 1935 and the presence of a national secretary of the Centrale des Mineurs in the Ministry of Labour was reminiscent of the 1918 to 1921 period, though in an even more emphatic manner. But in these years as a whole one can clearly observe (in 1924 and in 1932) the intervention of the political authorities in industrial relations.

---

57. B. Hogenkamp and H. Storck, 'Borinage', *Revue belge de cinéma*, 1983–4, 6–7. This volume devoted to the celebrated film by J. Ivens and H. Storck includes a good account of the strike by B. Hogenkamp.

Table 6 Strikes and working days lost in the Southern Belgian coalfields, 1935–1938

	1935	1936	1937	1938
Strikes	80	45	56	55
Working days lost in the southern coalfields	346,510	163,523	162,789	37,572

As a strategic sector, vital for industry and for the population, confronted with deep and long-lasting structural problems, the coal industry was under ever greater scrutiny. The profitability of the mines could only be maintained through political measures such as import controls, the abolition of concession fees, the restriction of taxes on enterprises, the participation of the state budget in social matters such as pensions and illness and, finally, the organization of the internal public market. The social actors well understood this and they regularly met the labour and finance ministers and even the head of the government. Coal was well and truly a special sector. The unions sought this intervention, which, in order to maintain public order or for purely political reasons, tended to be conciliatory.[58] By contrast, the employers continued to distrust it. The Coal Federation assemblies (1935–9) complained of the presence, as chairman of the National Mixed Commission for Mines, of a minister rather than a civil servant, and asserted that, in any case, they preferred to meet with a finance minister, who was usually a conservative, rather than a minister of labour who was a socialist . . . etc., etc.[59] The approach of war made the tripartite nature of the relationship even stronger, at least from 1935. As for the strikes, they diminished in intensity and length (Table 6).

The employers remained profoundly dissatisfied with the system

58. One thinks here of the role of Christian Democratic ministers in the centre-right governments which followed a liberal line between 1929 and 1935. With respect to unemployment, for example, they accepted increasing restrictions but maintained the existence of the system, in particular the role of the unions in the distribution of redundancy payments. They justified the preservation of this role by the need to reinforce the Christian unions in the face of the Socialist unions (Van Themsche, *De Werkloosheid*).
59. 'We have no more faith in the joint commissions . . . the workers sense that they are the masters', declared Baron Coppée to Prime Minister Pierlot on 30 December 1939 (FEDECHAR archives). Employers protested over the fact that workers waited for union directives before applying certain legal or regulatory measures and that, as a result, they no longer had the authority they needed.

of collective relations which had been set up in 1919. 'Handicapped by our coal deposits and our social legislation, it is really odd that it should be the country with the most deplorable economic situation that should go the furthest on the road of (meeting) workers' claims.'[60] The ILO report (1938) rather put this observation into perspective in establishing that the social charges on employers, measured per worker or per ton extracted, were lower in Belgium than anywhere else. If Belgian miners benefited from comparable, even better, social conditions in relation to those of neighbouring countries, it was because of state intervention and not because of charges weighing on the industry.[61] There was clearly a triangular relationship, but it was frequently based on bilateral ones between the state and employers and the state and the unions, rather than a direct employer–union relation, the importance of which declined compared to the growth of bilateral relations. The meeting of the socialist Prime Minister, P. H. Spaak, with the Coal Federation in 1938 perfectly symbolized this set of relationships. The Coal Federation wanted to take back a 5 per cent extra-conventional wage rise. The unions were not opposed to it. The Prime Minister responded, basing his position on the shortage of labour and the desire of the government not to reopen the path to the hiring of foreign workers (the measures of August 1935). It was necessary to improve the attractiveness of work in the mines by maintaining wage levels, the increased costs being divided evenly between the state and the companies, and by bringing about the necessary reforms: 'Let me give you a piece of advice: do things yourselves or the moment will arrive when they will have to be done and I do not want the state to

60. FEDECHAR archives, *Rapport 1926*, 21 January 1927. This complaint is repeated in numerous annual reports.
61. Average social charges on employers.

| | per worker | | per ton | |
	1929	1935	1929	1935
Great Britain	100	100	100	100
France	134.5	190.6	206.1	279.2
Holland	312.5	406	260.6	289.6
Belgium	84.5	76.7	139.4	106.3

The reduction observed in Belgium between 1929 and 1935 was a result of the policies of the centre-right governments.
Source: L'industrie charbonnière dans le monde, vol. 2, *Questions sociales*, BIT, Geneva 1938, p. 178.

intervene if you are able to do them. If not, the time will come when the state will have to do so.' He concluded: 'We do not have the power to prevent you from reducing wages but, if you do so, if you unleash a free-for-all over 7 million [francs; 50 per cent of the increased cost], you will need my gendarmes.'[62]

The Structural Crisis of the Coal Industry

Industrial Relations during the Occupation

The German invasion brought profound changes in industrial relations in the coal sector. The occupation authorities decided from the beginning to suspend meetings of the joint commissions. For its part, following the departure of the government for France, the governor of the Société Générale, with the support of the top-ranking leaders of industry and finance, set up the Galopin Committee which set out the ground rules for dealing with the occupying power. Its policy of 'the lesser evil' was one of continuing, or resuming, work in order to provide for the needs of the population and to keep the labour force together. The occupying power would not be provided with war materials.

On 12 June 1940, the committee of the Coal Federation met for the first time since the invasion at the Société Générale under the presidency of Edgar Stein, the person in charge of the coal division of the holding company.[63] In conformity with the Galopin doctrine, the committee issued a directive to the members of the employers' organization limiting production to the satisfaction of the needs of the civilian population. From that time, the Coal Federation became the intermediary with the German authorities over all problems in the coal sector.[64]

Very quickly the Galopin doctrine was shown to be very difficult to put into practice because the occupation authorities required combustible materials to be supplied. In addition, it was not possible to keep the workforce together because foreign workers employed in the Belgian mines were recruited to work in the

62. FEDECHAR archives, *Procès-verbal, Entrevue avec le Premier Ministre, mercredi 13 juillet 1938, 18 à 19h 30*.
63. FEDECHAR archives, *Procès-verbal, réunion du comité du 12 juin 1940*.
64. FEDECHAR archives, *Rapport sur l'activité de la Fédération des Associations charbonnières de Belgique pendant la période du 10 mai 1940 au 30 juin 1942. Procès-verbal des réunions du comité, juin 1940–avril 1944*.

German coalfields. The Federation engaged in numerous negotiations to limit the departure of workers and to obtain the repatriation of prisoners. Only mining engineers were freed.

While specifying that there was no 'question for the coal companies of making the least profit' but that for them it was a matter of 'making ends meet',[65] the leaders of the Coal Federation unsuccessfully tried to oppose the wish of the German authorities that wages should be raised and social advantages be given in order to contain worker agitation and to get miners to rally to the UMIW (Union of Manual and Intellectual Workers), the union organization set up in collaboration with the occupiers. On 1 June 1941 the German authorities decided to increase wages by 8 per cent, to put up family allowances and to institute attendance bonuses in the mines.

In order to deal with the rise in wages and social charges, the coal employers tried in vain to persuade the authorities to increase prices. At the most they were given compensatory subsidies, which only made up in part for the losses incurred. As far as coal sales were concerned, the cartel continued under a new name, the Belgian Coal Agency.

Nevertheless, the occupying forces did take into account complaints about the shortage of labour by favouring the transfer of miners to the Campine coalfield, whose output was of great interest to them, and, from May 1942, putting Russian prisoners to work. These measures did not in any way prevent the fall of output per man per day which, for the country as a whole, declined from 754 kg in 1940 to 479 kg in 1944. The decline in Campine was even more spectacular. It fell from 1,009 kg to 590 kg.

The Coal Federation called unceasingly for the food ration of mine-workers, whose health was suffering greatly from the shortages, to be increased. It did finally obtain from the German authorities on 1 May 1942 the granting of supplementary allocations which brought their ration of bread, meat and fats to the same level as that of German workers.

The suspension, during the Occupation, of the pre-war industrial relations system helped to accentuate concentration in the coal industry and the weight of the employers at the national level. None the less, in the localities, contact between employers and

65. See note 63 above.

workers was maintained by means of the local conciliation committees set up by the collective agreements of 9 February 1920.

The Battle for Coal

On its return from exile in September 1944, the Belgian government fixed as its priority goals the securing of adequate food supplies for the country, the elimination of wartime inflation and the establishment of social peace. A whole series of measures for restoring the post-war economy were considered. However, two factors had not been taken into account: Belgium remained a military zone until August 1945; and the coal problem asserted itself with unexpected seriousness. In September 1944, following mass desertion of the industry by the workforce, there was a sharp drop in output, which fell to 170,000 tons a month from a level of 1,980,000 tons a month in 1943. The employers declared themselves incapable of satisfying wage claims if they could not raise prices sufficiently. In addition, the obsolescence of the coal-workings was damaging to output.

From this time on, there was permanent confrontation between successive governments and the coal employers over the government's desire to ensure coal supplies at a low price for the nation's industry. This is the reason why, in January 1945, the government, chaired by the socialist Achille Van Acker, later nicknamed 'Achille Charbon', entered into 'the battle for coal'. Not content with sending the militia into the mines, it decreed civilian mobilization. To compensate for constraints intended to keep the miners at work, a special statute and numerous social advantages were granted to them. The failure of this policy forced the Belgian government, in agreement with the Allies, to set to work 45,000 German war prisoners and to recruit labour abroad. In autumn 1945, it concluded a 'miners for coal' agreement with Italy which obliged Belgium to deliver to Italy, at an agreed price, 5 tons of coal per month for every Italian worker supplied. In January 1947, another agreement was entered into with the Allies for the recruitment of 20,000 displaced persons from Central Europe. By the end of 1947, the German prisoners were freed and replaced by the Italian miners and the displaced persons.

As well as keeping an eye open to improve conditions for miners (from 1944 to 1947, there were nineteen decree-laws, thirty-one decrees of the Regent and fourteen ministerial decrees in favour of maintaining or recruiting the 'free, high quality labour force which was missing'),[66] the government imposed an average sales price on the coal employers of 320 francs a ton, a price considerably below the cost price. It compensated the loss-making enterprises by means of subsidies. This price was maintained until July 1946. However, the price rise agreed to at that time was insufficient to meet the rise in wages. Renouncing subsidies, the government in March 1947 set up a cross-compensation system between profit-making and loss-making mines, all within the confines of its policy of maximizing production and supplying at a low price. This system, energetically criticized by the owners as a brake on the modernization of the coal industry, was abandoned in 1949.[67]

The Question of Nationalization

The battle for coal and the subsidy policy which the state pursued after the war posed certain structural problems for the Belgian coal industry.

During the debate in the early twentieth century about how the coal deposits in Campine should be exploited, the leaders of the BWP had proposed that the state should take on their development, as was the case in certain German and Dutch coal enterprises. As early as 1886, G. De Greef, a socialist in the Proudhon tradition and a professor at the Free University of Brussels, had proposed the buying back of the coal mines. In 1922, A. Delattre had published a brochure devoted to nationalization of the mines: 'Only their exploitation without profit can save the coal mines.'[68] He suggested the setting up of a sales agency and a compensation fund. But, above all, the Labour Plan of H. De Man proposed the nationalization of the energy sector (1933).

It was after the Second World War that the idea of nationalization once again caught the attention of the trade union and parliamentary Left. At the end of August 1945, two bills, one Communist, the other Socialist, were put forward in the Chamber of Representatives within three days of each other. Neither of them was

66. G. Logelain, 'Problème de la main d'oeuvre dans les mines belges', in *Annales des Mines*, 1949, p. 365.
67. P. Wauters, '"De steenkoolslag". Krisis in de Belgische steenkoolindustrie na Wereldoorlog II (1945–1948)', unpublished dissertation, VUB, Brussels 1981; P. Sunov, *Prisonniers de querre allemands en Belgique et la bataille du charbon 1945–1947*, Brussels 1980.
68. A. Delattre, *La nationalisation des mines*, Brussels 1922.

considered. During spring 1946, the Minister for Reconstruction in the Socialist–Liberal–Communist coalition cabinet chaired by Achille Van Acker elaborated a scheme to reform the structure of the coal industry. Albert De Smaele, an engineer by training and a socialist and minister in various post-war cabinets, proposed in place of nationalization a reorganization aimed at dividing up the tasks and responsibilities by putting day-to-day management into the hands of the companies, overall supervision and general policy orientation into the hands of a national coal council and the defence of the general interest into the hands of the state.

In the eyes of Van Acker and De Smaele, it was important to remedy the deep causes of the malaise in the coal industry, in particular price instability and the extreme subdivision of the concessions. This is why they proposed the creation of five limited companies, one for each coalfield, led by a technical committee composed of managers from each enterprise, which would bring about the redivision of the mines into units which would be large enough to make them profitable. Even before it was presented in the Chamber on 27 June 1946, the project had been violently attacked by the Social Christian opposition and had brought about long debates within the government itself. The fall of the latter on 10 July put an end to this attempt.

The successor of De Smaele in the new Socialist–Liberal–Communist coalition cabinet, Paul De Groote, who was just as convinced of the need for structural reform but also aware of the opposition of industrialists expressed in the Liberal and Social Christian parties, brought about, though not without many changes from the original conception, the creation of a National Coal Council which began to function in October 1947. This council, composed of equal numbers of representatives of the employers, the unions and the government, was given the task of reporting, within eighteen months, on the main guidelines for reorganizing the sector and bringing about the necessary mergers.[69]

The failure of the nationalization proposals and the evolution towards economic consensus were part of the larger context of compromises arrived at between the employers as a whole and the unions at the end of the war.

69. Wauters, '"De steenkoolslag"'; R. Petre, 'Faut-il nationaliser les charbonnages belges?', *Clartés syndicales*, 31–2, Charleroi 1951; 'Les projets de réformes structurelles dans les mines belges', *Courrier hebdomadaire du CRISP*, 27 February 1958, 8, pp. 4–10.

The Emergence of the State as a Full-scale Participant in Industrial Relations

During the war, Socialist and Christian trade-unionists, employers' representatives and a high-ranking civil servant from the Labour Ministry conducted long, secret negotiations over a social solidarity pact[70] which laid down the basis for social relations after the war. Its principles were brought into being through the integrated social security system for wage-earners (1944) and the laws relating to the 'organization of the economy' (1947–8). The decree of 9 June 1945 legalized the joint commissions and their decisions. The National Labour Conference of 16 and 17 June 1947 established new institutions, namely elected union representatives in the enterprise (rather like pit representatives), Health and Safety Committees and Enterprise Councils. The last two were to be mixed and were created on a basis of collaboration and sharing of information. The first laid down union representation in the enterprise, but only for recognized unions.

If the political expression of the state, the government, was in theory generally of the Left through the presence of Socialists and Communists in most cabinets (1945–7), the policy followed took on through various checks and balances a perfect coherence. Until around 1950 the priority was production at any price in order to bring about the recovery of the country and of Europe. Afterwards, with the creation of the ECSC, supervision of production and producers was the main feature. Because of the general situation this last meant, in Belgian terms, the rationalization of the coal industry and the closure of the Southern pits, followed, from the beginning of the 1980s, by those of the North. Government policy was the social accompaniment of the battle for coal followed by the social accompaniment for its burial.

Within their organization the employers put on the same face as before, but the aims of the enterprises were not quite the same because at first they were being subsidized to produce and then they

70. 'The representatives of the employers and the representatives of the workers recognize that the satisfactory operation of enterprises, to which national prosperity is linked, requires their loyal collaboration . . . The workers respect the legitimate authority of enterprise managers and promise to carry out their work conscientiously. The employers respect the dignity of workers and promise to treat them with justice. They undertake not to do anything directly or indirectly to limit their freedom of association or the development of their organizations.' *Revue du travail*, 1945, p. 10.

were being subsidized to close down. Of the three partners, it was the unions who were most different after the Second World War. Having become independent of the BWP in 1938, the Socialist unions were dissolved along with the party in 1940 by H. De Man, who headed the organization of the Union of Manual and Intellectual Workers. This found more response among former mining trade-unionists than was the case in many other sectors. This peculiarity opened a large political space for Communist resistance in the mine through Comités de Lutte Syndicale (CLS) (committees for union struggle, CUS).

After the war, the Fédération Générale du Travail de Belgique (FGTB, Belgian General Federation of Labour) was the result of the merging of various unions, many of which had been set up in secret, such as the Syndicats Uniques which were influenced by the Communists. In the mines, the Communists maintained their autonomy within the new entity until their formal exclusion in 1950, when they were squeezed out as a result of three years of Cold War and Communist sectarianism. In 1948, the first elections to the Health and Safety Committees elected 958 candidates, with 583 candidates (61 per cent) from the Centrale des Mineurs, 236 (24 per cent) from the Christian Free Miners and 146 (15 per cent) from the Syndicats Uniques and from independent unions in the Southern coalfields.[71] But the Christians remained the majority in Campine (72 per cent). The relative size of unions close to the Communist Party had not played a major role despite its participation in various governments. Its withdrawal in March 1947 over the question of the price of coal had not brought about any particular social disturbances. As closures took place in the Southern coalfields, so the predominant Centrale des Mineurs lost ground to the Free Miners in the same way as the Christian union crumbled in the Northern coalfield. In 1971 the Free Miners overtook the others in the social elections when they obtained more than 50 per cent of the delegates.

Though the Socialist miners participated actively in the 1950 strikes against the return of the King, and the strike by a number of professions in 1960–1 against government measures, only the large strike of 1959 in the Borinage against closures and in favour of reconversion is worth noting.[72] In Campine, the Free Miners led

71. *L'ouvrier mineur* (Brussels), January 1949 (unverified results).
72. J. D. von Bandemer and A. P. Ilgen, *Probleme des Steinkohlenbergbaus. Die*

several limited strikes with precise aims related to working hours, and for a pension after twenty-five years. But in 1966–70, the Campine coal mines were shaken by unofficial strikes opposed to restructuring which escaped from union control. This type of strike reappeared in the 1980s, directed, with few exceptions, against closures.

The social movement had changed its nature after the war just as the coal industry had itself changed its nature. The unions had become bureaucratic organizations, well informed and effective in increasingly technical negotiations about professional training, accidents at work, work-related illnesses and, above all, pensions.

Decline of Industrial Relations at Sectoral Level and Birth of a Policy of Industrial Relations within the Enterprise

From the end of the Second World War, industrial relations were marked by the disappearance of the unions and their replacement by intervention by the state. Without abandoning the position of arbitration traditionally accorded to it, the state became, according to circumstances, the principal intermediary with and/or the adversary of the employers. Starting in 1952, the High Authority of the ECSC also played a part in industrial relations.[73]

Relations between the employers and the state   At the heart of the

---

Arbeiter- und Förderverlagerung in den Revieren der Borinage und Ruhr, Basel–Tübingen 1963.

73. For the most part, the literature dealing with this period is partisan. In addition to the archives and publications of the Coal Federation, the annual reports of the Société Générale and of Brufina and the works quoted above, see I. Orban, *Industrie des mines*, Brussels 1947; C. Hauzeur De Fooz, *Étude synthétique sur l'industrie charbonnière en Belgique*, Brussels 1951; M. Demasse, *L'industrie charbonnière*, Brussels 1951; J. Van Offelen, *Charbon cher et intérêt général*, Brussels 1951; *L'industrie charbonnière en 1952*, published by the Ministère des Affaires Economiques et des Classes Moyennes. Administration des Mines. Service d'études économiques de l'industrie charbonnière, 1953, p. 24; F. Baudhuin, *Histoire économique de la Belgique 1957–1968*, Brussels 1970. In addition, two personal accounts were particularly important to us: those of Max Dubois and Pierre Urbain who both started their careers as engineers in the coal mines after the war and eventually developed close connections, the one with Pierre Delville who was in charge of the coal sector of the Coppée group, the other with Paul Renders, director of the Société Générale, whom he succeeded in his various responsibilities in the coal companies and organizations on his retirement in 1971. We warmly thank them both and also M. Jacques Parent, Professor in the Mining Section of the Faculty of Applied Sciences of the Université Libre de Bruxelles.

relations between the employers and the state there existed a profound belief in the need to maintain coal as a vital sector in order to provide for the energy needs of the country. At the time, the Campine coal mines were considered to be the future of Belgium. Despite the instability of the coal market after 1948, it was the employers in the early fifties as much as the public authorities who were active supporters of the continued exploitation of marginal mines in order to preserve maximum flexibility in case of a severe downturn in the economic situation.

In the first phase, from 1947 to 1952, the game was played between the coal employers and the state. Three major problems were in the forefront: the labour force, the imbalance between the cost of production and the selling price of coal fixed by the government, and the lack of new investment.

The employers did not reject state intervention *a priori*, but rose up against a *dirigisme* which they considered harmful to the coal industry. By contrast with the pre-war period, they did not propose wage reductions because they were aware of the difficulties in recruiting a qualified labour force. Instead they sought help in constructing housing as well as a lightening of the tax burden and social charges. The devaluation of the Belgian franc following that of the pound sterling in September 1949 worsened the gap between the cost price in Belgium and abroad. According to the Société Générale, the cost of labour per ton of coal amounted to 461 francs in Belgium as opposed to 205 francs in Great Britain, 178 francs in Germany and 168 francs in Holland.[74] In addition, the employers did not hesitate to call on the state for assistance in recruiting foreign labour, while they remained mindful of conceding as little as possible in social terms.[75]

However, it was the coal price policy of the government which was at the heart of the debate. Once the battle for coal had been concluded, the Belgian government pursued its aim of providing combustible materials at a low price. In addition to the state fixing the sale price, there was free importation of foreign coal. This system, which imposed sacrifices on the coal industry for the benefit of its consumers, was considered to be unfair and the

74. *Rapport de la Société Générale pour l'exercice 1949*, p. 27.
75. M. Dumoulin, 'Les mineurs italiens en Belgique (1945–1957). Des relations bilatérales à la dimension européenne', *Relations internationales*, no. 54, Summer 1988, pp. 210–11.

employers reproached the state for slowing down the reshaping of the coal industry through its indiscriminate support for unprofitable mines. They denounced the compensation system for not providing any incentives for enterprises. In September 1949 the government gave up the compensation system and substituted for it a system of subsidies for loss-making coal mines, to be gradually reduced over a period of twenty months. The coal owners in the Borinage, the coalfield most affected by this new regime, obtained supplementary subsidies for five years and re-equipment credits to spread out the closures and to open up pits which could be exploited using up-to-date methods.

In the eyes of the employers, the *dirigisme* of the state over prices made it impossible for unprofitable pits to write off their losses. As for the others, it was difficult for them to find new investment resources. Apart from difficulties in financing their own investments, they could not enter into the capital markets because of the lack of confidence of the lenders. According to the employers, state intervention and the threat of nationalization had caused the shareholders to sell their shares. In addition, the fall in returns on coal shares compared to the general share index maintained the lack of confidence shown by the share-buying public.

Nevertheless, the fact remains that, starting in 1947, but above all in 1950, an effort at renewal was undertaken with the aim of increasing productivity with the financial support of the state and of credits obtained through the Marshall Plan.

The establishment of a policy of industrial relations in the enterprises   In the course of negotiations with the Economic Co-operation Administration, an American firm of consulting engineers, Robinson and Robinson, was called in. In May 1950 they presented a report with very harsh conclusions about the management and excessive costs of the Belgian coal industry, asserting that the average price of Belgian coal could be reduced by 150 francs a ton and by 250 francs in many particular cases.

The existence of this document, intended for use within the ECA, was made known to the press and they demanded that it should be published. For more than a year the coal employers succeeded in delaying publication while, at the same time, conducting a vigorous polemic against their detractors and denouncing the incompetence of the American consultants. Their experience, it was claimed, was based on the American coal industry, the conditions of development of which differed profoundly from those of European coalfields.[76] In any case, several missions made up of engineers, technicians and workers were sent to the United States in 1950. According to witnesses who were former engineers in the mines, the proposals of the Robinson report were not taken into account because they were not considered to be realistic. On the other hand, the missions to the United States led to the setting up of training centres in universities aimed at introducing directors and executives to American management methods. Their most essential contribution was the discovery of 'human relations' in the enterprise. In this way, in 1950, the Société Générale organized a special training course in Brussels for engineers in its mines which dealt with the 'social aspects'. The course was given by university staff and among them was L. Delsinne, a former worker militant and union leader, specialist in the trade union movement and professor at the Free University of Brussels.[77] This was a perfect symbol of the changes which were going on.

Up to the beginning of the 1950s, the conduct of discipline for staff and labour in the coal mines was spontaneously compared by the former engineers who had witnessed it to that in the army. In the words of one of them, the first instruction given to a young engineer on arrival at the mine was 'to bang on the table and swear'. Safety problems, the closeness of relations in the workplace and a very tough mental environment had sustained an extremely hierarchical system, based on both repression and solidarity, which was peculiar to the coal industry. After the end of the war, social relations within the mines were complicated by the tensions arising from the coexistence of workers of different nationalities. Falling more and more under American influence, the enterprises adopted a 'policy of industrial relations, the objective of which was to improve the climate of collaboration and the aspect of solidarity in the enterprise'.[78] A 'human relations department' was set up in the

76. E. De Vos, 'Les milieux industriels belges et les débuts de la CECA (9 mai 1950–5 février 1952)', *Lettre d'information des historiens de l'Europe contemporaine*, vol. 3, June 1988, nos. 1–2, pp. 84–7.
77. The course, entitled Social Studies in Industry, lasts for three years with one day of study per month. There are examinations at the end (evidence of mining engineer Parent, Professor at the Université Libre de Bruxelles.
78. *L'industrie charbonnière belge*, a pamphlet published by FEDECHAR, Brussels 1959.

enterprise with the task of introducing staff to new methods of command, helping new workers to settle into the job and organizing health and safety campaigns.[79] The establishment of this new policy was accelerated when Belgium joined in the Schuman Plan, which forced the Belgian coal industry to make an effort to reduce its costs.

The new policy of social relations in the enterprise, embodied in the satisfactory working of Health and Safety Committees, did not bring any corresponding softening of attitudes to the social movement. In the fifties, in the Société Générale mines, the personal signature of all workers concerned was required when a notice of strike was given, under threat of being sued for damages in the labour courts. The employers' representatives in enterprise councils changed constantly and the unions did not put in much of an appearance within the coal mines. The interests of employers and unions came together in the fight against unofficial actions. Relations between the employers and the unions were conducted more and more at the highest level, thus evading the control of coal mine management as well as that of workers and of local union branches. This development explains the reappearance of unofficial strikes in Campine in the 1960s and after.

The employers' opposition to the ECSC   Even though they had had to resign themselves to the setting up of the ECSC, the coal employers had fought energetically against the Schuman plan from the time that it was announced on 9 May 1950.[80] For two years, the employers' organization, the Coal Federation, was mobilized to produce propaganda in the press as well as through the publication of brochures and the organization of conferences. At the same time, the top-ranking leaders of the industry put pressure on the government and the administration to make their own point of view heard in negotiations with the other European countries. Their efforts among MPs to prevent the ratification of the treaty turned out to be in vain. Learning the lesson of this failure, Max Nokin, director of

79. 'Professional training . . . contributes to the growth of safety and productivity. It is a powerful factor in improving industrial relations and in the promotion of workers'; from *Les charbonnages belges et la formation professionnelle de leurs effectifs*, FEDECHAR, Brussels n.d. (1950). Also in *L'industrie charbonnière*, an annual publication of the Coal Federation.

80. De Vos, 'Les milieux'; P. Delville, 'L'industrie charbonnière devant le Plan Schuman', *Etudes économiques*, November 1951.

the Société Générale, took the initiative in creating a Study Committee of West European Coal Producers with the aim of organizing, at the European level, the defence of industrialists against the encroachment of *dirigisme* and nationalization.

During negotiations of the ECSC treaty, it was clear that the Belgian coal mines would have to be allowed a transition period to assist their integration into the Common Market.[81] With this end in view, a system of equalization was set up which favoured the Belgian industry for a period of five years. Its aim was to bring the price of Belgian coal down to the level of production costs foreseen for the end of the transition period and to establish a plan for deliveries to other countries under the supervision of the High Authority. The equalization consisted in imposing levies on German and Dutch coal production, whose average cost price was inferior to the weighted average cost price within the ECSC. The sums collected, of which the Belgian government had to pay half, were passed on to the Belgian coal mines in exchange for a lowering of the sales price from the time of the opening of the Common Market.

The adherence of Belgium to the ECSC exercised an undoubtedly stimulating effect on the efforts agreed to by the coal industry to increase its productivity and, thereby, reduce its costs. In addition to the measures taken in the social area and in the area of health and safety – in particular the battle against dust pursued by injecting water into the seams – the modernization of the mines and the search for new outlets were actively promoted.

As far as mining was concerned, a policy of concentrating coal-workings and companies was pursued. Production capacity was increased by extending and speeding up underground transport. The electrification and mechanization of the work process at the pithead and underground reduced employment through changing work conditions and reducing the required levels of qualification. The introduction of a system of non-vertical supports on the German model allowed a certain amount of catching-up in the mechanization of coal-cutting. In addition, aware of competition with oil, the coal-owners sought to develop new uses for coal. They developed methods of capturing and using pit-gas and oriented their efforts towards two main outlets, the manufacture of coke,

---

81. On relations with the ECSC, we were able to use a number of files which were made available to us by the ECSC archives in Brussels.

and the production of electricity through coal-fired power stations constructed in association with the electricity producers.

Relations with the High Authority of the ECSC remained no less difficult. In June 1955, the equalization system was revamped. A reduction in the sales price, conditional on an equalization subsidy, was decided on, but the progressive reduction of equalization for certain types of coal and for enterprises which were able to compete in the Common Market was also decided on. The reaction of the coal employers was not long in coming. The Coal Federation, and three mines in Campine affected by this measure, demanded the annulment of these measures in the Court of Justice in Luxemburg. The case was thrown out.

In January 1957, in agreement with the Belgian government, the High Authority distinguished three groups of coal enterprises in Belgium:

1 Twenty-one mines which could be integrated without assistance;
2 Thirty mines which could be integrated with the help of transitional assistance;
3 Four marginal mines all situated in the Borinage.

Only the second group would benefit from equalization.

At the end of 1957, the equalization mechanism ceased to function. Compared to the amount of subsidies given to the coal industry by the Belgian state, the proportion given by the ECSC was relatively small. From 1945 to 1957, they had received 42,927 million francs, of which 164 million had been given by the ECSC for marginal mines and 2,130 million in the form of equalization assistance.[82]

According to those in charge of the ECSC, the Belgian mines had not used the transitional period to full advantage by engaging in serious reorganization and had used the aid to pay off their debts rather than to re-equip. They reproached Belgium with having kept production at too high a level and with having imported too much despite the growth of stockpiles, in short with having displayed an excessive economic liberalism.[83]

The crisis of 1958 and the decline of the coal industry   The crisis of 1958

82. See note 78 above.
83. Note *Division du marché de la CECA, 1er décembre 1953*, ECSC archives, CEAB 7, no. 35.

hit the European coal industry with devastating force. Not only had the consumption of oil products grown significantly compared to coal and its derivatives, but the massive reduction of transport costs for American coal threw the European coal market, which until then had enjoyed protection from its geography, into complete disarray. In these circumstances, the Belgian industry, despite all its efforts, found itself in an even worse position than its partners in the ECSC because its costs remained higher.[84] From 1 January 1956 the unions had obtained a reduction in working hours. The Marcinelle disaster on 8 August 1956 unleashed vigorous attacks, within Belgium and abroad, against the unsafe conditions in the mines. The outcome was that the immigration of Italian labour was suspended and wages and social charges raised. With the end of the equalization system at the beginning of 1958, the Belgian industry entered the free coal market within the ECSC under very unfavourable conditions.

The disastrous consequences were not long in appearing. From 1959, the ECSC had to take measures to protect the market. For their part, under the supervision of the National Coal Council, the financial groups embarked on the reorganization of the Southern coalfields by merging their interests into two companies, the Charbonnages du Centre and the Charbonnages du Borinage, which were set up to organize pit closures.

The deterioration of the social atmosphere following strikes in the Borinage against pit closures, and the relaunch of proposals for the nationalization of coal induced the employers to seek an agreement with the unions over reorganization of the coal industry, on the consensus model perfected in 1955 for the electricity industry.

In 1961, at the instigation of the Social Christian and Socialist Lefevre–Spaak government, the National Coal Council was replaced by the Directory of the Coal Industry. The job of this new institution was to bring about closer co-ordination of the coal industry, to work with the main users in the elaboration of an energy policy and to guide the subsidy policy of the state. Under the chairmanship of the René Evalenko, an economist who was very active in Socialist circles, the Directory played a decisive role

84. See the diagram by J. Paelink and P. Markey, 'Impact des dépenses d'énergie et de main d'oeuvre dans l'économie belge', in *Bulletin mensuel de la direction générale des études et de la documentation*, 1963, 2, p. 36. Labour comprises 72.05 per cent of the value of output in mining, which is closer to that of white-collar workers and civil servants than to other categories of workers (25 per cent to 35 per cent).

in the policy of closure by succeeding in reconciling the interests of employers and workers. While the employers, the unions and the state were conducting an agreed policy for closing the Walloon coalfields, Campine and certain independent coalfields believed that better days were just around the corner. During the 1959 crisis, ten enterprises had left the Belgian Coal Agency to act independently on the market. Among them were some Campine coal companies whose cost prices were low enough to face up to competition, as well as some companies which were part of conglomerates, such as the Winterslag mine of the Coppée Group, which had a guaranteed market.

Nevertheless, the reprieve was a brief one. Independent of the formidable competition from America, the opening up of natural gas reserves in Holland gave the *coup de grâce* to the Campine coal industry.

In 1966, the Cockerill-Ougrée steel-making company was authorized to close its operation in Zwartberg. This unleashed a wave of violent strikes, in which the influence of the Christian union was dominant, and for the first time several deaths were caused in the Campine coalfield. From then on, a process similar to that in the south started up. At the initiative of the state, the coal companies merged into the Kempense Steenkoolmijnen company created by royal decree in 1967. Two years later, the state took over 10 per cent of the company, becoming the majority shareholder in 1981 and organizing the closure of the Campine coalfield where, at the present time, there is only one mine still in operation.

6
Industrial Relations in French Coal Mining from the Late Nineteenth Century to the 1970s

Joël Michel

The French coal mining industry, like its counterparts in other developed countries, has an image derived from figures for miners' strikes of being a focus of social conflict. In fact miners' disputes were not as frequent as is generally thought. In the French context, the essential feature of the industry was rather the precocity and modern style of its industrial relations which, at the beginning of the twentieth century, were several decades ahead of French practice in general. For instance, the 1891 agreement in the Pas-de-Calais coalfield is regarded as the first collective agreement, and it was only in 1936 that collective agreements were signed in other major industrial sectors, and even then they were imposed by law. After 1945, miners still enjoyed an official status which was relatively privileged compared to the rest of the industrial labour force. This modern style implied the existence of social partners who were well organized and prepared to come to an agreement. But were trade unions and mining companies so different from the rest of French industry, and from heavy industry, in particular, where authoritarian, paternalist or very tense relations prevailed?

The optimistic point of view needs to be strongly qualified. First of all, the relative 'social harmony' that existed was no more than an equilibrium in power relations which were sometimes brutal and which could even develop into violent confrontation. It was not only in 1893 in the Nord or in 1900 in Montceau that strikes turned into all-out wars. The miners' strike of 1948, which involved acts of sabotage and pitched battles with the army resulting in casualties among the workers, was the most brutal that the mining industry, and probably French industry as a whole, has experienced this century. Secondly, it is misleading to consider the question only in national terms because negotiations never took place at the national level. Power relations between workers and

271

employers varied considerably from one region to another.
Thirdly, with an output of 41 million tons in 1913, France did not
rank among the major producers and cannot be compared with
Great Britain or Germany. But, above all, the geographical distribution of French mines was very uneven. There was a large
number of scattered coalfields but the large concentration in the
Nord–Pas-de-Calais (the North field)[1] came to equality with the
major producing area, the Loire coalfield, in 1860, then accounted
for more than half of national output in 1886, rising to three-
quarters in 1908–12 with nearly 100,000 workers.[2] In this period
the second most important coalfield, that of the Loire, only produced one-tenth of national output. The other coalfields were
dotted around the Massif Central and, with the exception of the
Gard, they often consisted of just one isolated company such as
Carmaux, Decazeville or Blanzy which employed a few thousand
or sometimes only a few hundred workers. Therefore the North
field of the Nord–Pas-de-Calais region was the only French coalfield of international significance and, within that group, the more
recently developed Pas-de-Calais coalfield accounted for two-
thirds of total output, that is approximately half French production
as a whole. Prior to the annexation of Lorraine in 1919 and its late
development after the Second World War, the Pas-de-Calais was
the only really dynamic axis of economic, and therefore social,
growth. Consequently, the awareness of employers and workers
was more pronounced in the ten or so companies producing
between two and four million tons a year each, which is more than
most whole coalfields elsewhere. In the Pas-de-Calais, and to a
lesser extent in the Nord, trade unions and employers' organizations were set up on a firm foundation. This enabled relations
between the social partners to establish themselves before the end
of the nineteenth century. By comparison, in coalfields of the
centre and the south, and later on in Lorraine, workers' organizations were always fragile and employers traditional. As a result,
they remained areas of conflict where authoritarianism dominated.

However, because of its size and its leading role, the Pas-de-
Calais gradually set the style for industrial relations in the whole
coal mining industry. Everywhere employers imitated, more than

1. We will use the word 'North' when we refer to the two coalfields of the Nord
and of the Pas-de-Calais and the word 'Nord' when we refer to the *département*.
2. Marcel Gillet, *Les charbonnages du nord de la France au 19ème siècle*, La Haye 1973.

they approved of, its collective agreements. Similarly, the mining industry as a whole was nationalized a short time after that of the North. How can one explain this relationship to the dynamic axis? It was necessary for economic reasons. For instance in the Nord the old and powerful company in Anzin had to follow its more competitive rivals or lose its labour force.[3] But as far as the coalfields dotted around the Massif Central are concerned, economic competition from the mining industry in the North did not extend beyond the Paris region and did not have any influence on the local market. All the coastal areas and the southern half of the country were the almost exclusive preserve of British coal, while Germany worked its way into the east. The real reason for this development was, perhaps, the pressure from the state which intervened in the mining industry in the light of what was appropriate in the North. One essential conclusion which can already be drawn is that from the very beginning of the period under examination, industrial relations in the mining industry did not involve two partners but three, the presence of the French state being particularly prominent. It is all the more necessary to take this into account in view of the fact that after 1945 the state became the only employer. Thus, in order to study industrial relations in the mining industry, it is necessary to distinguish two periods, that of private employers and that of nationalization.

There was no unity among employers during the first period. The wide dispersion of the coalfields, the differences between large private companies in the North, small enterprises in the Loire and often family-based enterprises elsewhere resulted in the fact that the only common feature of their attitude towards the labour force was an unfailing paternalism which took a large variety of forms. In general the other coalfields did not have the same need as those in the North to attract their labour force through the provision of housing, nor did they have the means to undertake large-scale building projects, in the Loire in particular. Paternalism in the companies of the North was anonymous but heavy, that of La Grand' Combe in the Gard was oppressive, that of La Mure near Grenoble patriarchal. Often the paternalism of dynasties such as the Solages in Carmaux or the Chagots in Blanzy was strongly opposed by the workers. This was even more true as far as

3. Odette Hardy, *Industries, patronat et ouvriers du Valenciennois pendant le premier vingtième siècle*, Lille 1985.

Decazeville was concerned.[4] Generally speaking, and particularly in the North where reconstruction was necessary, paternalist institutions became even stronger in the interwar period.

But the whole range of possible attitudes towards workers can be found, ranging from recognition of trade unions in Lens to the 'Herr-im-Haus' attitude in Lorraine. Employers were local rather than national. Shareholders in the north mostly came from the bourgeoisie of this highly industrialized region, in particular from the textile industry. Similarly, the Lyons bourgeoisie had large financial commitments in the Loire, while elsewhere families, sometimes aristocratic ones, were deeply entrenched. It was these longstanding investors, together with the steelmakers in Lorraine, who were to share among themselves the seats on the various boards of directors in Lorraine after its annexation. Some companies in the Gard such as Blanzy, and, to an even greater extent, in Anzin, which were more deeply integrated into national and, therefore, Parisian capitalism were the exception. As for salaried engineers and managers, even though they were trained in national institutes, they often settled down for life in coalfields where they made a career and were adopted, as it were.

The dispersion of the coalfields, the local implantation of the ruling elites and the absence of any competitive national market account for the fact that employers did not have strong economic organizations like those of their opposite numbers in Germany. The Comité Central des Houillères de France (CCHF, Central Committee of French Collieries) which was set up only in 1887, mainly on the initiative of companies of the centre and south coalfields, was not a powerful body such as the Comité des Forges. However, no matter how open-minded or narrow-minded they were, employers in the mining industry shared the same social creed in two respects. First of all they refused to make links between separate disputes between the CCHF and the miners' federation. There was no question of any negotiations on a national scale; they dealt only on a regional level and, even then, only with organizations with which they chose to negotiate. Anyway, the national federation of coal miners set up in 1892, and transformed into the 'federation of underground workers' in 1910 by taking in

4. Rolande Trempe, *Les mineurs de Carmaux, 1848–1914*, Paris 1971; Donald Reid, 'Decazeville: company town and working-class community 1826–1914', in J. M. Merriman (ed.), *French Cities in the 19th Century*, London 1982, pp. 193–207.

slate quarry workers, was never a strong body. Workers' unions, too, remained regional. Secondly, a constant effort was made to limit as much as possible the interventionist activities of the state. The position expressed by Darcy, the creator of the CCHF, giving evidence to the Senate commission on the length of the working day in 1902 was valid for the whole period: 'Employers are far from hostile to all reform. They have already done a great deal for their workers but they absolutely refuse anything which enables the state to interfere in the regulation of working conditions.'[5] Therefore the CCHF was primarily a parliamentary lobby, as was the workers' federation, because one of the major battlefields was legislation. It can be said that between 1900 and 1945 employers were relatively successful in achieving these two objectives: the two short periods of industrial unrest (1917–20 and 1936–7) were followed by periods of almost complete return to 'normality' (1921–35 and 1938–45), that is to the type of industrial relations favoured by the employers and established prior to the First World War.

Industrial Relations up to the Beginning of the Twentieth Century

The paternalist tradition made few adjustments to worker expression, which was still very weak in the 1880s. The miners of Saint-Étienne, who were the first workers to become organized, were crushed during a bloody strike in 1869. During the economic crisis and also in the early 1880s, violence still dominated industrial relations in Montceau, Decazeville and above all Anzin where, in 1884, the company dismissed nearly a thousand workers following a long dispute which Zola was to use as a model when he wrote *Germinal*. However, a few years later, the climate had changed. This was not the result of any change on the part of employers, even though, from the end of the nineteenth century, some general managers often took precedence over the boards of directors in a spirit of economic innovation and of rationalization of industrial relations. In 1889 it was primarily thanks to a return to prosperity that, as in the other European coalfields, miners in the Pas-de-Calais organized and won a major strike. From it emerged the first worker organization to become instantaneously powerful. Follow-

5. Archives of the CCHF (in the National Archives) 40AS39, 28/2/1902.

ing a second strike in the autumn of 1891, the signature was
obtained of a delegation of employers on the first 'Arras agreement'
of 29 November. Even though its contents were very disappointing for the workers, this agreement marked a decisive stage in the
development of industrial relations. However, the mining
companies, taken by surprise by the strike, were resigned to the
outcome rather than agreeable to it and it took the intervention of
the prefect, and above all of the government, to force them to sign.
In any case, the mining employers pulled themselves together
quickly. In the Nord they had been cool towards the formation of
the CCHF in 1887 and it only became effective in 1892, while the
regional Chamber of Collieries almost disappeared in 1889 as a
result of the withdrawal of the companies from Lens and Anzin
who were in disagreement on the issue of transport. However, in
March 1891, the companies in the area created a Union of Collieries
on the German model to resist strikes.[6] This initiative was followed
in the Loire and in the Centre, but it did not succeed on a national
level and even in the North, Anzin and Lens refused to take part in
it. Following the signing of the Arras agreement in November
1891, the employers felt trapped and decided to fight back. In order
to prevent it being used as an intermediary by the workers, as had
happened in 1891, the regional Chamber of Collieries was dissolved in 1892 and a new organization of the North was not set up
until 1897. But the anti-strike association remained.

The deliberate elusiveness of employers in the mining industry
was obvious during the dispute of the autumn of 1893, which was
brought about by a failure to enforce the collective agreement.
During that strike the CCHF clearly defined the strategy which it
was to adopt from then on:[7]

> If our committee were prepared, through a feeling of good relations, to
> contribute to the expenses which the defence of the interests under threat
> might involve (to an extent which is left to the discretion of the bureau
> or its delegates) it would be dangerous – from the point of view of our
> relations with the public authorities, the workers and also from the point
> of view of the impression it would make on public opinion concerning
> issues of this nature – for our intervention to appear, officially and
> openly, to be a collective act.

6. Marcel Gillet, 'Aux origines de la première convention d'Arras. Le bassin
houiller du Nord–Pas-de-Calais 1880–1891', *Revue du Nord*, April–June 1957; Joël
Michel, 'Syndicalisme minier et politique dans le Nord–Pas-de-Calais', *Le Mouvement Social*, April–June 1973, pp. 9–33.
7. CCHF 40AS267, letter from Darcy to Bollaert dated 29 October 1893.

Our committee must take action in conditions of absolute discretion through indirect methods so that our name will never appear. Our committee has always thought that, in the eyes of the workers' unions,
• we should never appear to constitute a union of employers, a new enlarged version of the Committee of Collieries of the Nord–Pas-de Calais . . .

The companies in the Nord–Pas-de-Calais shared this attitude. Having restored their authority without destroying the workers' organization, they started signing agreements again in 1898. In practice, it was informal but actually very stable employer delegations which negotiated with the trade union. They had the same power as an official association but they enjoyed greater room for manoeuvre; they could maintain a low profile when it suited them as well as refuse any attempt to institutionalize joint discussion. They presented a united front, despite the lack of willingness displayed by Anzin, and did not commit themselves too publicly.

The most original feature of the model developed in the North was the signature of collective agreements with worker organizations from 1891 until the First World War. At the turn of the century this practice became more widespread in the Loire, but, as power relations there were more unequal, it was more a concession made by the employers than a victory won by the workers so that government arbitration, with all the difficulties it involved, was often required. Elsewhere, outbursts of authoritarianism were still in evidence in large companies, particularly Montceau and La Grand' Combe, and wage agreements, usually reached after intervention from the prefect, remained much more informal.

Yet it can be said that this major step forward which was unique in France and comparable only to practices in British coalfields, was achieved in conditions which remained favourable to the employers. On one hand, the recognition of trade unionism which seemed to have been granted was always disputed within the companies, as shown by workers' complaints when a parliamentary enquiry was held in 1902. Although employers agreed to recognize reformist trade union organizations, they still frequently refused to meet militants – many of whom were protected by their independent position as safety officers, elected by the men and paid by a levy on the owners through the agency of the state – and persecuted trade union members, for instance in Béthune or Anzin. Even the trade union in the Pas-de-Calais did not achieve the creation of a permanent conciliation organization, although it often proposed one.

Even though they did not always make it possible to avoid strikes, agreements actually played a moderating role by reducing the range of claims. As was the case in Great Britain, the negotiation of agreements in the Pas-de-Calais was limited to wages, 'excluding any claim not included in the debate', and, on top of that, to a fairly artificial notion of the wage compared with the actual wages received, namely the coalfield average wage. They recognized the relationship of wages to prices, but only to prices, not to production or profit, and in that respect they operated like a disguised sliding scale. According to the well-known study by F. Simiand,[8] the various Arras agreements were closely subjected only to the fluctuations of the economic environment and to the law of the market – and this, incidentally, was accepted by most elected trade union representatives even though it provoked growing irritation among miners.

However, although in the short term the workers' discontent was probably justified, in the long term, which was the chief concern of workers' organizations, the benefit that the beginning of a system of parity meant for trade-unionism justified the sacrifices made. In any case, this recognition needed to be constantly reinforced because employers were vigilant and kept trade unions within narrow limits. In 1900, 1902 and 1906, it was only the pressure exerted by strikes which forced employers to come to an agreement with a workers' organization which was also anxious to end the dispute. After 1906 it was the existence of an anarcho-syndicalist union which pressured the companies into definitively recognizing the 'old trade union' of the Pas-de-Calais, finding it useful after all. Through this alliance, employers in the mining industry succeeded in maintaining divisions among workers and weakening the national federation – which was essentially run by miners from the centre – to the advantage of local organizations. As they benefited from specific advantages as a result of regional agreements, miners in the Pas-de-Calais were not particularly inclined to go on strike on behalf of fellow workers in the south and, between 1900 and 1902, the only major attempt by the miners' federation to gain national status failed as a result of partisan manoeuvres. As a result, employers in the Nord–Pas-de-Calais

8. François Simiand, *Le salaire des ouvriers mineurs de France*, Paris 1907.

reached a regional consensus in the mining industry against recognition of the trade union elite, which was composed of republican notables who relied more on their political offices than on wide trade union membership. After 1900 a new spell of vigorous paternalism with economic as well as social targets, resulting primarily from a shortage of labour, contributed to the creation of a spirit of corporatism and of co-operation in the defence of the mining industry.[9]

For the companies whose authority and profits were not threatened by conciliatory trade-unionism, the fly in the ointment was state intervention. The miners' deputies put into legislative action all the energy which they did not devote to developing their organizations. On the issues of safety, the working day and retirement pensions, French miners won significant victories. As early as 1890 they obtained elected, paid and independent safety officers. In 1894 they obtained independent funds for mutual assistance, and then in 1914 a national pension fund. In 1905 the eight-hour day was decreed by law for coal-hewers and for all underground workers in 1913. Employers were able to carry the fight to the mining commission of the National Assembly and managed to delay many measures, thanks to their vigilant defenders in the Senate. They were also clever enough to prevent the adoption of legislation by their own action. For instance, the 1902 agreement increased the level of retirement pensions for mineworkers in the North in order to prevent a law on this issue from being passed. Even better, in 1912, in exchange for local concessions, the president of the miners in the Pas-de-Calais voted at the Chamber of Deputies for the law on retirement pensions to be regional rather than national in scope. On this occasion the division among workers resulted in the collapse of the national federation. Even so, the archives of the CCHF reveal how employers were frightened of projects for compulsory conciliation and arbitration, especially from 1908 on.[10] Even the agent of the most socially advanced company – in Lens – was worried: 'It is clear that in no other industry have employers done as much for their workers as managers and engineers in the coal industry. And it is sad to observe

9. J. H. Porter, 'Wage bargaining under conciliation agreements 1860–1914', *Economic History Review*, 1970, pp. 460–75; Joël Michel, *Le mouvement ouvrier chez les mineurs d'Europe occidentale (Grande-Bretagne, Belgique, France, Allemagne). Etude comparative des années 1880 à 1914*, thesis, University of Lyon II, 1987.
10. CCHF 40AS72.

that all the reward they get is to be subjected to exceptionally severe obligations.'[11]

As a result, between 1906 and 1913, employers' organizations were strengthened and the co-ordinating role of the CCHF was accepted with better grace. After the major strike of 1906, the national anti-strike fund which had been talked about in 1893 was set up.[12] It was renewed just before the First World War, even though it had been of little use. The reason was that employers in the mining industry were deeply worried. Certainly they retained their authority, coming to agreements with trade unions only when they chose to do so and to their advantage, but the increasing rationalization of industrial relations led them to fear that they might progressively lose control.

1914–1935: First Alarms and the Return to Normal

The First World War represented a huge upheaval for the French mining industry. For four years the front line cut across the Pas-de-Calais coalfield in its most productive part, around Lens. The major part of the coalfield under German control ticked over until it was finally destroyed by the occupiers. Production in Anzin was around 30 to 40 per cent of the 1913 level and, as the German administration kept a more watchful eye on staff administration than on the running of the operations, miners had to endure very severe conditions. In the region of Lens, where there was heavy fighting, pits were flooded and mining villages destroyed. A large part of the population fled further west to the few concessionary coalfields still managed by Frenchmen. They amounted to 47 pits out of the 160 of 1913, employed 40,000 workers and produced 8.7 million tons as against 27 million. Following the stabilization of the front line in November 1914, production was pushed to the maximum, especially in Bruay, at least until the German offensive of spring 1918, to such an extent that this company alone produced 4.5 million tons, that is to say more than the second French coalfield (that of the Loire) in 1917. In the other coalfields, production also increased, often with the help of the labour force who fled from the North, but it only increased by 45 per cent, whereas in the non-

11. CCHF 40AS39.
12. CCHF 40AS72.

occupied Pas-de-Calais it increased by 90 per cent.[13]

Inevitably the French mining industry was integrated into the war economy, essential in such a vital and exposed sector. The state intervened first of all to fix prices and direct production, then to distribute the labour force and finally to organize industrial relations. As early as July 1915, selling prices were controlled and the law of 22 April 1916 organized the taxation of coal. The National Coal Office introduced more and more stringent distribution regulations, then total requisition and finally, in 1917, parity with British prices by imposing a tax on French coal.[14] Rapidly, the state called back reservists to produce coal, directed the requisitioned labour force to the coalfields where it was most in demand, such as Saint-Étienne, and organized production by imposing a nine-hour working day as early as 1915.

Thus the state was gradually drawn into intervention in industrial relations as well, even more than before 1914. None the less, it was not a case of solving a social crisis. French miners were very patriotic, even nationalistic, and were prepared to make many sacrifices. Along the front line, in Noeux or in Béthune, they worked in pits at night under bombardment and suffered heavy casualties. In any case, before 1914 real trade union power had been confined to the coalfields in the North where it was wiped out by the invasion. Most leaders, the mayors of towns situated in the German zone, remained there with the population. The miners' national federation, which was transferred to Paris, had very little power. The beginning of the war marked the lowest ebb for trade-unionism in the mining industry, which came to rely essentially on Saint-Étienne, since the small coalfields scattered around the Massif Central were in no way capable of playing a leading role. In any case, disputes were almost non-existent – thirteen strikes for the whole duration of the war (two in 1915, one in 1916, four in 1917, six in 1918). All were very small and there were none in the North. The miners' representatives who stayed on the spot continued to defend workers vigilantly. In the Pas-de-Calais, three pay settlements were signed in 1916 and the momentum increased in 1917. During the first years of the war, basic pay was still at the level which had been determined by the Arras agreement of 1891, even

13. M. Georges, 'L'effort des mines du Pas-de-Calais non envahies de 1914 à 1919', *Revue de l'Industrie Minérale*, 1921, pp. 339–60.
14. CCHF, report to the 1921 annual general meeting.

though it proved increasingly inadequate. In the centre and south coalfields, which did not have this yardstick, they turned more and more to the CCHF, which circulated information about increases agreed in the Pas-de-Calais and concessions which might be granted. In this way the employers' federation gained more authority as a result of the war and the better co-ordination of pay policies it required.[15] However, the increase in the cost of living made it necessary gradually to work out a complicated, but unsatisfactory, system of bonuses and allowances which had to be constantly renegotiated.

In such a context – where miners were perhaps better paid than other workers but had the feeling that their earnings were not in proportion to the sacrifices made, the increase in productivity and most of all the increase in prices – state intervention was almost entirely in favour of the workers, even though it took the form of arbitration. Its role was decisive in coalfields which had been badly organized before the war and where, under its protection, the real beginning of trade-unionism and the process of collective bargaining took place. In the Gard, for instance, the company of La Grand' Combe had been totally uncompromising before 1914, but pressure from the prefect forced it to become more docile and to concede pay increases. In 1917, miners who fled from the Pas-de-Calais organized a trade union at La Grand' Combe, and on 25 June 1917, under the aegis of the prefect, companies in the Gard signed, for the first time, a contract with the workers' organization. However, this type of conciliation remained artificial as it was assisted by the state which, in fact, gave up the idea of institutionalizing these meetings because it considered the workers' trade union to be too weak. But at least employers had had to come to terms with it. In that coalfield, the running of the joint commission set up in February 1918 showed that employers knew how to assert themselves over the workers, for instance when they succeeded in maintaining divisions by company and by category. In order to prevent any excesses, the trade union turned regularly to the prefect to ask for his arbitration. But in those fragile coalfields, as soon as the state became less favourable, the negotiating power of the workers decreased.[16]

15. CCHF 40AS50, I, in particular Albi 1916.
16. Raymond Huard, 'Les mineurs du Gard pendant la guerre 1914–1918', in *Economie et société en Languedoc–Roussillon de 1789 à nos jours*, Montpellier 1977, pp. 275–94.

State intervention was also extremely valuable for the national movement which was reconstituted in the course of 1917. The federation of underground workers was weakened by the invasion of the largest coalfield but benefited from new assets. Trade-unionism had split in 1913. As a result of the war the issue which had caused the split – whether the law on retirement should apply regionally or nationally – disappeared and unity was restored, even though it only became official after the war. Also, as the centre of gravity of the federation shifted towards the south, it acquired a new dynamism for its demands. Between 1917 and 1920, the workers' federation seemed to rush from one success to another. The wave of new members was impressive. With 130,000 members in 1919 against a few thousand at the beginning of the war, miners' organizations became mass trade unions at last. Therefore, the national federation, and the trade union of the Pas-de-Calais which led parallel action, obtained satisfaction on many claims with or without resorting to strike action but with vital support from the government. Employers in the mining industry, no longer able to resist, could only adopt a relatively flexible attitude, and the economic environment allowed them to do it without too many problems.

From the end of 1917 the federation of underground miners announced its major claims for the following years: pay rises, of course, but also shift work and round-table joint commissions where the partners had equal status. The government was entirely in favour of this last point, particularly Albert Thomas, the Minister of Munitions, and Loucheur, the Minister of Labour. Even though, in September 1917, the companies rejected the proposal by the miners' deputy Bouveri that the competence of the joint commissions responsible for assigning mobilized workers to various coalfields should be extended to include work organization, hygiene and discipline, in practice the government showed its support for the miners by favouring compulsory arbitration without actually saying so. Thus workers obtained without any difficulty, under the aegis of prefects, the upward reassessment of their wages and the determination of a cost-of-living bonus which came on top of previous bonuses. In 1918 they even tried to obtain a minimum wage which would be determined by the round-table joint commissions. Both reforms caused a lot of concern among employers, as shown by the large amount of mail sent by frantic employers to the CCHF in an attempt to 'avoid this peril'.

Fortunately for the employers, the lack of experience of the new
workers' representatives, the continuation of a system of separate
negotiations for each individual coalfield, the sometimes contradictory actions of the prefects and the fierce resistance of the
employers' negotiators, made it possible to maintain some confusion as regards the nature of the payslips which were issued. At the
end of the negotiations, which had already dragged on, the regional
joint commissions which had been set up had proved to be disappointing. In autumn 1918, workers in most coalfields had accepted
ministerial arbitrations which set aside the minimum wage by
category, but granted significant increases and the incorporation
into the basis wage of the various bonuses accumulated during the
war, which went against the employers' policy of slicing up wages.

The detailed account, coalfield by coalfield, of these lengthy,
interrupted, resumed negotiations can be read in the abundant
archives of the CCHF.[17] The omnipresence of the authorities in the
negotiations is quite obvious. Sometimes the minister had private
meetings with employers' representatives and then workers' representatives in order to get them to come to an agreement before
the meeting of the commissions; sometimes he telephoned the
prefect to give him instructions on the current negotiations; sometimes he exerted direct pressure on the employers, because there
was no doubt that the underground federation and, to a larger
degree, the trade unions in the Nord–Pas-de-Calais which had not
yet re-entered it, were looked on favourably by the government.
The same archives also reveal the overwhelming activity of the
CCHF and its dynamic secretary, H. de Peyerimhof, aimed at
preventing its members from being routed or the employers' front
from breaking up in this sensitive period. It was a matter of
organizing a retreat, but with the utmost discretion. Except for a
few meetings with the minister, the CCHF never appeared as such
for fear of becoming involved in negotiations at national level,
although the workers' side never really put this on the agenda. A
conductor who performed in the shadows, de Peyerimhof was
anxious to maintain a united front and to give in only to threat,
particularly the threat of legislative intervention. In June 1917, he
told worried employers: 'This is not the time to weigh up these
new demands. We are going through an abnormal period and we
are threatened by imminent legislative intervention which we must

17. CCHF 40AS50.

prevent at all costs.' The CCHF was particularly annoyed by Loucheur, who was 'hypnotized by the wartime methods used in the factories and the British experience' of joint committees. But for many employers, for instance those in the Gard, frustration was intense. For instance, the managing director of Bessèges wrote to de Peyerimhof on 4 February 1918:

> I acknowledge receipt of the 'protocol' presented by the permanent secretary of the Minister of Labour and accepted by the representatives of the national council of the federation of the underground and, this evening, of your confidential communication of the text of the Loire agreement decided upon by the prefect between the companies of the Loire and the federation of miners.
> The hardest aspect of this agreement is not so much the financial sacrifices made as the way in which they were extorted.
> Our colleagues in the Loire can say that they made their sacrifice on the altar of the country. I feel sorry for them and I admire them.
> As far as we are concerned in the Gard our calvary has not yet begun. I hope it will not be as hard to climb as was that of 26 June 1917 . . .

But the workers' offensive had only just begun. Supported by a multitude of new members and by the social movement developing in the whole country following the return to peace, it achieved new successes in 1919. In April, the federation of the underground put forward a maximum programme, demanding, for instance, the representation of miners on the boards of directors of concessionary companies. Employers took refuge behind the government, alleging that commercial freedom had not been restored. In order to force the government to move to a vote before 12 June and to enforce, before 16 July, the law on the eight-hour working day, the federation of the underground threatened to declare a general strike, which actually broke out in the Nord–Pas-de-Calais where there was great mistrust over possible backsliding. The agreement of 6 June introduced the net eight-hour working day as well as significant wage increases which were ratified by ministerial arbitration. On 11 June, the Chamber voted a declaration of intent which introduced a working day of seven hours and twenty minutes for underground workers, of which six hours fifty minutes would be actual work. Even though the employers were preparing to enforce it without waiting for the vote of the Senate, the miners' federation declared a general strike in order to get its text voted on and forced the government to capitulate. On 19 June the government agreed that there had been a 'misunderstanding' and on 24 June the law was voted without anything being taken away.

Following this first surrender, the federation of underground workers was triumphant and in December 1919 it demanded a bettering of retirement pensions and once more forced the government to modify its bill, which was introduced in January 1920 to provide a strong reassessment. In fact it was feared that this series of workers' victories might lead to excesses. In March 1920, the miners in the Pas-de-Calais went on strike for a wage increase. The government proceeded immediately to arbitration. The companies promptly accepted it and the trade union leaders came round to it without succeeding in stopping the strike. The negotiation brought new evidence of the powerlessness of employers. On 18 March 1920, the Minister of Labour and Minister of Public Works modified their arbitration award, following a very insistent request from the workers, in order to make it easier to standardize wages between companies.[18] Similarly, in the spring of 1920, by threatening to start a new strike, the federation of the underground succeeded in forcing the government to vote a law which put slate quarrymen on the same footing as miners despite resistance from the Mines Commission.

However, although power relations remained favourable to miners, these victories were achieved by engaging in confrontation with the government and no longer through its support. Since the victory of the National Coalition at the 1919 elections, the government was far less anxious to support the miners and, on the contrary, was ready to confront an increasingly virulent Confédération Générale du Travail (CGT). Employers, who had been passive spectators able only to ratify the concessions (except in Lorraine where their intransigence sparked off the strikes of September 1919 and April 1920), were well aware of this change. Already in the second half of 1919 they flatly refused to apply the law on the eight-hour working day and repressed militants in the weaker coalfields, though not in the Nord–Pas-de-Calais.[19] Employers were biding their time and so was the government. The decisive test was the attempted general strike on 1 May 1920 supported by the Federation of Railwaymen. The Federation of Miners reluctantly accepted in order to show solidarity. In fact

18. CCHF 40AS71.
19. National Archives, F7 13791.

railwaymen in the north and in the east did not go on strike and this nullified the miners' efforts. Relatively short and patchily supported, the strike, which was marked by incidents in the Loire and the Gard, failed rapidly. On 20 May, without waiting for the order of the national council of the CGT, the Nord–Pas-de-Calais resumed work and on 26 May there was a general return to work. During the strike, employers reopened the offensive. The CCHF sent a circular letter to companies to propose a poster threatening strikers, and another circular letter dated 18 May revealed its new frame of mind:[20]

> The continuation among mineworkers of a strike devoid of any corporative interest and which in all the other sectors has rapidly proved to be a failure, shows, in comparison with unreasonable trade union instructions, a deplorable drop in national and employers' authority, in common sense, in free enquiry and in the freedom to work.

The CCHF observed that:

> the certain failure of this strike and its impending end can provide a double occasion to arouse individual thinking in every worker and to restore principles of order and authority in so far as this is possible. A few days ago the coalfield representatives held talks about the situation . . . A common feeling emerged regarding the need for a limited number of well thought out sanctions in mining as well as in other industries. The public authorities and public opinion are waiting for these sanctions to become final although they should not take on the character of reprisals.

Indeed employers co-ordinated their action and circulated among themselves the list of the 210 dismissed workers so that they could not spread the contamination to any other region.[21] Even the intervention of the federation of the Nord–Pas-de-Calais, whose moderation was acknowledged, did not succeed in getting the dismissed workers reinstated.

In such a context, the offensive launched in October 1920 by the national federation – which, in addition to the cancellation of the sanctions resulting from strike activities, also requested a review of payslips by local and regional joint commissions according to a

20. CCHF 40AS77.
21. 40AS2, 6/8/1920.

national system in order to obtain wages five times higher than in the prewar period – ran into the employers' intransigence. Faced with the threat of a general strike, the government convinced the CCHF to accept a meeting at national level, but very soon the Minister of Labour and the Minister of Public Works condemned the excessive demands of the workers' federation and appealed to public opinion against it. The national federation of miners had to put off its call for a general strike until the discussions in the regional commissions had been held. In general the commissions only granted an increase in allowances, which the trade unions reluctantly decided to accept in the spring of 1921.

At this time, a drop in the cost of living together with the general industrial situation enabled companies to resume their offensive against wages. Wage cuts were extorted in the centre, then the south and eventually in the Nord–Pas-de-Calais in January 1922. The CCHF congratulated itself on the disapperance of the 'artificial procedures' which had emerged during the war. In his report to the 1922 annual general meeting, de Peyerimhof observed that:

> these wage readjustments marked the abandonment by tacit, if not by stated, agreement of the institutions of the payslip and of the individual minimum wage, which had not been customary in the mining industry and were introduced as a result of the war regime . . . The end of state intervention in the determination of our prices should, and did, result in the abandonment of the extraordinary procedures of joint and arbitration commissions that, in your anxiety to preserve social peace, you decided to accept during the war. Alone with your workers you have solved, peacefully and sensibly, problems which would not have gained anything by being brought up by people other than those directly concerned. In this way, you have regained one of the essential features of your authority as employers at the same time as your workers redis-covered the meaning of the close relationship between their interest and yours.[22]

Fear gave way to relief and even satisfaction. The cost of the war period had been too high. While wages increased, the minimum wage was abandoned. Although the law fixed the length of the working day, the mining industry had already been ahead of its time on this issue in 1914. The institutionalization of the parity system was dropped in favour of the more informal relations which

22. CCHF, report to the 1922 annual general meeting.

had existed before the war. In spite of the apparent unity of the national federation, trade unions remained divided between the organized 'excited and dangerous neophytes' of the south and 'our old trade unions of the Nord–Pas-de-Calais which, with twenty-five years of strikes and class struggle behind them, are far more sensible'. For instance, the latter forced regional rather than national negotiations at the end of 1920. The *union sacrée* with workers' leaders such as Basly, the mayor of Lens, was secured in order to rebuild the economy of the devastated region and demand reparations.[23] It was a return to the pre-1914 corporatist situation cemented by a shared patriotic concern. The future looked bright.

Until the Popular Front, employers benefited from this restoration of order and from very favourable power relations. On one hand, the state wanted to become much less interventionist. The CCHF opposed the tradition established in the industry concerning any scheme for compulsory arbitration which it regarded as 'an unreliable, harmful and impossible solution. No such solution achieves anything.'[24] On the other hand, where the government continued to intervene in pay negotiations, it was done in a more neutral way or even in favour of the employers, as when Laval gave an arbitral decision in favour of the owners, when the men's case was obviously better, in the North in 1925.[25] The position of employers had grown stronger without any real change. At the head of the major companies, the same men as in 1914 had benefited from the additional asset of good engineers who had been promoted to general managers and of general managers who had become members of the boards of directors. After the war, many private companies became public limited companies, but one cannot talk about capitalist restructuring. For instance, there were no mergers. Perhaps the only novelty was the increased importance of the steelmakers from Lorraine. Their breakthrough in the North, which had begun before 1914, became more marked. In particular they took part in administering Lorraine after its annexation, although the Saar and Moselle were in the hands of a group of mining capitalists from Blanzy and, especially, the North.

The strengthening of these links does not allow one to conclude that a nucleus of heavy industry was forming. The involvement of

23. On the symbolic actions of the *union sacrée*, see National Archives, F7 13791.
24. See CCHF 40AS75 on the numerous projects for 1924–7.
25. CCHF 40AS50.

'national' capitalists increased in boards of directors and this is perhaps shown in the growing role of the CCHF in spite of some regional tensions. In 1925 the death of Darcy, the creator of the CCHF, did not diminish in any way de Peyerimhof's untiring activity. The CCHF actively co-ordinated negotiations but made sure that they took place on a local basis. However, the CCHF worked in agreement with the Nord–Pas-de-Calais where employers were more flexible and open-minded. Employers in the Loire and in the Gard who came up against economic difficulties and more virulent communist trade unions were sometimes dissatisfied to see their interests being equated with those of the North. Above all, the CCHF – even though Darcy had been first and foremost president of the Confédération Générale du Patronat Français (CGPF, French Employers' Federation) – was not on the best of terms with steelmakers such as de Wendel and never came out victorious from a confrontation with the Comité des Forges which dominated the CGPF. It remained, therefore, slightly on the fringe of French employers and its unity was strengthened, though on pre-1914 lines.

Another aspect of its restored power was the development of paternalism, encouraged by the reconstruction of 50,000 houses in the North where companies were also closely involved in the creation of allotments for workers. In 1913, employers in the Pas-de-Calais provided housing for 49 per cent of their 33,847 workers. In 1931 they provided housing for 72 per cent of their 65,677 workers. Also in 1931, employers in the Nord provided housing for 64.7 per cent of their workers, and in Lorraine for 51.8 per cent.[26] Traditionally coalfields in the south provided housing for fewer workers. However the study of the paternalistic activities of a large firm such as Blanzy[27] reveals a moralizing and often narrowly Catholic character which weighed perhaps more heavily on the labour force than before 1914. The 1920s were pre-eminently the years of mining villages closed in on themselves.

Besides, this carefully supervised labour force was less coherent than in the past. It is true that rural origins were more distant but the war scattered the old mining communities in the North. In order to reconstruct quickly it had been necessary to call massively on Polish workers, at first in the North and then everywhere. In

26. CCHF 40AS63.
27. CCHF 40AQ24–5, meeting of the board of directors in Blanzy 1914–34.

1927, foreign miners accounted for 58 per cent of the labour force in the North, nearly 20 per cent in Lorraine, the Loire and the centre. Even though they may have brought some traditions of organization with them from Westphalia, these workers were supervised by the consulates and the companies and could hardly take part in the workers' movement without incurring the threat of deportation. In addition, the workers' movement was severely weakened by the division between socialists and communists. Even though the communist CGTU did not make a big dent in the already-unionized membership, 20 per cent at the very most, fratricidal hatred between militants sprang up in these years. In any case, as far as unionized members were concerned, the leaders were rather pessimistic with respect to arbitration. They set themselves the objective of 'winning by convincing public opinion, since we are incapable of being a match for the mining plutocracy which is extremely well organized and benefits from the support of financial organizations and politicians of all parties who are prepared to sell each other to the highest bidder'.[28]

In these circumstances, disputes were rare. The CGT and employers called them 'communist strikes' as most of the time they were led by the CGTU alone. The CGTU acted without taking local conditions into account, and often it was directly following instructions from the International Federation of Trade Unions. As a result it failed systematically. The 1923 strike in the Pas-de-Calais provoked dozens of convictions, and the strikes which took place at the same time in the Saar and Moselle were crushed by a real 'white terror', with hundreds of dismissals.[29] Communist harassment was particularly noticeable in the coalfields in the centre and the south in 1928–9.[30] These sectarian actions did more to harm workers' organization than to strengthen it.

Employers had therefore succeeded very well in recreating pre-1914 conditions and in taking advantage of this period of 'social peace'. First of all, they took advantage of the regional character of workers' trade-unionism and avoided entering into any national agreements. The instructions given by the CCHF in 1925 were followed to the letter: 'It would be desirable that, in the various coalfields, we should not content ourselves with purely and simply

28. C. Bartuel and H. Rullière, *La mine et les mineurs*, Paris 1923.
29. National Archives, F7 13903.
30. CCHF 40AS78.

generalizing the solution started in the Pas-de-Calais as this would help to confer on it a national character, which we must avoid at all costs.'[31] Informed by the CCHF of the concessions made in other regions, employers were very successful in maintaining diversity of wages according to locality and category on the grounds of differing economic conditions. As a result it was difficult to make a comparison between wages on the eve of the 1930 crisis. Of course, negotiations were constantly taking place but only with reformist worker representatives. Employers showed great reluctance to call them collective agreements. When the quarterly bulletin from the Ministry of Labour published in 1933 made that mistake, the CCHF issued a correction in the following terms: only the agreements in the Nord–Pas-de-Calais were actually collective agreements in the legal sense, signed by leading representatives from both parties and sent in to the Conciliation Board for Labour Disputes. Everywhere else they were arbitrator's declarations or agreements tacitly accepted following discussions with the workers or, more frequently, unilateral decisions from the coalfield management.[32] In the Loire and the Gard, the minutes of meetings were only exceptionally signed by workers' representatives, and did not commit them; and they were never sent to the Conciliation Board for Labour Disputes. In Lorraine, Blanzy and Carmaux, workers were informed of wage alterations by means of posters or circulars even though workers' representatives, with the exception of communists, were commonly allowed to present claims and to discuss the wages notified by employers. Finally, the CCHF insisted on the fact that neither the CCHF itself, nor even the Committee of Collieries of the Nord–Pas-de-Calais, should sign as such. Only company representatives could do so.

These agreements rapidly followed on from one another. Eighteen alterations were made between November 1919 and April 1932 in the North, often on the same model. Following fruitless meetings, the Minister of Labour or the Minister of Public Works intervened unofficially at the request of the reformist trade union which was disavowed by the unitarians. Another employers' victory, from 1922 on, was the return to the selling price as the sole criterion, whereas workers, of course, wanted the cost of living to

---

31. CCHF 40AS51, letter from the general secretary to the Committee of Collieries of the Loire and of the Centre dated 8 June 1925.
32. CCHF 40AS75.

be taken into account in the fixing of wages. Therefore the results were very predictable. Wages followed the economic situation and did not increase in relation to increases in production or in profits. However, the trade unions fended off the threat to the eight-hour working day – in spite of severe attacks on it in 1922, 1925 and 1927–8 – and the introduction of work in groups in 1926–7. The trade unions also won a major victory with respect to the increase in retirement and disability pensions. But these were legal victories, not the result of direct trade union action.

This consensus, or the weakening of worker trade-unionism, therefore gave employers a free rein to carry out the most import-ant change of the period, the rationalization of labour which started at the very beginning of the 1927–8 recession. It has been particularly thoroughly studied with respect to Anzin.[33] The rationalization concerned human labour, not technology. It con-sisted in the specialization of tasks, the precise co-ordination of individual labour through time-keeping, increasing the role played by foremen (who doubled in number in Anzin between 1927 and 1929) and most of all in the adoption of longwall mining. The new method cut down heavily on administrative services, made it possible to extend the mechanization of underground transport to all sectors and broke up the old work team. From then on, the remuneration of labour could become individual. For workers, the consequences were significant – loss of autonomy at work, an increase in productivity, deskilling, an increase in discipline and a reduction in employment. Between 1927 and 1929, Anzin laid off 13 per cent of its workers. However, the trade union response was not in proportion with the workers' discontent. The unitarian trade unions were alone in committing themselves totally against this transformation of labour, but, because of their low membership, they were hardly capable of launching strike actions, which in any case were made difficult by dismissals. As a matter of principle, reformist trade unions were favourable to technical progress and therefore, *a priori*, to rationalization, which, incidentally,employers asked them to explain to the workers in the press. They were disturbed by the social consequences, but in 1927–8 they accepted the employers' argument according to which rationalization was

---

33. Odette Hardy, 'Rationalisation technique et rationalisation du travail à la compagnie d'Anzin 1927–1928', *Le Mouvement Social*, July–September 1970, pp. 3–49.

connected with the deterioration of the market. Therefore, they turned essentially towards the government and contented themselves with its promises to help the coal market.

The economic crisis of the 1930s presented employers with the opportunity to intensify their programme. Collieries were hit by the crisis just like the rest of the French economy – which, it must be remembered, was affected later, less deeply but for a longer period than the world economy. After a relative lack of anxiety in 1930, production began to decrease in March and especially in November 1931 following the British devaluation. Stagnation was interspersed with phases of improvement in the winters of 1932/3 and 1933/4. The depression reached its lowest point between April 1934 and September 1935. Only after that did collieries see some precarious improvement. As far as French companies were concerned, the analysis of the crisis was simple and dictated by the particular situation of France. As an important producing country, France was also one of the main importers of coal in the world. In their opinion, overproduction allowed dumping, against which French producers were inadequately protected. Therefore their answer was first a Malthusian reaction. As France produced two-thirds of what it consumed itself, the protection of the national coal industry was possible and the industry requested that from the government before it contemplated a reorganization of the market. At the same time, it was necessary to cut production costs, and therefore labour costs, as much as possible.

Under these circumstances, industrial relations became tense without reaching breaking point. The negotiating power of trade unions was weakened, despite the small number of lay-offs, which were avoided by short-time working and the expulsion of foreign workers. Most of all, trade unions were taken by surprise. In 1930, still satisfied with wages, the trade unions organized a 24-hour strike on 6 October for holidays and pensions. The agreement reached on 29 November made many miners unhappy. In return for six days' holiday, the trade unions accepted the principle of overtime. The agreement was never enforced, but it was a bad omen for trade union policy. Already short of inspiration to fight against employers' initiatives at the end of the 1920s, the trade unions did not have any proper analysis of the crisis to offer and limited itself to a defensive action which did not fundamentally question employers' policies.

The only bastion of resistance was the CGTU, which put forward a proposal for seven-hour working day and did not hesitate, for instance, to launch the strike of November 1931. But the CGTU had to follow the instructions of Internationale Syndicale Rouge (ISR, Profintern), regardless of local conditions. Consequently its action was not important. However, it won a certain amount of sympathy among the miners, which it was able to draw upon at a later date when the CGT trade-unionists, who refused to engage in strikes for fear of losing them, earned the reformist union a reputation as a lackey of the bosses. In fact, reformist trade unions were not only defensive in outlook. Sharing the analysis of the crisis as one of overproduction, they sided with the employers to demand protectionism from the government. In 1931, their common action obtained, first of all, the application of quotas on British imports, and then a duty on import licences. Trade unions in the North boasted about this and the press talked about the *union sacrée*. However, in the autumn of 1931, the Mining Association of Great Britain succeeded in obtaining the postponement of these decisions. It soon became obvious that the French government was not prepared to put the interests of the mining companies ahead of those of importers, dockers and, in last place, consumers. Although the mining pressure group did manage to secure a reduction in railway freight costs, the commercial agreement of 1 July 1934 between France and Britain gave guarantees to British exporters. In the end, the actual fall in imports did not have much of a regulating function.[34] As a result, companies were forced by the government to organize themselves. Although they did not favour the German-style cartel (which incidentally existed in Lorraine), they signed, as early as December 1931, an 'agreement on the rate of output' which lasted until 1936. It divided France into five zones and limited the production of the various coalfields.[35] The enforcement of this agreement was put into the hands of a collieries statistics bureau close to the CCHF which was given access to the account books. Trade-unionism remained dependent on the goodwill of government until 1932, when it proposed the creation of a National Coal Bureau, which was unacceptable to the employers, and sided with the Plan put forward by the CGT. However, companies and trade unions met

34. Robert Laffitte-Laplace, *L'industrie charbonnière de la France*, Paris 1933.
35. J. H. Jones, G. Cartwright and P. H. Guénault, *The Coal-Mines Industry. An International Study in Planning*, London 1939.

again in January 1935 in order to obtain the integration of coal from the Saar in the quota of imports from Germany.

But the most important issue was the attempt by the employers to cut production costs. Trade-unionism found it difficult to defend wages on a day-to-day basis. The crisis made possible a return to harassment, fines and abuses, particularly after 1933. In fact, in that year, relations with employers in the North were broken off for six months, and in the autumn the trade union organized demonstrations which had a certain amount of success. It was also at the end of 1933 that the Anzin company, which had always found it difficult to put up with being 'enrolled' in the regional Chamber of Collieries, tried to recover its wage freedom, without success. Otherwise relations between workers and employers were maintained during the crisis. In fact employers took advantage of the powerlessness of workers in order to draw maximum profit out of the old-style relations. From the beginning of 1931 until the beginning of 1935, history repeated itself. When they got a warning of a forthcoming drop in wages or extension of short-time working, the unions in the Nord–Pas-de-Calais condemned the principle, met with the employers, called on the authorities, led a campaign aimed at public opinion protesting about injustice done to the miners, and then they accepted the measures taken by the employers, which were usually slightly improved in order to save their face. The slightly more radical national federation of miners threatened employers with a general strike, but it was weak and disarmed by the reformist syndicalism of the North and had to give up its threat. As for the unitarians, they systematically prepared their own general strike and the reformist compromise gave them the opportunity to proclaim loudly that they had been betrayed.

Under these circumstances, nominal wages fell ineluctably and the purchasing power of miners was also severely eroded by short-time working. At least the trade union did obtain the creation of an unemployment fund from March 1934. This was a rare achievement in France. It is, perhaps, true that employers had, as usual, given in less because of pressure from the workers than out of fear of seeing the state impose such a fund by law. Nevertheless this victory had an unfortunate consequence. As a result, companies preferred to lay off workers; only the CGTU reacted, with a strike which was rather poorly supported by the miners.

Essentially, the trade union failed to prevent employers from increasing productivity by developing measures of rationalization.

Except in Lorraine, the production process was improved without mechanization of the hewing but by concentration in large workings where working conditions were dictated by the speed of mechanical conveyors.[36] In the Nord, for instance, between 1931 and 1932, the number of coal-faces dropped by 47 per cent. The major innovation was the Bedaux system, which enabled wages to be determined in proportion to individual effort so as to obtain the maximum effort possible from every worker. 'Bedaux points' represented the amount of work carried out in one minute. As one point was worth one-sixtieth of the hourly wage in the workers' category, they had to gain sixty Bedaux per hour. If they did less for more than two days they were downgraded, whereas points above sixty were only remunerated at three-quarters of the hourly rate, a system which was therefore regressive. At Roche-la-Molière, for instance, in the four years after the introduction of the Bedaux system the underground cost of production fell by 25 per cent while wages had increased only by 15 per cent. The system was introduced slowly. But the whole array of constraining methods used made a general increase in productivity possible without much wage compensation. The trade union could only cite its position of principle: rationalization must not be achieved at the expense of the workers but through better equipment of the mine to everyone's benefit. In fact profits remained sufficient to pay substantial dividends and these made relatively small inroads into the reserve funds only in 1937–8. In Anzin itself the whole decrease of trade was paid for by depreciation.[37] Companies were, therefore, much less affected by the crisis than the miners, whose working conditions had deteriorated badly and whose skill had been devalued. Social peace had been maintained, but the results were far more unbalanced than in the past. In the absence of reform, the tensions which had accumulated by 1935 could only lead to a social explosion.

Workers' Upsurge and Employers' Revenge – 1936–1944

The will to put an end to the sacrifices of the crisis was evident before June 1936, partly as a result of the change in the economic

36. Aimée Moutet, 'Une rationalisation du travail dans l'industrie française des années 30', *Annales ESC*, September–October 1987, pp. 1061–78.
37. Hardy, *Industries, patronat et ouvriers*.

environment, partly as a result of the awakening of the workers' movement. The reunification of the unions decided on during the summer of 1935 was only achieved by the end of that year in the mining industry. The communists, who gained many seats as miners' representatives during the crisis, retained an important place in the regional committees which, from then on, became more aggressive. Following a local success in February 1936, there was a threat of strike action, first of all in the North and then on a national scale, for 1 May. After a dispute which lasted six weeks, miners in the small coalfield of La Mure obtained a guaranteed minimum wage. In the North the government intervened and the agreement of 28 April put an end to harassment and introduced a wage scheme for unskilled labourers based on length of service. In June 1936, miners started a general strike in which the most badly treated day-time workers played a leading role. The Matignon agreements on 7 June reversed power relations. This time the wave of new trade union members came from the 'forgotten' groups – surface workers and Polish workers. Gains for workers were very significant and those achieved in Matignon were quite significant for lower-paid workers. Company agreements became compulsory and trade union rights were recognized. Employers' organizations, in this case, regional chambers not the CCHF, had to agree to be the real negotiators with workers' organizations.

Two points were particularly important for miners. On the one hand, collective agreements[38] abolished downgrading, time measurement, extraction demonstrators, individual payment and, thereby, the Bedaux system. On the other hand, miners returned to a week of forty hours or rather thirty-eight hours forty minutes over five days.

These benefits were less important that the general feeling of liberation. The 1936 strikes, which incidentally were brief, were joyful.[39] But behind the celebrations all the accumulated frustrations opened up into a general crisis of authority. Miners gained fresh confidence. Those in Anzin burned a straw dummy representing 'Bedaux'.[40] In Béthune a miner declared to his foreman: 'With the Popular Front we will make all of you sweat blood.'

38. See CCHF 40AS75 for details of the agreements by coalfield.
39. R. Hainsworth, 'Les grèves du Front Populaire de mai et juin 1936. Une analyse fondée sur l'étude de ces grèves dans le bassin houiller du Nord–Pas-de-Calais', *Le Mouvement Social*, July–September 1976, pp. 3–30.
40. Jean Dewaulle, *Le chancre de l'idéal*, Wallers 1973.

Throughout the summer, there was no real authority. The foreman, more so than the absent engineer, was held responsible for the
accumulated sufferings. While employers paid the economic price
for 1936, the supervisory staff bore the brunt of the rejection of the
oppressive hierarchy and of authoritarian order, whereas representatives, especially communist ones, formed real trade union
counter-power.

Confronted with this mini-revolution, employers in the mining
industry shared the fear experienced by all French employers. It
was not so much the financial losses they deplored but the 'spirit of
laxity', the 'immense drop in social values' and the 'intolerable
anarchy' which were created.[41] This imbalance increased, at least
until the spring of 1937.

The fact is that the political environment had become an essential
element of the social climate. The fall of the Blum government in
April 1937 gave the signal for an employers' reaction which was
based on facts. It is beyond question that production costs increased
and productivity in 1937 was 90 per cent of that in 1935. A board of
enquiry on production concluded at the end of 1937 that, under
these circumstances, rising demand had benefited imports.[42] In the
employers' minds, boosting productivity and putting an end to the
crisis of authority were one and the same problem. It was because
discipline was no longer enforced that production stagnated and, as
the company in Ostricourt emphasized, 'that productivity decreases. As long as we tolerate replies such as "you are not the boss
here!", "I will get you sacked" etc. (and we tolerate them since we
do not impose any sanctions in order to avoid incidents), there will
not be any real discipline underground and the problem of productivity will not be solved'.[43]

The only conceivable solution seemed to be to revert to the 1935
situation, namely to attack the social gains made in 1936. As a
result, companies did not invest in new equipment such as mechanical hewers,[44] but proceeded to a new concentration of coal-
faces. Of course, miners were violently opposed to this, but trade

41. J. Masson and E. Staniec, *1936 dans les mines du Pas-de-Calais*, thesis, University of Lille III, 1976.
42. Maurice Hamon, 'La loi des 40 heures dans les mines et la métallurgie', in
Actes du 98ème congrès national des sociétés savantes, Paris 1975.
43. Masson and Staniec, *1936 dans les mines du Pas-de-Calais*.
44. Aimée Moutet, 'La rationalisation dans les mines du nord à l'épreuve du
Front Populaire. Etude d'après les sources imprimées', *Le Mouvement Social*,
April–June 1936, pp. 63–99.

unions found themselves in an awkward position. On one hand, they absolved workers of any responsibility. The solution, they said, consisted in the creation of more jobs, which was done too slowly, and modern equipment, which was non-existent. But on the other hand, in order that the government's experiment of industrial recovery should succeed, it was necessary to develop production. Consequently trade unions tried to convince their members to accept a return to discipline. They did not have much success. The disputes which broke out at the end of 1937 demonstrated the bitterness felt by workers, who considered that employers had sabotaged the social laws in a spirit of revenge.

Companies attacked especially the law on the forty-hour week – 'which is responsible for all the current difficulties', said the president of the Chamber of Collieries of the Nord–Pas-de-Calais. A decree dated 21 December 1937 imposed two-and-a-half additional working days per month if necessary. The trade unions accepted this in exchange for a promise to employ more workers. Employers also tried to reintroduce time measurement. The failure of the second Blum government in April 1938 signalled the break-up of the Popular Front and enabled employers increasingly to resume control. Up to then the government had been acting as referee and had been enforcing the essential points of the advantages gained in 1936. Daladier, whose objective was to increase production, now accepted a return to harassment, dismissals, time measurement – at the cost of an increase in fatal accidents – and, a major challenge, the dismantling of the law on the forty-hour week.

As it happened, the communists won major victories at the elections of representatives in 1938. Egged on by its grass roots, the federation of underground workers again started to threaten a general strike, but employers were not intransigent. On 1 September they imposed a national agreement involving eleven extra days of work before 1 March 1939, an agreement which the workers had no intention of complying with. In November 1938 the national confrontation dragged miners into the collapse of the entire workers' movement. On 12 November Paul Reynaud introduced provocative orders-in-council which, among other things, planned to abandon the forty-hour week. The CGT had no choice but to declare a general strike for 30 November. Miners, some of whom started to strike as early as 24 November, supported it relatively well, in spite of requisitioning of their labour by the government. The national failure had severe consequences. The Anzin company dismissed 759 workers, representing 6 per cent of its personnel.[45] On 4 January 1939, the Committee of Collieries of the Nord–Pas-de-Calais summoned the trade union representatives in order to denounce various clauses included in the 1936 agreements so that they could reintroduce the working and pay conditions which existed during the years of economic crisis. The arbitration of the Minister of Labour, de Monzie, on 28 January was hardly less harsh. He reintroduced time measurement, payment by small work-teams (though not individual wages) and procedures for downgrading workers. A decree dated 22 February allowed 186 hours of overtime per year. In the spring of 1939, the power of employers was completely restored, while trade unions became divided and their membership declined. But this victory opened up an enormous gap between employers' and workers' representatives and permanently alienated miners as a whole. The last vestiges of the friendly relations built up since the end of the nineteenth century were shattered and power relations were entirely exposed.

But employers did not content themselves with this victory. The reaction, which began in 1937/8, developed during the 'phoney war' between September 1939 and May 1940 because of the requirements of national defence, and it continued after the defeat. The CCHF was dissolved as well as the CGT, but R. Fabre, the general secretary of the CCHF since 1931, and his team took command of the Mines Organization Committee in order to suppress all constraints on the working of the mines, a task in which the occupiers were soon to assist. British imports stopped, Lorraine was annexed. In order to meet the demands of the Wehrmacht, the Bundesbahn and of a shivering France, Vichy and the German army insisted that the Nord–Pas-de-Calais should increase its output as quickly as possible.[46] In the absence of technical resources, they renewed their drive for high productivity through authoritarian practices and extended working hours. Economic necessity and social revenge linked up with the requirements of the occupier in order to sweep away the last gains of 1936. From December 1940, seven companies reintroduced the Bedaux system. Down-

45. C. Boileux and C. Carlier, *La grève générale du 30 novembre 1938 dans le département du Nord*, thesis, University of Lille III, 1974.
46. Étienne Dejonghe, 'La reprise économique dans le Nord–Pas-de-Calais', *Revue d'Histoire de la Seconde Guerre Mondiale*, January 1970, pp. 83–112. Also by the same author, 'Pénurie charbonnière et répartition en France 1940–1944', *Revue d'Histoire de la Seconde Guerre Mondiale*, April 1976, pp. 21–55.

gradings and sackings multiplied. The employers'offensive was at its peak in the months following the defeat, before the Germans became involved.[47] The latter attributed the ill will of the miners to the toughness of the French capitalists! In addition the miners of the Nord–Pas-de-Calais, under the influence of the Communist Party, threw themselves into an audacious strike from 27 May to 6 June 1941, against the occupier and also against the companies who were trying to extend the individual wage system.[48] National aspirations and class struggle fused together. The occupation forces took the situation in hand and the repression was ferocious. The companies acted as assistants to the occupier. The Germans arrested undesirable elements pointed out to them by companies. Only one, Drocourt, refused to denounce war prisoners released on parole who had gone on strike.

The Germans quickly went beyond this collaboration. From May 1942, they put deported Russians to work and in November 1942 they imposed Sunday working. They intervened directly in industrial relations but the fiction of agreement was maintained with respect to the Vichy unions. Also, the arbitration of 20 January 1943 authorized the companies to downgrade any worker who earned less than 80 per cent of the average wage for his category for two fortnights in succession. The Oberfeldkommandant modified the arbitration in order to accelerate the downgradings.[49] For the miners, the collusion between the enemy and the employers was obvious. However, from 1943, the latter showed themselves lukewarm and even relatively passive during the strike of October. The Oberfeldkommandant, aware of this reluctance, went so far as to threaten them.[50] From 1944, the Germans named an official with total power over production and distribution in mines in the occupied areas.

By this time it was too late. The companies were definitively compromised in the eyes of the population; they were reproached with having forced up production for Hitler in order to enrich themselves. The employers replied that they produced primarily for France and that, in any case, they scarcely had a choice. They

47. Étienne Dejonghe, 'Problèmes sociaux dans les entreprises houillères du Nord–Pas-de-Calais pendant la seconde guerre mondiale', *Revue d'Histoire de la Seconde Guerre Mondiale*, January–March 1971, pp. 124–47.
48. CCHF 40AS78.
49. CCHF 40AS68, letter dated 3 December 1943.
50. CCHF 40AS78, letter to the prefect dated 17 October 1943.

were also accused of having relied on foreign support in order to obtain revenge for 1936. The toughness of the employers tended to confirm this.

As though they had learnt nothing, in September 1944 the companies again refused to grant a payment to the families of those who had been shot or to reintegrate those who had been sacked in 1936, because the agreements which had been signed related only to the reintegration of those who had been sacked in 1939. At the Liberation, the crisis of authority and social hatred were at their peak. As usual it is necessary to point out regional differences. In Lorraine, the Wendel family, for example, were stripped of their possessions, but the Moselle had been absorbed into the economic region of Saarpfalz and the French employers had disappeared, and thus were not the object of any rancour.[51] In the south of France, certain employers and engineers had had links with the resistance although this was not significant in the Loire or the Gard. In the north, which had been united to Belgium, the tensions were such that the explosion could not have been anything other than brutal.

The Mines After Nationalization

After a half-century of effort by the workers and, by and large, victorious resistance by the employers, the state took over from the employers. And nationalization, in the relatively confused period from 1944 to 1947 when it was being set up, bred illusions among the workers.

The violent explosion of social as well as national liberation translated not only into political purification but also into the eviction of employers who had lost respect and of many engineers. There was a lengthy crisis of authority which affected the whole of management. In the centre and the south, Liberation was sometimes difficult. At Montceau, in early September, a master miner was murdered and the miners' representative, who was also the trade union secretary, was killed in his home. At Bessèges, at La Grand' Combe, the engineers were arrested and held in poor

51. P. Gérard, 'Le protectorat industriel allemand en Meurthe-et-Moselle', *Revue d'Histoire de la Seconde Guerre Mondiale*, January 1977, pp. 9–28; H. E. Volksmann, 'L'importance économique de la Lorraine pour le Troisième Reich', *Revue d'Histoire* 1980, pp. 69–93.

conditions.[52] But, above all, in the Nord–Pas-de-Calais, 3,000 official complaints were lodged with the purge committees against engineers and foremen.[53] The miners wanted, in one fell swoop, to rid themselves of the employers, who seemed to be a good target for the slogan 'Punish the traitors! Confiscate their goods', and to free themselves from the methods of increased exploitation which had been imposed on them during the war, simultaneously getting even with the supervisory staff who had enforced them. The Commissaire of the Republic in the Nord gave an excellent description of this explosive climate:[54]

> Before the war the mining employers were capitalist employers in the strongest and most odious sense of the word . . . at the coal-face their short-term conception meant that coal had to be produced at any cost and at the lowest possible price, even with poor equipment and mediocre techniques; each time it appeared to be preferable to inflict hardship on the workers rather than to invest capital . . . However, if the system of control in the mines was barely tolerable before the war, it became totally unacceptable during the Occupation.
>
> In order to carry out the employers' orders, the embittered engineers, separated from their social context, and brutal foremen used methods which made them odious to the bulk of the workers.

The most serious aspect was the questioning of authority. The technical staff acted like company officials.[55] Many of them, unhappy with the Malthusian attitudes of the management, welcomed nationalization as a step towards the realization of their technical ambitions. But they were completely cut off from the workers in social terms, in the same way as the foremen, who were not from the masses. They tended to be compliant: 'the bosses who were clumsy or mediocre, brutal, loudmouthed and disdainful towards the workers, men who despised the people, were considered to be collaborators of necessity.'

As Étienne Dejonghe insists, 1944 was a new 1936 for the mines. In the conditions of the time, nationalization was undoubtedly the only option. In addition, conservative measures were very rapidly

52. CCHF 40AS78.
53. F. L. Closon, *Commissaire de la République du général de Gaulle. Lille septembre 1944–mars 1946*, Paris, 1980.
54. Ibid.
55. Étienne Dejonghe, 'Ingénieur et société dans les houillères du Nord–Pas-de-Calais de la Belle Epoque à nos jours', *Le Mouvement Social*, 1985, pp. 173–89.

adopted by the departmental Committees of Liberation in order to prevent anarchy and a social explosion. In Montpellier and Toulouse, mines were requisitioned from the end of September 1944. In the Nord, the committee provisionally suspended the colliery owners of the region and the decree of 13 December 1944 nationalized the collieries of the Nord–Pas-de-Calais. Elsewhere, mines were either requisitioned or the status quo was maintained. The law which finally nationalized all collieries was passed, after much hesitation, on 17 May 1946. From the outset, nationalization did not imply worker management. From January 1945, the eighteen concessions in the North were reorganized into seven units of production. At the mercy of the vagaries of national politics, the debate on the direction nationalization should take dragged on for a year; as a result, the employers were able to turn circumstances to their account. At the end of 1944, Henri de Peyerimhof and the reactivated Chamber of Collieries were mobilized.[56] In spring 1945, the boards of directors in the North and the association of shareholders in mining companies set up in Saint-Étienne were mobilized to salvage as much as possible, that is from activities which were not directly part of mining operations, and to obtain the largest possible compensation. Against the 'totalitarian attempts' of the communist Minister for Production to 'skin the investors' they obtained full compensation inflated in relation to the actual value of shares.[57]

Above all, the final set-up in the coal industry showed that the workers' ideas had not really won out. If the autonomy of the separate coalfields had been safeguarded, prolonging the regionalism so dear to the companies, it was because the CGT wanted it to be so. Charbonnages de France (CDF, French Coal) only coordinated the activities of the separate coalfields, which produced and sold on their own initiative. The unions forced the election of the chairman of each board of directors but they did not have a majority of members. The boards were composed of representatives of three groups. The CGT had six out of eighteen seats on the national board of the CDF and seven out of nineteen on each

56. Rolande Trempe, 'Les charbonnages, un cas social', in Claire Andrieu, Lucette Le Van and Antoine Prost, *Les nationalisations de la Libération: de l'utopie au compromis*, Paris 1987, pp. 294–309. R. Fabre, the former secretary of the CCHF, wrote many brochures, CCHF 40AS2.
57. Contributions by Annie Lacroix-Rix and Jean-Charles Asselain in Andrieu *et al.*, *Les nationalisations de la Libération*.

regional board, the remainder being divided between the state and representatives of consumers and other interested parties.[58] But the state retained significant controls over prices and levels of output as well as naming the managing directors and having a veto over nominations to the boards.

None the less, the social gains were immediate. The Communist Party tirelessly repeated that 'the miners had gained more in a year than in fifty years of trade-unionism' and the miners' statute appeared to give them good grounds for this assertion.[59] This statute, negotiated with the CGT, took the place of the collective agreement. Worker representation was strengthened and miners' representatives, the keystone of trade-unionism, were elected for pithead workers as well. The local joint committees had more power than in other sectors. They established the conditions of work, which were approved by the chief engineer, and they supervised the taking on of labour (including the appointment on a permanent basis of executives) and, most important, sackings, which were only agreed to in the event of grave misdemeanours. The statute also allowed for some upward mobility between the categories. It allowed for supervisory staff to be recruited from among workers and engineers from among the supervisory staff, making provision for study grants to be made available. It created a single salary scale for all personnel based on 112.5 percent of the rates in the well-paid steel industry for pithead workers and 132.5 per cent for underground workers. Material progress was undeniable in all areas, with the exception of arbitration.

None the less, from spring 1945 until the end of 1948, worker disenchantment grew and the 'social dynamite' of the mines ended up by exploding. This came about because the social aspirations which had been pent up during the war came into collision with the exigencies of reconstruction. In December 1944, coal production was 60 per cent of the 1938 level, and in December 1945 at 87 per cent, but exports were only 39 per cent. The battle for production was essential if plans for modernization were to be realized. In the absence of equipment, the increase in production depended on

58. Étienne Dejonghe, 'Les houillères à l'épreuve 1944–1947', *Revue du Nord*, 1975, pp. 643–60.

59. G. Levasseur, 'La situation du personnel dans les entreprises nationalisées', in M. Boiteux and G. Bouquet, *Le fonctionnement des entreprises nationalisées en France*, Paris 1956; Pierre-Marie Delesalle, *Le statut du personnel dans les entreprises nationalisées*, Paris 1953.

redoubled efforts on the part of miners. The despised methods of the interwar period, accentuated during the war, were reintroduced – rationalization, individual salaries, Bedaux points. The collieries 'wore the same clothes as the companies'[60] and even used German methods like the staggering of paid holidays and Sunday working. As the only force capable of imposing such bitter sacrifices on the miners, the CGT and the Communist Party were all-powerful and turned the regional production and headquarters committees into weapons in the battle for output. Their logic was simple – to produce more was an extension of the struggle of the resistance to save the country. To succeed in this battle, with the Communist Party in the government, was to demonstrate the managerial capacity of the unions and to establish a necessary economic foundation to satisfy the justified social claims of the miners. The fervour of the Communist Party for this task was great, but so were the difficulties. On 21 July 1945, Maurice Thorez, speaking in the communist fiefdom of Waziers, launched his famous appeal to miners to produce and to take into account the solidarity of the miners with the rest of the working class, whose fate depended on them. Vigorously denouncing the obstacles in the way of the 'rebirth of France', Lecoeur, the communist Minister of Mines, wrote in the CGT newspaper in January 1946:

I say in all tranquillity, no one can dispute that the authority of the foreman must be exercised . . . Is it not true, comrades, that in your pits you know at least one idler, one disorganizing influence? . . . The supervisors, who are afraid of goodness knows what, say nothing. This can only discourage those who put their backs into their work. Shouldn't the foreman have the unanimous support of honest workers if he takes justifiable action against the disorganizer, against the idler? . . .

While this was going on, the CGT had to struggle constantly to preserve its dominant role in the nationalized industries, a struggle in the course of which it was forced to give up more and more territory, particularly when the socialist Robert Lacoste replaced the communist Marcel Paul as Minister for Production in December 1946. Even before this, many political forces, and the engineers' unions, had denounced the 'communist colonization' of the CDF. Of the eighteen members of the board of directors in May 1946,

60. Dejonghe, 'Les houillères à l'épreuve 1944–1947'.

308 *Joël Michel*

twelve belonged to the CGT, of whom ten were communists. A decree of 5 August reduced the number of CGT representatives to five. The president and the managing director of the collieries in the Nord–Pas-de-Calais, elected in the autumn, belonged to the CGT. In July 1947, Lacoste sponsored a series of decrees hostile to the CGT. The state was only to nominate representatives having at least five years' experience in government service. He reserved to himself the right to delegate powers to the managing director, revoke administrative appointments, dissolve the board of directors and approve conditions of work in the enterprises. Numerous decisions of the boards of directors, particularly in financial matters, could only be carried out with the approval of the appropriate minister.

From the beginning of January 1947, because of the financial crisis in the collieries and a new fall in living standards, the social malaise grew. In March 1947, while remaining 'constructive', the Communist Party renounced its productionist slogans. After the ejection of the communists from the government in May, Lacoste launched a new offensive with a decree in June which enacted that no member of the board of directors could belong to any other group than the one he represented. For instance, a representative of the consumers could not be a member of the CGT. In addition, the term of office was reduced from six years to three, and one-third would be replaced at a time. All was ready for the CGT to be winkled out. In August, certain wage concessions were suspended. This was resented as a provocation. In November, the government subsidy was ended and the price of coal went up by 60 per cent. This seemed to mark a return to liberal economics. After the resignation of V. Duguet, the president of the CGT and of the CDF, on 10 November, and above all the dismissal on 16 November of Delfosse as one of the representatives of the state on the board of the CDF, the miners engaged in a general strike from 17 November to 9 December. It was a show of force between the Communist Party and the government, but it was also intended to defend nationalization now that it was under threat. They demanded wage concessions, the annulment of Delfosse's dismissal and weekly meetings with the union in each region. The government offered a wage rise and organized strong repression. The socialist Minister of the Interior, Jules Moch, sent in the forces of order to intimidate the miners rather than out of necessity. At one time, the CGT had had twelve members out of eighteen on the board of directors of the CDF. Lacoste's decree of June 1947 had reduced this to six. In December, three went over to the Force Ouvrière (FO) union and the other three were excluded for taking part in the strike. The strike precipitated a split between the CGT and the reformist CGT–FO. After a new strike in spring 1948, the government was determined finally to break the communist CGT and re-establish its own authority. Just at this time, in June 1948, a commission of enquiry on the coal industry denounced the crisis of authority resulting from the excesses of the purge process, the slow setting up of the boards of directors and the fact that wage earners were in a majority in them. The Lacoste decrees of 19 October and 2 November 1948, aimed at combating absenteeism, constituted a declaration of war. They ordered the sick and those with silicosis to go back to work again. The powers of miners' representatives were reduced. After a ballot, the miners went on strike on 4 October 1948.[61] The recapture took place after eight weeks of social war. Against this 'insurrectional' strike, the government ordered the CRS, the special police forces, to use their weapons if necessary, and the pits, notably in the area of Valenciennes, were retaken by tanks. Six miners died in the battle, 2,000 were imprisoned, 6,000 sacked. The enterprise committees were suppressed, the conditions of eligibility for miners' representatives were modified, managerial authority was taken in hand. In the board of directors of the Nord–Pas-de-Calais, the nine CGT members of 1946 were reduced to four in 1948. All were dismissed for participating in the strike. Former administrators of the 1930s and the Vichy regime, such as E. Marterer, made their reappearance. In general, however, it was technicians who replaced the militants. Audibert, the new chairman of the CDF, was able to feel highly satisfied with this governmental energy, which had allowed 'the social climate to improve and authority to be exercised' in such a way that 'the politics of the mining industry after the strike could no longer inspire narrowly sectional perspectives'.[62] For the miners, frustration was enormous. The hopes of nationalization had collapsed under the difficulties. The enemy, from then on, was the govenment, that is

61. Françoise Saliou, *La SFIO et les grèves du Nord–Pas-de-Calais en 1947–1948*, thesis, University of Paris I, 1973; Darryl Holler, *Miners against the State: French Miners and the Nationalization of Coalmining 1944–1949*, Ph.D. thesis, Michigan, 1980.
62. E. Audibert, *Cinq ans de nationalisation*, Paris 1951.

the state–employer, which, in the final analysis, was no better than
'the companies', the name which still survived in the language of
the workers.

If, later on, miners recognized any virtue in nationalization, they
remained very hesitant about and critical of state management.[63] It
was quite different from that of the concessionary companies
because it grew out of centralized economic policy. However,
contrary to the expectations at the time of Liberation, the energy
sector was not part of the harmonious execution of the plan,[64] but
joined in the return after 1949 to the free market. This meant both
rising output and a fall in production costs. Coal output rose until
1953, but with rising productivity there was a loss of 70,000 jobs,
that is 20 per cent in five years. Coal subsidies to the industry
disappeared, to the benefit of customers. Above all, the signing of
the treaty setting up the European Coal and Steel Community had
important implications for the reorientation of the coal industry.[65]
The differences between groups of coalfields were erased in the
interests of national strategy. The new coal policy meant closures,
which brought about social conflicts in the Gard and Aquitaine.[66]
But the crisis was analysed as a temporary one, without any overall
reaction.

From the social point of view, the state proved itself to be a
somewhat more tolerable employer, even though it was able to
insist on having its own way. For one thing, despite the restoration
of authority after 1948, the benefits of the miners statute, which
private employers would have dismantled, were, for the most part,
preserved. Given the abandonment of really equal relations, the
state took its lead from the statute of public employees. Also, the
state made considerable investment in housing for the ageing
workforce, more from social than production considerations,
something that private employers would never have done. But on
one essential point, wages, the miners' situation changed for the
worse. With the ending of wage controls in the private sector in

63. Serge Moscovi, 'Les mineurs jugent la nationalisation', *Sociologie du Travail*,
1960/2, pp. 216–29.
64. Jean-Paul Thuillier, 'Les charbonnages et le Plan 1946–1962', in *De Monnet à
Massé. Objectifs politiques et objectifs économiques dans le cadre des quatre premiers plans*,
Institut d'histoire du temps présent, Paris 1986.
65. M. Vilain, *La politique énergétique en France de la deuxième guerre à l'horizon
1985*, Paris 1969.
66. Serge Moscovi, 'La résistance à la mobilité géographique dans les expériences
de reconversion', *Sociologie du Travail*, 1959, pp. 24–36.

February 1950, wages in the steel industry, on which miners' pay rates were indexed, shot up. In November 1950 the state decided to fix wage rates in the public sector for itself and to abandon any reference to the steel industry. In contrast to its impact on Renault and the electricity industry, this decision was damaging to the interests of miners.[67] In 1951, wage rates were fixed separately for each coalfield. This opened the way for economies to be made in the less profitable ones. Legal action by the unions over this decision came to nothing. In 1955 the government linked wages to the price index and to the growth of productivity. But in 1958, reference to prices disappeared and, with it, all guarantees of improvement in miners' incomes, which clearly fell back from then on.

In addition, miners became alarmed that capitalism had taken the offensive to recapture its lost positions when, on 11 May 1953, the managing director of one of the former companies was nominated by decree to the board of directors for the Nord–Pas-de-Calais. But if links with the private sector were perhaps of some significance in Lorraine, this was not generally the case. But trade unions did not increase their influence as a result. No matter what the social policy of the state, equality did not exist and contact with unions was limited to the FO and the Confédération Française des Travailleurs Chrétiens (CFTC, French Confederation of Christian Workers). The CGT, to which the majority of unionized miners belonged, was ignored. This was a period of emptiness in industrial relations. And yet there were hardly any local strikes, only general movements.

This impotence is much more easily understood if one considers that alongside this organizational division between trade unions, which had never been so well defined, there was a corresponding anaesthesia, something completely new in the mining communities. Apart from miners' sons, no new workers were taken on, so that, when the need arose, it was impossible even to find enough applicants. Above all, technological change brought a profound alteration in work methods, something which rationalization in the 1930s had not done. This time there was a real modernization of equipment. The coal-hewer gave pride of place to the mechanic or electrician. The number of foremen doubled in ten years. The

67. Monique Maillet-Chassagne, *Influence de la nationalisation sur la gestion des entreprises publiques*, Paris 1956.

unions were slow in adapting to these changes. It was only in the 1960s that they set up sections for different categories. Finally, action in these increasingly demoralized communities was not made any easier through the disappearance of employers present on the spot. They were replaced by technocrats who managed from afar, and even took away the essence of their wage power from the supervisory staff, who were as disoriented and bitter as the miners themselves.

With the European turning point of 1959, recession became widespread and entered a critical phase. There was no longer a coal policy, only an energy policy which depended on the price of oil products. Thanks to centralization, the state was able to engage in an inflexible policy of closures, but the first Gaullist governments carried them out in a clumsy, even provocative, way. This time, the mining communities did react.

Because of a shortage of space, and also because, in essence, this action moved further and further away from the types of industrial relations we have been considering up to now, these events will only be glanced at briefly.[68] In Decazeville, the 2,000 miners occupied the pits from December 1961 to February 1962. At the outset, the CGT called for the mine to be kept open, but, in practice, it negotiated over pensions, social security and retraining bonuses. The national leadership discouraged local bodies from continuing the fight against closure. This first dispute provided a social model for future closures. At Faulquemont, in Lorraine, in 1971, 1,500 miners went on indefinite strike against the closure announced for 1973. The unions began by defending the profitability of the mine but, as at Decazeville, quickly slipped into discussing conditions for retraining and transfer. When this conflict flared up again, in 1974 at the time of the closure, it was a fight of honour in order that the mining should survive, albeit elsewhere.

A short time after the conflict at Decazeville, in 1963, the mining

68. There is a wide variety of works on the 'tragedy of the collieries'. Concerning Decazeville, besides D. Reid, see: Rolande Trempe, 'Un combat d'avant-garde. La grève des mineurs de Decazeville 19 décembre 1961–21 février 1962', *Economie et Société de Languedoc-Roussillon*, Montpellier 1978, pp. 295–309; Jean Tibi, *Les houillères de la Loire 1960–1980. La mine foudroyée*, Saint-Étienne 1973; C. Mazauric and J. Dartigue, *Ladrecht*, 1982; Joël Michel, 'Les réactions syndicales à la récession charbonnière en Europe occidentale', *Revue Belge d'Histoire Contemporaine*, 1988.

community conducted its last great national strike (that of 1968, which brought important social improvements, had little to do with the mining industry as such). But this strike came about as the result of wages having fallen behind in the 1950s rather than from a desire to ensure the future of the coal industry. The miners, essentially, defended the gains they had made. But there was one new element in the dispute. For the first time, white-collar workers, engineers and supervisors gravitated to the side of the workers in the face of centralized, anonymous, technocratic power. For the first time in many years, the entire mining community came together in self-defence against an outside threat. This coming together was a sign of the extent of the danger. In fact, it was a prelude to an even greater enlargement. From this time on, defence of the mines became less a question of industrial relations than of regional struggle for employment or survival. These 'ghost town' operations counted for as much as the occupations of the pits. In the light of this, industrial relations in the strict sense went into the background. Unions operated under difficult conditions. Membership fell as retirements became more numerous and many miners gave up the struggle. The CGT campaign to defend coal at the national level depended mainly on the Communist Party. The state, which temporized over particular cases, did not take money into consideration. The French case was certainly a defeat for mining redeployment seen from the regional point of view, but from the purely corporative point of view, the miners came out well because the state gave or maintained important social guarantees. Starting from the mid-1970s, even though the saving of the pit at Ladrecht in the early 1980s went against the current, miners thought less about saving pits than about coming out of the situation as well as possible. From then on, the real conflict was to be found among Moroccan workers who had been recruited in ever-increasing numbers and who agitated to obtain the guarantees embodied in the statute of miners.

Appendix

There are so few statistics concerning the number of trade union members in France that theses have been written on this difficulty. French trade-unionism embraced only a minority of workers. It was unstable, deeply divided and only asked for low subscriptions because it did not provide any services. Often it did not keep any accounts. In any case, membership was less important than the audience. This was true for miners' unions, except in the numbers of their members – in the coalfields of the North a majority of miners belonged to a union as early as the beginning of this century; and for coalfields in the rest of France, unionized members were in a majority from the inter-war period.

Table 7 French coal production 1889–1984 (millions of tons)

1889	24.3	1915	19.5	1941	43.8	1967	50.5
1890	26	1916	21.3	1942	43.8	1968	45.1
1891	26	1917	28.9	1943	42.4	1969	43.5
1892	26.1	1918	26.2	1944	26.5	1970	40.1
1893	25.6	1919	22.4	1945	35	1971	35.7
1894	27.4	1920	25.2	1946	49.2	1972	32.7
1895	28	1921	28.9	1947	47.3	1973	28.4
1896	29.1	1922	31.9	1948	45.1	1974	25.6
1897	30.7	1923	38.5	1949	53.4	1975	25.6
1898	32.3	1924	44.9	1950	52.5	1976	25
1899	32.8	1925	48	1951	54.9	1977	24.3
1900	33.4	1926	52.4	1952	57.3	1978	22.4
1901	32.3	1927	52.8	1953	54.3	1979	21
1902	29.9	1928	52.4	1954	56.3	1980	20.7
1903	34.9	1929	54.9	1955	57.3	1981	21.5
1904	34.1	1930	55	1956	57.3	1982	19.9
1905	35.9	1931	51	1957	59	1983	19.6
1906	34.1	1932	47.2	1958	60	1984	19
1907	36.7	1933	47.9	1959	59.7		
1908	37.3	1934	48.6	1960	58.2		
1909	37.8	1935	47.1	1961	55.2		
1910	38.3	1936	46.1	1962	55.2		
1911	39.2	1937	45.3	1963	50.2		
1912	41.1	1938	47.5	1964	56.2		
1913	40.8	1939	50.2	1965	54		
1914	27.5	1940	40.9	1966	52.9		

Nord–Pas-de-Calais: production (millions of tons), and as a percentage of national production

1890	14.2	54.6%	1960	28.9	50.7%
1913	27.3	66.9%	1966	25.2	47.6%
1938	28.2	59.3%	1975	7.7	30.3%
1949	27.6	52.2%	1984	2.5	13.1%

Source: Annuaire Statistique de la France, INSEE.

7

Goodbye to Class War: The Development of Social Partnership in Austrian Coal Mining

Franz Mathis*

In the field of industrial relations Austria has for several decades held a special position. Rightly or not, the social partnership built up in the first years after the Second World War is referred to again and again with approval as an exemplary system, as a partnership between employers and employees whereby workplace strife is, as a rule, supposed to be solved by way of negotiations, without strikes or the like. This sort of conflict resolution has, according to the majority view, often been successful. Critics of this point of view do exist.[1] What follows consists of a more precise examination of the development of social partnership in Austria, with one specific branch taken as an example.

First of all, the framework for the analysis must be set up with the help of several details concerning the size and extent of Austrian coal mining (see table 8):

- Coal production and the number of mining employees have both been of only modest proportions, the former peaking at, at the most, seven million tons in 1957, and the latter at somewhat more than 23,000 mining employees at the beginning of the 1920s.
- The long-term development in annual production saw a phase of stagnation lasting from the turn of the century to the end of the First World War, with an annual production of between two and three million tons, followed by an initial upturn to about three to

---

* Translated by John Toal.
    1. Cf. e.g. Anton Pelinka, *Modellfall Österreich? Möglichkeiten und Grenzen der Sozialpartnerschaft*, Vienna 1981.

316

Franz Mathis

Table 8 Production, employment and productivity in Austrian coal mining, 1900–1987

Year	Production in 1000 t Brown coal	Black coal	Total	Employment in 1000 Brown coal	Black coal	Productivity (Production per employee in t) Brown coal	Black coal
Average 1900/12	2411						
1913	2621	87	2708	12.1	0.6	216	149
1914	2361						
1915	2463						
1916	2493						
1917	2186						
1918	2241	95	2336	12.6	1.9	178	50
1919	2217	90	2307	16.5	1.4	134	67
1920	2697	133	2830	18.5	2.1	146	64
1921	2797	138	2935	20.9	2.5	134	56
1922	3136	166	3302	21.1	2.3	149	71
1923	2685	158	2843	18.6	1.9	144	82
1924	2786	172	2958	16.7	1.8	167	94
1925	3033	145	3178	15.0	1.6	203	92
1926	2958	157	3115	14.1	1.2	209	135
1927	3064	176	3230	12.0	1.0	255	177
1928	3263	202	3465	10.7	1.1	304	190
1929	3525	208	3733	11.2	1.1	314	192
1930	3063	216	3279	9.95	1.1	308	193
1931	2982	228	3210	9.4	1.2	318	196
1932	3104	221	3325	9.6	1.3	322	171
1933	3014	239	3253	9.95	1.3	303	180
1934	2851	251	3102	9.15	1.45	312	172
1935	2971	261	3231	9.2	1.5	322	172
1936	2897	244	3142	9.1	1.4	320	171
1937	3242	230	3472	9.5	1.3	341	175
1938	3340	227	3567	10.5	1.3	317	179
1939	3533	217	3750	11.5	1.3	308	167
1940	3614	228	3842	11.6	1.3	312	171
1941	3537	226	3762	11.5	1.3	307	171
1942	3523	225	3748	12.7	1.3	278	175
1943	3651	214	3866	13.0	1.3	280	170
1944	3676	195	3872	12.9	1.2	285	159
1945	2066	72	2138	11.6	1.1	178	66
1946	2407	108	2515	13.2	1.4	183	76
1947	2839	178	3017	15.1	1.9	188	95
1948	3338	181	3518	15.6	1.6	215	106
1949	3816	183	3999	15.3	1.5	247	119

Table 8 *continued*

Year	Production in 1000 t			Employment in 1000		Productivity (Production per employee in t)	
	Brown coal	Black coal	Total	Brown coal	Black coal	Brown coal	Black coal
Average							
1950	4308	183	4491	16.5	1.4	262	130
1951	4989	196	5184	17.6	1.4	284	139
1952	5179	190	5369	17.0	1.5	305	129
1953	5574	162	5736	16.5	1.4	338	111
1954	6285	177	6462	16.5	1.3	382	134
1955	6619	175	6794	16.8	1.3	394	130
1956	6730	166	6896	16.8	1.4	401	118
1957	6877	152	7029	16.8	1.5	409	102
1958	6494	141	6635	16.85	1.5	385	95
1959	6221	134	6355	15.8	1.2	394	112
1960	5973	132	6105	14.7	1.0	407	127
1961	5661	106	5767	13.3	0.9	427	116
1962	5712	99	5811	12.4	0.9	459	113
1963	6053	104	6157	12.1	0.85	499	121
1964	5761	103	5864	11.2	0.8	513	129
1965	5450	59	5595	11.1	0.7	491	88
1966	5283	20	5303	10.0	0.2	529	112
1967	4604	14	4618	8.7	0.1	531	113
1968	4177		4177	7.2		581	
1969	3841		3841	6.5		589	
1970	3670		3670	5.6		659	
1971	3770		3770	5.9		636	
1972	3756		3756	5.6		676	
1973	3634		3634	5.4		670	
1974	3629		3629	5.4		668	
1975	3397		3397	5.2		654	
1976	3215		3215	4.8		672	
1977	3127		3127	4.5		689	
1978	3076		3076	3.9		789	
1979	2741		2741	3.6		762	
1980	2865		2865	3.5		815	
1981	3061		3061	3.4		913	
1982	3297		3297	3.3		986	
1983	3041		3041	2.85		1065	
1984	2901		2901	3.2		917	
1985	3081		3081	3.2		973	
1986	2969		2969	3.1		949	
1987	2786		2786	2.9		971	

Total production includes coal allowances in kind given to the employees.

Table 8 *continued*

Sources: The details have been collated from: *Österreichisches Montan-Handbuch*, which appears periodically; the earlier annually published *Mitteilungen über den österreichischen Bergbau*; Leopold Weber and Alfred Weiss, *Bergbaugeschichte und Geologie der österreichischen Braunkohlenvorkommen*, Vienna 1983, p. 11; Karl Bachinger and Herbert Matis, 'Strukturwandel und Entwicklungstendenzen der Montanwirtschaft 1918 bis 1938. Kohlenproduktion und Eisenindustrie in der Ersten Republik', in Michael Mitteraur (ed.), *Österreichisches Montanwesen*, Vienna 1974, pp. 115ff. and 128; *Kohle. Unseres Landes Hauptquelle für Wärme und Kraft*, ed. Fachverband der Bergwerke und Eisen erzeugenden Industrie, Vienna 1963, pp. 64f.; and from *Rot-Weiß-Rote Kohle*, ed. Kohlenholding GmbH, Wien, Vienna 1956, tables II and III. On the measurement of productivity cf. Österreichisches Institut für Wirtschaftsforschung (ed.), *Beschäftigung und Produktivität im österreichischen Bergbau von 1913 bis 1950*, Vienna 1950, p. 7.

almost four million tons, and then a rapid increase to the level of seven million tons in the 1950s. Annual coal production sank again in the following years, and since 1969 it has hovered around the three to four million tons mark.

- Mining's productivity, which was unstable for a long time, started a continuous rise during the 1950s and prevented a parallel rise in manning figures. These, apart from a few exceptions, varied between 10,000 and 20,000 right up till the 1960s, and fell afterwards to 3,000 in 1985.

- By far the greater part of Austrian coal production – it was, in fact, almost always more than 95 per cent - was brown coal; black coal only ever played a minor role, and its production was stopped in 1965.

- In the long term, the most important centres of Austrian coal production lay in the Upper Austrian Hausruck; in the Upper Styrian mining region, Leoben; in the West Styrian mining region around Köflach; and in the only black coal mining region of major importance, the Lower Austrian Grünbach/Schneeberg. Of minor or only transitory importance were the brown coal mining regions north of Salzburg; in the Carinthian Lavant valley; in South Styria; and in the northern part of the Steinfeld, on the Lower Austrian–Burgenland border.

Class strife and the Beginnings of the Balance of Interests up to 1914

Because the start of modern coal mining in Austria goes back to the nineteenth century, its development has to be briefly described. Early on, the brown coalpits of Styria were run by several smaller firms, most of them being amalgamated into two major firms during the course of the second half of the nineteenth century. The Graz-Köflacher Eisenbahn- und Bergbaugesellschaft (GKB), founded in 1855, was joined in the Voitsberg–Köflach and more southerly Wies area by a series of pits already in operation. It was an unparalleled period of expansion. These pits were then extended and, by the end of the century and on the eve of the First World War, the GKB employed over 2,000 miners.[2] The Österreichisch-Alpine Montangesellschaft, on the other hand, was founded as a complex, bringing together in 1881 not just a number of iron works, including the ore mines at Erzberg, but also several coal mines, the more important of them being those at Fohnsdorf, Seegraben and Köflach, with about 4,000 employees in all.[3] The third large concern in Austrian brown coal mining was built up in the Hausruck mountains, where in 1855 two medium-sized enterprises were amalgamated to form the Wolfsegg-Traunthaler Kohlenwerks- und Eisenbahngesellschaft (WTK). From the beginning of the twentieth century, about 1,500 people worked here.[4] As a contrast, the only black coal mine worth mentioning, that in Grünbach, where black coal had been mined since the 1830s, employed only about 450 people just before the First World War.[5]

From the 1850s onwards, mining activity expanded progressively, but not just that – a fundamental change took place as regards industrial relations in mining. The mining decrees which had been passed or issued by various governments at intervals ever since the Middle Ages, prescribing, among other things, an eight-hour shift and a week of five-and-a-half days, were withdrawn as a result of the increasing liberalization of economic life, and replaced in 1854 by a new General Mining Act.[6] The latter contained no rules about

2. Franz Mathis, *Big Business in Österreich. Österreichische Großunternehmen in Kurzdarstellungen*, Vienna 1987, pp. 128ff.
3. Ibid., pp. 23ff.
4. Ibid., pp. 367f.
5. Ibid., pp. 133f.
6. Cf. Max Metzner, *Die soziale Fürsorge im Bergbau unter besonderer Berück-*

the length of shifts, about women's or child labour, nor about
Sundays and feast days as days of rest. In the spirit of freedom of
contract, all such rules were left to the discretion of the employers
and employees involved.[7] As a result, the employees had to work
much longer hours, up to twelve or fourteen hours a day; women
and children were employed in mining; and Sundays and feast days
were treated as ordinary working days.[8] New regulations introduced on support for miners and their dependents in case of
accident, death, invalidity or sickness were to the employees'
disadvantage: even though the mine-owners did not have to pay
any contributions, they controlled the obligatory Bruderladen
(Friendly Societies).[9]

The government could not ignore the clear decline in the miners'
fortunes and the occasional resistance put up by those immediately
affected. As a result, it felt obliged to pass stricter regulations on
working conditions in mining. In 1884 the General Mining Act was
amended for the first time. Female employees were no longer to be
put to work underground, and children under fourteen were not to
be employed in mining at all. Sundays and feast days were fixed as
days of rest. A shift could last no more than twelve hours, including the trip to and from the face, with an effective working time in
coal mining of nine hours.[10] The reasons given for the amendment
to the Act – which allow some inductive thinking as to the miners'
situation at the time, as well as to the government's motives – are
the physical damage wrought by too-long working hours, and the
conservation of the miner's working energy in the interests of the

---

sichtigung Preußens, Sachsens, Bayerns und Österreichs, Jena 1911, pp. 4f. and 20.
    7. Karl Pribram, *Der Normalarbeitstag in den gewerblichen Betrieben und im Bergbau
Österreichs*, Vienna 1906, p. 39; Kohlenholding GmbH Wien (ed.), *Rot-Weiß-Rote
Kohle*, Vienna 1956, pp. 97f.
    8. On the situation of the Styrian miners in the nineteenth century, cf. Karin
Maria Schmidlechner, 'Arbeits- und Lebensbedingungen der steirischen Industriear-
beiterschaft im vorigen Jahrhundert', in Paul W. Roth (ed.), *Glas und Kohle. Katalog
zur Landesausstellung 1988*, Graz 1988, pp. 167–78, or cf. Karin Maria Schmidlechner, *Die steirischen Arbeiter im 19. Jahrhundert*, Vienna 1983. On the situation of
women in particular, cf. Ruth Ellen Bader, 'Die Frau im Revier', in Roth (ed.), *Glas
und Kohle*, pp. 179–87.
    9. Metzner, *Fürsorge*, p. 114; *Rot-Weiß-Rote Kohle*, p. 102; Alois Adler, 'Die
soziale Lage der Berg- und Hüttenarbeiter in der Steiermark ab 1848', in *Der
Bergmann. Der Hüttenmann. Gestalter der Steiermark*, Graz 1968, pp. 297ff. Cf., too,
Herbert Hofmeister, 'Die Rolle der Sozialpartnerschaft in der Entwicklung der
Sozialversicherung', in Gerald Stourzh and Margarete Grandner (eds.), *Historische
Wurzeln der Sozialpartnerschaft*, Vienna 1986, pp. 281ff.
    10. Pribram, *Normalarbeitstag*, p. 40; Metzner, *Fürsorge*, p. 27.

state's defence needs, while general humanitarian aspects turn up as well.[11]

Soon after, in 1889, a new 'Bruderlade' Act was passed, splitting the contributions fifty/fifty between employers and employees, and reducing the voting strength of the former to one-third of the board.[12] The 'Bruderlade' was entrusted with payments to the sick and pension payments to invalids, widows and orphans, although 'even those with the most modest demands could not manage to make ends meet' with them.[13] The accident insurance, also handed over to the 'Bruderlade', was just as inadequate, so that the miners repeatedly demanded to be included in the regular Workers' Accident Insurance. This was conceded to them in 1914.[14]

Despite the steps taken by the government to improve industrial relations in mining, workers' demands, such as that for the eight-hour day, had only been partially fulfilled, and the practical application of the new laws left much to be desired. Not surprisingly then, the miners carried on trying to back up their demands by means of strikes. Naturally such strikes also served their central demand, that for wage rises. Wages were, as before, to be negotiated directly between employer and employee and were, therefore, left out of the mining Acts. Whereas few strikes are heard of earlier on, they increase quite clearly in number and intensity after the late 1880s. The miners in Seegraben, for example, had, according to contemporary reports, enjoyed a relatively good position in the 1870s,[15] but in July 1889 they went on strike and demanded, among other things, a wage rise and the eight-hour day. The same year saw a strike by the Köflach miners, who laid claim to a minimum wage.

In both cases, the management, which no longer consisted of the owners of the Alpine or of the GKB, but of managers without any capital investment in the companies worth mentioning,[16] tried to

11. Michaela Gindl, *Der Kampf der österreichischen Bergarbeiter um den gesetzlichen Achtstundentag 1900–1919*, doctoral thesis, Faculty of Arts, Vienna 1982, p. 16. Cf., too, Gerhard Pferschy and Gerald Gänser, 'Sozialrechtliche Entwicklungen bei den Glasmachern und Bergknappen', in Roth (ed.), *Glas und Kohle*, pp. 189–97.
12. Metzner, *Fürsorge*, pp. 114f.
13. Franz Aggermann, 'Die Arbeitsverhältnisse im Bergbau', in F. Hanusch and E. Adler (eds), *Die Regelung der Arbeitsverhältnisse im Kriege*, Vienna 1927, p. 172.
14. *Rot-Weiß-Rote Kohle*, p. 98; Aggermann, 'Arbeitsverhältnisse', pp. 174f.; Metzner, *Fürsorge*, p. 115. Cf., too, Hofmeister, 'Rolle', pp. 303ff.
15. Eduard Staudinger, *Die Bildungs- und Fachvereine der Arbeiter in der Steiermark von 1848 bis 1873*, D. phil. thesis, Graz 1977, p. 238.
16. Mathis, *Big Business*, pp. 24f. and 129.

stand firm. The regional authorities helped with military intervention, arrests and banning orders, in order to cow the strikers – unsuccessfully this time, as it turned out. After barely two weeks, when the Seegraben pits could no longer be maintained by the supervisory authorities alone, the employers gave in and agreed to give the workers a 10 per cent wage rise and introduce the eight-hour day, and the latter was also introduced into the Alpine's pits in Fohnsdorf.[17] The GKB miners in the Köflach district also obtained the minimum wage demanded, and, after further waves of strikes in 1890, 1891 and in January 1892, the eight-hour day as well.[18] On the other hand, the managers of the firms involved – apart from a few exceptions, such as the mine-owner Ludovika Zang – kept up their repressive behaviour by trying to circumvent concessions already made, which included the granting of a holiday on the 1 May, and either to sack potential and actual leaders of the workers or to secure their removal in general.[19]

The managers' anti-worker stance did not stop them later on from exploiting the introduction of the eight-hour day, forced on them as a piece of social policy on the part of the Alpine since the law referring to it was only passed 30 years later.[20] They emphasized the allegedly obliging attitude of the employers and underlined their repeatedly stated opinion that the workers had little reason to be dissatisfied and that their protest originated in the subversive activity of Social Democratic workers' leaders from outside, who were less interested in improvements than in political power.[21]

17. Ernst Karl Hinner, *Arbeit und Leben des Bergmannes in Fohnsdorf in volkskundlicher Sicht im 19. und 20. Jahrhundert*, Leoben 1978, p. 121; Felix Busson, 'Sozialpolitische Entwicklung', in *Die Österreichisch-Alpine Montangesellschaft 1881–1931*, Part I, Vienna 1931, pp. 144 and 146.
18. Busson, 'Entwicklung', p. 148. Cf., too, Pribram, *Normalarbeitstag*, p. 42.
19. On strike activity in the Styrian mining districts around 1890, cf. especially Eduard Staudinger, 'Die Anfänge der Gewerkschaftsbewegung im weststeirischen Kohlen- und Industrierevier', in Roth (ed.), *Glas und Kohle*, pp. 204ff., as also Michael Schacherl, *30 Jahre steirische Arbeiterbewegung 1890 bis 1920*, Graz 1979, reprint, pp. 20, 22f., 30f. and 36ff. Also Julius Deutsch, *Geschichte der österreichischen Gewerkschaftsbewegung*, vol. 1, Vienna 1929, pp. 310ff. Helmut Konrad reports a strike in the Hausruck area in March 1888 in his *Das Entstehen der Arbeiterklasse in Oberösterreich*, Vienna/Munich/Zurich 1981, pp. 298f.
20. Busson, 'Entwicklung', p. 145.
21. Ibid., p. 154. The agricultural minister, who was responsible for mining at that time, was of the same opinion and suggested that therefore union rights be restricted, a suggestion that the Ministry of Justice rejected. Kurt Ebert, 'Die Einführung der Koalitionsfreiheit in Österreich', in *Wurzeln der Sozialpartnerschaft*,

In fact, Austrian mining in this period saw the foundation of local workers' clubs as precursors of the trade unions set up later in the wake of the repeal of the combination prohibitions still fixed in the General Mining Act. The representatives of the working men's clubs met in December 1890, on the initiative of Viktor Adler, at the First Austrian Mine- and Steelworkers' Congress in Vienna.[22] In 1894, a first Central Federation of the Mine- and Steelworkers of Austria was formed. Just like its successor organizations, it was dominated by the important coal mining district of north-west Bohemia.[23] The employers' federations, on the other side, were amalgamated in 1897 into the Central Association of the Mine-owners of Austria, which took over the programme of the Montanunion, which had existed since 1874.[24]

Both the increased strike activity before and after 1890 and the growing unionization of the miners must have contributed greatly to the passing of a further law in 1896, after several years of protracted negotiations. This law provided for the setting up of mining Genossenschaften.[25] Such Genossenschaften are not to be regarded in the same way as the co-operatives (the usual trans-lation), but as professional joint organizations for mining em-ployees and employers, an institution that only existed in mining. In so doing, the government was not so much interested in satisfy-ing the demands made by the miners at their congresses and elsewhere as in the avoidance or the harmonious regulation of indus-trial strife through negotiation. Co-operation between employees and

---

p. 116. Cf., too, Karl Stocker, 'Bergbau und Arbeiterbewegung', in Ernst Hinner *et al.* (eds), *Fohnsdorf. Aufstieg und Krise einer österreichischen Kohlenbergwerksgemeinde in der Region Aichfeld-Murboden*, Graz 1982, p. 232.

22. Norbert Englisch, *Braunkohlenbergbau und Arbeiterbewegung. Ein Beitrag zur Bergarbeitervolkskunde im nordwestböhmischen Braunkohlenrevier bis zum Ende der österreichisch-ungarischen Monarchie*, Munich/Vienna 1982, pp. 108 and 114; Deutsch, *Gewerkschaftsbewegung*, vol. 1, p. 263; *Rot-Weiß-Rote Kohle*, p. 97. On club foun-dings in general, cf. Staudinger, 'Anfänge', pp. 199-208. Cf., too, the foundation dates recorded in the 'Stammbaum der Gewerkschaft Metall-Bergbau-Energie', as referred to in Fritz Klenner, *90 Jahre Kampf. 1890–1980. Gewerkschaft Metall-Bergbau-Energie*, Vienna 1980, pp. 200ff.; Stephan Philippovich, *Arbeitsrecht im Bergbau. Von den alten Bergordnungen zum modernen Sozialrecht Österreichs*, Innsbruck 1935, p. 133.

23. Englisch, *Braunkohlenbergbau*, pp. 126ff.

24. Wilhelm Denk, 'Die Unternehmerverbände des Berg- und Hüttenwesens im letzten halben Jahrhundert', in the *Österreichischer Berg- und Hüttenkalender 1969*, pp. 23–6.

25. Cf. for the following, Aggermann, 'Arbeitsverhältnisse', pp. 175f.; Philippo-vich, *Arbeitsrecht*, pp. 130ff.; Metzner, *Fürsorge*, pp. 159ff.

employers was to take the place of the class struggle. The emphasis
was to be laid not on the divisive, separating aspects, but on concerns
common to everyone occupied in mining, 'the care of the common
interest and the mining spirit, as also the maintenance and improvement of the branch's sense of honour' among its members.[26]
Regional Genossenschaften were to be formed by both sides together,
employers and employees, and the actual arbitration function was
to be fulfilled by a committee appointed on the basis of full parity
for both sides. Furthermore, the law provided for special local
workers' committees, which were to present any wishes and complaints on the part of the workers and to work toward the avoidance of any workplace strife.

Although the harmonization of the working climate in the mining districts must have been in the employers' interest, and although they, therefore, came to see the law in later years in a
positive light, their first impulse was to oppose its implementation.
They attempted to stop the setting up of Genossenschaften, and,
since it was for the parties to decide whether or not to submit to the
Genossenschaft arbitration committee, the Genossenschaften were
not able to take a very active role in conflict resolution.[27]

The employees' representatives rejected the law from the start.
They recognized the law's fundamental intentions for what they
were, especially as regards the policy connected with it of, on the
one hand, driving back the Free Trade Unions, which were almost
exclusively social democratic in their orientation and stood at that
time outside the established system, and, on the other hand, bringing about peaceful settlements of industrial disputes with the help
of the Genossenschaft delegates, who were partially dependent on
the employers and stood inside the established system. The delegates' dependency resulted from their lack of protection. They had
to accept their election explicitly and risk being given notice if their
activities were deemed 'anti-employer' in character.[28]

The free unions suffered further disadvantages: the workers saw
less need to join them, since they had legal representation in the
Genossenschaften; and the Genossenschaften were used as organizational bases by the more employer-friendly 'independent' or 'yellow' unions, enabling them to exert a greater influence over the

26. Stocker, 'Bergbau', p. 227.
27. Metzner, *Fürsorge*, p. 162.
28. Ibid.,

miners.[29] In fact, a comparative study of the situations in Prussia, Saxony, Bavaria and Austria, published in 1911, stated that the Austrian Miners' Union, founded in 1903, was, as opposed to the equivalent German workers' organization, relatively ineffective.[30]

However, the free unions tried, despite their rejection of the Genossenschaften, to gain as much influence over them as possible by way of the elections. For this purpose they regarded it as necessary to strengthen their own movement. As a result of the combined activities of the Free Unions and the Genossenschaften – such is the opinion of the comparative study mentioned above – the interests of the Austrian miners as a whole 'were promoted much more than those of the German miners, to whom such compulsory Genossenschaften were unknown'.[31] Integration in the existing legal establishment meant that there was a formal basis on which to stand up for the interests of the workers when it came to governmental White Papers. The Genossenschaften acted as a basis for the finalizing of collective bargaining agreements, they made it easier to join in the foundation of educational institutions and, finally, they enabled workers and bosses to engage in closer contact with one another.

The latter mentioned this last function with especial approval, when they looked back on developments in industrial relations. Just this direct contact between the management and the legal representatives of the labour side resolved the tension so often felt earlier on, contributed to the clarification of minor differences and led to more balanced discussions of wage questions.[32]

Even if the Genossenschaften did not have any spectacular results to show for their efforts, their introduction – and this is the main point – was a first step on the path from class warfare to (social) partnership politics. To begin with, they had a pacifying effect within the confines of present-day Austria. Strike activity of a far-reaching nature was mostly avoided, although smaller conflicts were not unknown;[33] at the time of a wider ten-week strike, the

---

29. Englisch, *Braunkohlenbergbau*, p. 171.
30. Metzner, *Fürsorge*, p. 151.
31. Ibid., p. 165.
32. Busson, 'Entwicklung', p. 151.
33. They are covered, among other things, in the reports of the inspectors of mines: K.k. Ministerium für öffentliche Arbeiten, *Die Bergwerks-Inspektion in Österreich*, Vienna 1892ff. Cf., too, diverse material from the company archives of the Alpine Montangesellschaft, now part of the Austrian State Archives, 'Alpine-Montan (VOEST)', Karton 198; and Schacherl, *Arbeiterbewegung*, pp. 118ff. and 168f.

workers in the Köflach mining district, around 3,000 in number,[34] took part for only a few days, till they obtained wage concessions, recognition of the legally elected representatives and a no-punishment deal for the strikers. Apart from all that, the eight-hour shift, the main aim of the approximately 80,000 miners in Bohemia, Moravia and Silesia, had already been achieved in the districts belonging to the Alpine – at first excluding, but gradually including the time for getting to and from the face.[35]

There was a strategy of the employers which quite possibly contributed to the calming down of industrial relations over and above the foundation of the Genossenschaften. It was not a new strategy, but towards the end of the century it became more and more apparent. The employers had realized that repressive measures against and to the discredit of the social democratic workers' movement were not sufficient in order to establish a workfloor climate favourable to the enterprise and mostly free of conflict. On the contrary, the dismissal of 'disturbing elements' often made martyrs out of them and, instead of curbing the influence of the workers' movement, gave it new impetus.[36] The strategy's purpose was to counteract the influence of the workers' leaders, which was seen by the managers to be negative, with the influence wielded by the managers themselves. Management considered that it had bothered too little about the psychological attitudes of the workers, so that now it aimed at the planned education of the workforce in the spirit of the works community.[37] The managers reminded the miners of the traditional *Knappschaft* bodies, their own confinement to small, narrow areas and of the work hazards specific to mining, in order thus to imbue them with a feeling of solidarity in belonging to a community separate from and above the rest of the industrial workers, a community welded together by fate and fortune.

Social benefits supplied by management and meant to promote positive influence over the miners were also to be seen in Austrian coal mining.[38] There was not even a pretence that a feeling of

34. According to Staudinger, 'Anfänge', p. 207, there were only 2,175; according to Gindl, *Kampf*, p. 112, over 3,000.

35. Gindl, *Kampf*, pp. 24, 62 and 112f.

36. Busson, 'Entwicklung', p. 142.

37. Ibid.

38. Cf. for details on this the reports of the inspectors of mines as well, in *Die Bergwerks-Inspektion in Österreich*, 1892ff.

responsibility for the workers' welfare was the reason for such measures. For example, the management of the Seegraben mine owned by Baron Drasche admitted openly in 1872 that the provision of housing, fuel and, partly, of small plots of land, as also the financing or co-financing of a works infirmary, works school and the Bruderlade (Friendly Society) were all aimed at cutting down the increasing agitation of the workers and at 'keeping the latter away from disturbing organizations'.[39] Decades later, the management of the Alpine was also the motor behind the provision of free or cheap works housing, baths, works infirmaries, factory medical services, as also of free or reduced-cost fuel; its intentions had little whatsoever to do with social motives, but much more with the economic interests of the enterprise, which wanted to create for itself a good foundation for an expansion of its plants without unnecessary friction. For this purpose it was necessary to ensure that the workers were healthy and contented.[40] The design and the equipment of the dwellings, which consisted mostly of kitchen-bedsit flats for married couples, with an area of thirty to forty square metres each, and of dormitories with up to twenty beds plus the requisite kitchens for groups of up to forty single workers, all with fixed, exactly formulated house rules,[41] were based on 'the principle of maximal economy' and had to be suitable for service as 'mass quarters'.[42] Whatever the motivation behind the social benefits bestowed by the respective works, they were without doubt yet another aspect of management's efforts to substitute the values of partnership for those of class struggle, without the concessions being too great.

Class Struggle or Partnership – 1914 to 1933

The latter years of the First World War and especially the 1920s showed how unstable and how insecurely founded in fact was the partnership idea that state and bosses were striving for. The attitudes of the two sides of employee and the employer have, however, to be examined with a greater degree of differentiation.

39. Staudinger, *Bildungs- und Fachvereine*, pp. 237f.
40. Busson, 'Entwicklung', pp. 138 and 154.
41. Ernst Hinner, Helmut Lackner and Karl Stocker, 'Bergarbeiterkultur', in *Fohnsdorf*, p. 300.
42. Hinner, *Arbeit und Leben*, pp. 153f.

Management, for its part, carried on with its attempts at harmonizing industrial relations, increased them even, while at the same time intensifying repressive measures against strikers and the social democratic or communist workers' leaders. The working class itself appeared split not just between the free and the independent 'yellow' unions, but also, within the free unions, between the representatives of the more moderate and those of the more radical wings.

The first quarrels started up during the war, caused to begin with by the food distribution question. Fohnsdorf miners were already demonstrating in spring 1915 and the Grünbach workers in November 1916 against the inadequate food supply. From 1917 on, restlessness, semi-uprisings and strikes increased more and more as a result of supply difficulties and other negative effects of the war such as the reintroduction of the nine-hour shift and Sunday working. The fact that wages were falling behind inflation did not help either.[43] As in previous decades, the Alpine's management held the Social Democratic Party alone responsible for the workers' protests, rather than the bad situation as such. It reacted by implementing the same repressive measures that had been used before, seeing to it that this time they were put into practice by the military arm. The latter arrested the strike leaders in Fohnsdorf in July 1917, at short notice, and despatched them to the front line – although the military authorities did not always comply with the expressed wishes of management in this respect. How little management cared for social aspects of industrial relations and how much priority was given to economic matters could be seen by the fact that 'the priority position given to nutrition in pay questions as opposed to that given to work merit or achievement' was regarded as 'a mistake with dire consequences', since 'growing inflation would finally lead to the last common labourer getting, as long as he was blest with children, higher wages than the specialist trained worker, who carried the burden of production on his shoulders'.[44]

After the war, as well, there was repeated strike activity, and strike leaders were dismissed.[45] Although wage and salary demands were adjudged to be 'not unjustified' by management in the face of

43. Aggermann, 'Arbeitsverhältnisse', pp. 180f., 188ff., 195f., 198, 204ff.; Busson, 'Entwicklung', pp. 156f. and 160f.
44. Cf., for example, Busson, 'Entwicklung', p. 186.
45. Ibid., p. 156.

galloping inflation,[46] any say by the employees in things such as employment practice was rejected out of hand. A strike for this right in the Köflach district failed in spring 1923 because of the lack of solidarity from salaried employees (office workers), who were ready to work, and of the intervention by the army and armed police.[47]

In the meantime the short-lived participation in government by the Social Democratic Party had led to the passing of a series of Acts of Parliament, which did not secure the domination of the works councils aimed at, but still brought about a definite improvement in the miners' lot.[48] The eight-hour day, including the trip to and from the face, and the six-day week were finally fixed; child labour in mining and women's labour underground were forbidden again; May Day and 12 November were declared state holidays; the social insurance system was extended and a special workers' insurance introduced. Collective agreements regulating rights, obligations and other matters to do with work contracts were made legally binding and, in analogy to the Chambers of Commerce and Trades, chambers for workers and for salaried employees were introduced. Last, but not least, the works councils were set up. These took the place of the local workers' committees, after the dissolution of the Genossenschaften, and at first consisted almost entirely of members of the social democratic free unions, the latter using them as connecting links to the works themselves. The works councils had, among other things, the right and the duty to speak up for and promote the interests of the workers and the salaried employees, to watch over the execution of the Worker Protection Act and other Acts relevant to the employees, to make a case at the arbitration office against those dismissals which had occurred for political reasons, and also to make collective bargaining agreements with the bosses and their organizations with the help of the unions.[49]

The points in need of regulation in collective agreements included wages and salaries, payday dates, working hours, Sunday working, work on public holidays, night work, overtime pay, the length of dismissal notice, breaks and holidays – in other words,

46. Ibid., p. 175.
47. Ibid., pp. 178f.
48. Cf. *Rot-Weiß-Rote Kohle*, p. 102; Busson, 'Entwicklung', p. 166.
49. Gerald Wurm, *Gewerkschaftliche Zielvorstellungen in Richtung Partnerschaft*, diploma thesis of the Social and Economic Science Faculty, Linz 1974, p. 17.

some of the most important aspects of a work contract. Special agreements deviating from the collective agreement were to be valid only if they were more favourable for the employee involved. Such agreements had existed in mining for some time – the so-called Work and Shift Rules, which were supposed to take the interests of the employees into account and had to be approved by the state mining inspectorate. At first they were just arbitrary rules fixed unilaterally by the employers. Only in 1896 did the employees get the right to a hearing, and in 1919, to a certain amount of co-determination.[50] Unlike the old local workers' committees, which were supposed to encourage co-operation and ideas of partnership with the owners, albeit in the interests of the workers and within the Genossenschaft framework, the works councils were primarily there to represent the interests of one side alone, pushing them through not so much together with as against the other side's interests. It is therefore quite understandable that the employers' side complained about the friction in the mutual relationship, since, instead of the peace they wanted, unrest entered the plants.[51]

Management's reaction was not at all confined to verbal criticism of the Works Council Act. Apart from the repressive measures already mentioned, which also included the occasional dismissal of so-called disturbing elements, the blacklisting of free union workers, practised since 1927 by the Alpine, and the systematic discrediting of social democratic workers' leaders by holding them responsible for the economic difficulties,[52] there were also attempts to gain influence over the works councils themselves.[53] At the works or local level, all non-socialist or non-marxist movements were promoted, especially by maximum support for the foundation of owner-friendly, independent or 'yellow' unions.[54] These latter stood

50. Aggermann, 'Arbeitsverhältnisse', pp. 187f.; Philippovich, Arbeitsrecht, pp. 20ff.
51. Busson, 'Entwicklung', pp. 166ff.
52. Stocker, 'Bergbau', p. 253; Busson, 'Entwicklung', pp. 154 and 166f.
53. Cf., too, Robert Hinteregger, 'Die steirische Arbeiterschaft zwischen Monarchie und Faschismus', in Gerhard Botz et al. (eds), Bewegung und Klasse. Studien zur österreichischen Arbeitergeschichte, Vienna/Munich/Zurich 1978, pp. 280ff.; Karl Stocker, 'Arbeiterschaft zwischen Selbstbestimmung und Unternehmerkontrolle – Einige Aspekte über Disziplinierung, Machtverhältnisse und Widerstand in Eisenerz', in Otto Hwaletz et al. (eds), Bergmann oder Werksoldat. Eisenerz als Fallbeispiel industrieller Politik, Graz 1984, pp. 34 and 140.
54. As early as 1919 the Alpine founded a 'yellow' office workers' union in Fohnsdorf. Deutsch, Gewerkschaftsbewegung, vol. 2, p. 112; Stocker, 'Bergbau',

ideologically near to the bourgeois Heimwehr (Home Defence) and replaced 'class struggle and marxist rabble rousing' by 'community feeling and independence of every political party'. The 'idea of the works community, the common interest of everyone in the expansion of the works, of everyone co-operating to everyone's advantage' – all this enjoyed the highest priority. The managing directors were not 'enemies and exploiters' for them, but 'friends and leaders to these goals'.[55] Whether these slogans fitted the actual circumstances was not questioned – what mattered to management, at least to that of the Alpine, which took over the GKB in 1928,[56] was that the independent union from the start declared its aim to be the most favourable relationship possible with the employers, something quite different from the intentions of the Metal and Mining Workers Federation, which adopted the role of an opposition.[57]

Besides the promotion of independent unions, which included giving the latter's members preferential treatment at the workplace and in the distribution of works accommodation, the firms carried on with their well-known nineteenth-century strategies for pacifying the workers. Workplace social benefits, such as accommodation, infirmaries, swimming baths and food distribution centres were extended and complemented by playgrounds, gymnasiums, libraries, dental clinics, child welfare offices and the like. All this was done in order to intensify the employees' sense of belonging to their firms and to reduce their readiness for conflict.[58] On top of all that, the practices of the German Institute for Technical Work Instruction (DINTA) were imitated in Austria as well – since, after all, the Alpine majority had been sold to the Vereinigte Stahlwerke in Düsseldorf in 1926 – and works schools and newspapers were introduced.

By 1930 works schools had been set up in Fohnsdorf, Köflach district and at the Grünbacher Steinkohlenwerke, the latter belonging

p. 248. In 1896 a local chapter of the Christian Unions had been set up in Fohnsdorf as well, but the Christian Unions did not play any significant role in Austrian mining. Busson, 'Entwicklung', p. 152.

55. Busson, 'Entwicklung', p. 187.

56. Mathis, *Big Business*, p. 129.

57. Busson, 'Entwicklung', p. 188.

58. Robert Pohl, 'Die Kohlenbergbaue der Österreichisch-Alpinen Montangesellschaft', in *Die Österreichisch-Alpine Montangesellschaft*, Part II, pp. 20f., 53 and 69; Busson, 'Entwicklung', p. 172; Wilhelm Richter and Franz Kirnbauer, *Der Bergbau Seegraben 1606–1726–1964*, Vienna 1964, p. 22; Hinner, Lackner and Stocker, 'Bergarbeiterkultur', pp. 297ff.

at that time half to the industrialist family Schoeller and half, via the Hirtenberger Patronenfabrik, to the Creditanstalt or, later on, to the industrialist, Felix Mandl.[59] They were at first concerned with the theoretical and practical sides of the miners' apprenticeship training and, on the side, with sports and arts activities, which were part and parcel of the firm's soft soap strategy to influence the staff in favour of management policies.[60] Work was supposed to be brought nearer to the people spiritually, to become an experience for them 'that would not be rejected with disgust, but instead be regarded positively as a duty both necessary and welcome'; for the employer side, this incorporated the key to the solution of the social problem.[61]

All these efforts to increase the miners' emotional commitment to their firms showed up fairly obviously, and especially in the works newspapers published by management. These papers, too, had been initiated by the Austrian Association for Technical Work Instruction, a junior branch of the DINTA in Düsseldorf. They were published with a large run and mostly free, at first in the Alpine, but after 1929 also at the GKB and the Grünbach collieries.[62] A main aim of the works newspapers, evident not only in the papers' own programme manifestos but also in the actual content of the published numbers, was to seek the 'way to the (workers') hearts'. The efforts, trials and sufferings of the miners were not overlooked, but pushed into the background by the elevation of the 'beautiful, noble and the sublime on earth' to a prime position,[63] in other words, bourgeois values came first. As part of their evasion strategy, management clearly wanted to make the miners forget their real situation and turn them instead into a nice little works community, happy and content with its lot and not expected to offer much opposition or criticism.

Over and above this, the aim was to tie a noose around the works personnel, which would then belong together like 'members of a family'.[64] References to the 'ancient' tradition of the miners' collec-

59. Mathis, *Big Business*, p. 133.
60. Stocker, 'Bergbau', p. 248.
61. Fritz Erben, 'Die fachliche Ausbildung der Arbeiter', in *Die Österreichisch-Alpine Montangesellschaft*, Part I, p. 209.
62. Stocker, 'Bergbau', p. 250; Hans Kreiner, *Der Grünbacher Steinkohlenbergbau und seine Zeit 1823–1965*, Grünbach/Schneeberg 1985.
63. *Werkszeitung der Österreichisch-Alpine Montangesellschaft*, vol. 1 (1927), no. 15, p. 242.
64. *G.K.B. Zeitung für Eisenbahn und Bergbau*, vol. 1 (1929), no. 1, pp. 2ff.

tive work were meant to further the complete self-identification of the miners with their mines. For that reason, the newspapers were supposed to report on factors that bound people together and not on those that kept them apart. They were there to help 'tear down the wall of mistrust and enmity that had been erected, to the disadvantage of both sides, between employers and employees'.[65] Pursuing these aims meant that as a rule one did not bother with objective or even polemical criticism of the Social Democrats, since 'everyday politics and party political strife' were to be excluded,[66] which did not stop the Alpine's management from preventing the distribution of a workers' newspaper, if necessary by force.[67] Any reports about strikes were pitched in a generalizing and at the same time patronizing tone, more or less saying that they did not pay for the workers in the end and that taking part in them was something that required long and hard thinking.[68]

The principle of preferring to mask whatever counted as negative led to a virtual news block on anything to do with social problems inside the respective works. Besides purely factual information about what was going on inside the firm, about the geological situation in mining, about historical, ethnological and general economic matters. There was also advice on accident prevention and special columns for publishing messages, personnel changes, marriages and births; there were spots for women, for gardening, small animal breeding and numerous other such interests. The articles on jubilees, anniversaries and deaths, which were often illustrated with photographs, promoted the feeling of unity with the firm. Various stories, poems, comic articles and puzzles were included to keep the readers amused. Whole series of articles of a German nationalist character, written in an instructive moralizing tone against the social conflict and pro-firm unity and community feeling, were published. Reports were composed about the works schools as well as about the firm's sports events, about new books and – rather less often – about the results of the works council elections.[69]

Besides promoting the independent unions, expanding the social

65. *Werkszeitung der Österreichisch-Alpine Montangesellschaft*, vol. 1 (1926), no. 1.
66. Ibid.
67. Stocker, 'Arbeiterschaft', pp. 42ff.
68. *Werkszeitung der Österreichisch-Alpine Montangesellschaft*, vol. 1 (1927), nos. 24 and 25, and vol. 2 (1927), no. 4.
69. Cf., for example, the *G.K.B. Zeitung*, vol. 1 (1929), nos. 1 and following.

benefits system in the works, setting up works schools and publishing works newspapers, management saw to the reintroduction
of the feastday of Saint Barbara, which was meant to deepen the
workforce's feeling of belonging to the works and to help improve
industrial relations. An old tradition that had not been observed
since the start of the century and was now being reactivated by
management,[70] this feast of the miners' patron saint was celebrated
from the mid-1920s on in Fohnsdorf, Köflach and Grünbach,
revealing where the most fruitful ground was for this professional
rank-based employer-friendly idea. While the office employees and
senior managers took part in the St Barbara celebrations, the great
majority of the workers joined in the annual May Day festivities.[71]
     This separation into an employer-friendly and a less employer-
friendly, even anti-employer, group, showed up in other areas. In
Fohnsdorf, for example, there existed after 1898 a club, the *Berggeist*, with social democratic orientations, beside and separate from
the older music club supported by the Alpine, the *Berg-Capelle*. The
Knappschaftsverein des Kohlenbergbaus Fohnsdorf, founded in 1926, to
which mostly officials and office employees belonged, but hardly
any miners,[72] applied itself to promoting trades or craft culture
with the accent on a person's station in life, stressing among other
things the reactivation of old traditions or usages. Similar clubs
existed in other mining districts of Austria as well.[73]
     Did it pay for the Austrian mine-owners? Did they manage to
intensify the miners' sense of belonging to the firm, to promote the
growth of the independent unions and reduce the influence of the
free unions, which were simplistically called 'marxist' by management? If one can agree with the words of the Alpine manager
responsible for social policy, Felix Busson, then it was the case that
'in 1929 the majority of the workers in all works belonged to the
Heimatschutz [Homeland Protection Society] or to the independent unions, so that these organizations now became the collective
bargaining partners'.[74] In fact, the results of the workers' elections
for the works councils did show, from 1928 on, a certain change in

    70. Stocker, 'Bergbau', p. 252, and Hinner, Lackner and Stocker, 'Bergarbeiter-
kultur', pp. 315f.
    71. Hinner, Lackner and Stocker, 'Bergarbeiterkultur', pp. 313ff.
    72. Ibid., pp. 311ff.
    73. Cf., for example, 'Die Bergknappenvereine und Bergkapellen des Hausruck-
bergbaues', in *Festschrift zum 7. Europäischen Knappentag Ampflwang, 16. bis
18.9.1977*, pp. 67f., Richter and Kirnbauer, *Seegraben*, pp. 46f.
    74. Busson, 'Entwicklung', p. 192.

favour of management. Although records of the election results differ in part from one another,[75] the free unions suffered large losses in both Seegraben and Fohnsdorf. In Seegraben, where in 1927 social democrats and communists had still managed to occupy all the places on the works councils, the independent union gained ground from the following year on – and from 1932 on, it had the majority of the seats. Despite raising its share of votes and seats, it seems not to have managed a majority in Fohnsdorf.

We have no information on the levels of worker representation on the Grünbach works councils, but the very narrow majority of the independent unions among the office workers alone (36 to 25 seats) in 1929 and the continuing dominance of the Social Democrats in the local elections of the same year (769 for the Social Democrats, 157 for the Communists and 373 for the Christian Social and the Great Germany People's Parties) suggest that the free unions managed to keep their end up in the Grünbach colliery till then.[76] Even in the new lignite pit, Hart, near Gloggnitz, which from 1930 on belonged to the GKB or to the Alpine, the works council elections of 1931 gave only 33 out of 261 valid votes to the independent union.[77] Nevertheless, the Styrian area at least showed a certain success for the strategies pursued by management.

The free unions had not appeared nearly as united as one would have expected them to have been if one had gone by the common uniform term for them used by management – 'marxists'. Two examples from the coal mining sector – which could easily be duplicated in other sectors – ought to be sufficient to show that more radical groups had been working against the Social Democrats since 1918 at the latest.

After the so-called January strike of 1918, when about 40,000 Styrian industrial and mining workers gathered in Graz for a grand demonstration, unsatisfied strike demands, increasing supply jams and shortages and various arbitrary decisions made in the factories and mines led to increasingly revolutionary feelings, so much so that the Social Democratic Party leadership thought it necessary to

75. Cf., for example, Karl Stocker, 'Akkumulationszwang und Arbeiterinteresse', in R. Hinteregger *et al.* (eds). *Auf dem Wege in die Freiheit*, Graz 1984, p. 259, for the results collated from the *Bergmann* and other newspapers and published in the works newspaper of the Alpine, vol. 4, no. 21.

76. *Werkszeitung der Grünbacher Steinkohlenwerke A.G.*, vol. 1 (1929), nos. 5 and 6.

77. *Österreichischer Metall- und Bergarbeiter. Organ des österreichischen Metall- und Bergarbeiterverbandes*, 2 May 1931, p. 3.

try and calm the workers down, in order to keep the movement under control.[78] Conversely, this line of action shook grass roots confidence in the party leadership to such an extent that only a few of the 60,000 Styrian industrial and mining workers who stayed away from work on 1 May went to the various party events held on that day.[79]

The Communists were the most radical of all, yet they were regarded, in Upper Styria especially, as no more than the left wing of the Social Democrats, which led to common demands being made by both groups, at least at the start. One of the claims made, for instance, was for the immediate nationalization of the Seegraben coal mine, a demand made without waiting for the result of the Austria-wide nationalization negotiations being held in Vienna. A protest against excessive flour and fat prices was, in spring 1919, occasion enough for such a worsening of the situation in Donawitz and Seegraben that the Alpine managers there were dismissed and the workers set up their own management instead. A workers' council was elected, consisting of an engineer, an accountant, a pit overseer and twelve workers. It was supposed to fix the conditions for the nationalization of the coal mines.[80] The party leadership, which was now entrusted with the tasks of government, described this action by the grass roots as 'naive' and after a while it was called off.[81] This led to a further loss of confidence in the Social Democrats in Upper Styria and to a clearer drawing of the line between them and the Communists there.[82] This must be regarded as an early sign of the Social Democratic Party distancing itself from the idea of class struggle in the coal mining community.

The second example of this type showed up around a decade later in the mining district of Grünbach.[83] Wage tariffs had been allegedly kept under the level of the collective agreement for years, despite a doubling of productivity, and a strike broke out at the end of November 1930 in which about 1,100 miners took part, of whom only 150 were union members. The strikers, with whom

78. Robert Hinteregger, *Die Steiermark 1918/19*, D. phil. thesis, Graz 1971, pp. 10ff.
79. Ibid., p. 15.
80. Ibid., pp. 169ff.
81. Cf. on Otto Bauer and the efforts towards nationalization, Judit Garamvölgyi, *Betriebsräte und sozialer Wandel in Österreich 1919/1920*, Vienna 1983, pp. 82ff.
82. Hinteregger, *Steiermark*, p. 189.
83. For the following cf. Franz Honner, *Streik in der Krise. Die Lehren des Grünbacher Streiks*, Vienna (n.d.), pp. 3ff.

the unemployed showed solidarity, demanded higher wages, improved work security measures and a better collective agreement. The strike did not originate with the union, dominated by Social Democrats, but with the workfloor and it was led by the Communist-orientated revolutionary union opposition (RGO). The free union, described as reformistic by the RGO, had even tried to prevent the strike, but the direct vote had produced an overwhelming majority of 82 per cent (817 to 166 votes) in favour of pressing on with the strike. Nevertheless, the strikers still had to cope not just with the owners' resistance and with pit overseers and officials, who together with a few blacklegs carried out the most basic maintenance tasks, but also with a lack of solidarity on the part of the union officials, who prevented the strike from spreading to other districts and undermined various supporting activities for the miners involved. After six weeks, in January 1931, the strike, which had not been mentioned once in the works newspaper, was finally called off by the union officials, acting on their own initiative and without anything having been achieved. On the contrary, several of the sacked workers were not reinstated, the wages of many others were cut and all five RGO works councillors disciplined. A possible effect of the strike and of the attitude of the Social Democratic union leadership is perhaps to be seen in the fact that the RGO was able to gain more than half the votes in the Hart coal works council elections held the following year.[84]

The departure from the ideals of class struggle had clearly been passed on as policy from the party leadership in Vienna to the regional party officials and the socialist works councils in the enterprises, but not yet to the miners at the grass roots level. Despite this, the partnership orientation or bias of the Social Democrats – which many had not even registered, veiled as it was by a form of verbal marxism – did not suffice to prevent them being forbidden and driven into illegality after the so-called self-dissolution of Parliament and the civil war clashes in February 1934.[85]

---

84. *Österreichischer Metall- und Bergarbeiter*, 2 May 1931, p. 3.
85. Cf. Hans Prader, *Probleme kooperativer Gewerkschaftspolitik. Am Beispiel der Politik des ÖGB im Wiederaufbau 1945–1951*, D. phil. thesis, Salzburg 1975, p. 90; Gindl, *Kampf*, p. 448. On the anticipated social partnership of 1919/20, cf. Garamvölgyi, *Betriebsräte*, pp. 245ff., and Prader, *Probleme*, pp. 87ff.

The Disqualification of the Miners in the *Ständestaat* and Under the Nazi Dictatorship

After the setting up of the authoritarian *Ständestaat* (corporate state) in 1933, the partnership idea in Austrian mining, aimed at for years by management and more and more accepted by some of the workers, was confirmed by decree from above. Strikes and lockouts were forbidden. Instead of the Works Council Act of 1919, a class- or rank-oriented law on the installation of 'works communities' was introduced, while, instead of works councillors, there were so-called trustees, who were nominated by the workers' chambers, which in their turn were administered on an interim basis by appointed officers. Together with the bosses they formed a body in which the grouping of the latter, with its absolute veto, had a controlling vote.[86] The free unions were dissolved and replaced by the Union Federation, which was to be newly formed as a uniform, non-sectional union for Austrian blue-collar and white-collar workers. The federation was also to take over the tasks of the chambers for workers and salaried employees. An Industrial Confederation was also to be instituted for all Austria, with relevant specialist sub-federations, such as the one for the mine- and steelworks, which had a further subdivision for the mines and within that a subdivision for the coal mines.

The brief of the Confederation and its subdivisions differed from that of their predecessors in that they were now responsible for wage negotiations with the employees' organizations. In general, the making of collective bargaining agreements, one of the most important achievements of the unions after the war, was now exclusively the province of the Union Federation and the Industrial Confederation, both institutions being subject to state law.[87]

The governing body of the Union Federation was not elected by the employees in any way – which goes to show their loss of rights – but appointed by the Minister for Social Administration.[88] Besides making collective agreements, the Union Federation was, as before, responsible for the interests of the employees in labour law, economic and social matters – but 'in a Christian, patriotic and

86. Stocker, 'Bergbau', p. 255; Emmerich Talos, 'Sozialpartnerschaft: Zur Entwicklung und Entwicklungsdynamik kooperativ-konzertierter Politik in Österreich', in P. Gerlich *et al.* (eds), *Sozialpartnerschaft in der Krise*, Vienna 1985, p. 53.
87. Philippovich, *Arbeitsrecht*, p. 18; Denk, 'Unternehmerverbände', p. 25.
88. Philippovich, *Arbeitsrecht*, p. 135.

social spirit, to the exclusion of all party politics'.[89] In order to make sure of partnership-orientated work in the union, the law allowed the rejection of membership applications on the grounds of 'class struggle or political attitudes'.[90]

Just how much such laws suited the interests of the firms involved in Austrian coal mining does not have to be emphasized, after what has been stated above. Reports about them in the works newspapers tended as a result to be positive and showed up at the same time the attitudes of the employer side in general. In the eighth issue of the GKB newspaper in 1934, the class struggle and class struggle tendencies were first of all presented as wrong, something that was concealed by the 'higher maths of socialist science as interpreted by the political party and the union belonging to it.'[91] At the same time, the opinion was ventured 'that in the actual firm itself "capitalist" and "work-slave" have it equally good and bad' because of the unsettled economic situation, an opinion that in the face of such different working conditions can only be regarded as dubious.[92] As before, it was denied that socialism and the unions made any genuine effort to improve the workers' lot and alleged that they had 'a programme aimed at destroying the economy and social life as a means to bring in Socialism and Bolshevism'.[93]

In contrast, the new works community with its ideal of 'social pacification' was celebrated as an achievement of the then Federal Chancellor, Engelbert Dollfuss, since a 'harmonious, peaceful cooperation between the several success bases in the works was the best guarantee for the success of the works as a whole'.[94] The writer just 'omitted' to mention that the fruits of the same works' success were distributed very unevenly among the people involved in its achievement, which accounts for the special interests of the miners and their families. The costs of the works social benefits were even to be piled on the shoulders of the workers, with the result that the trustees were given the right 'to set up and administer welfare institutions for the sake of the workers and salaried employees of the works and to impose a works charge for this

89. Ibid., p. 134.
90. Ibid., p. 136.
91. *G.K.B. Zeitung*, vol. 5 (1934), no. 8, p. 115.
92. Ibid.
93. Ibid., p. 116.
94. Ibid.

purpose on all employees'.[95] Elsewhere the same author admitted
the opposing interests of employees and employers and the resulting necessity of an institution which would be outside the works
sphere and in a position to 'mediate and balance things out in its
intervention activities.'[96]

Thus the introduction of the new Union Federation was approved of – which might appear surprising at first, but can be
understood once one knows how the leaders of the new federation
were appointed and how workers' representatives who were awkward were kept out of the new structure. As long as the official
representatives of the employees were dependent on a government
biased towards employers, or at least oriented to law and order, it
was easy for management to accept a parity-based board for the
professional trades, where 'neither entrepreneurs nor workers were
to occupy a dominating position'.[97]

Although the Ständestaat period was short, it counted without
doubt as a further important step on the way from class struggle to
partnership ideas. The rolling back of class struggle attitudes in
favour and partnership-oriented harmony was, so to speak, decreed
by law and, to a certain degree, comparable with the 'Genossenschaft' Act of 1896. Its effect on later decades was quite considerable.

During the seven years of Nazi rule in Austria, from March 1938
until the military collapse in May 1945, coal mining also saw the
employees being more and more deprived of their rights in the
interests of employers. The task of direct representation of the
workers at the various workplaces was handed over to a trustee
council, which had no powers of decision. Only the owners, or the
management in the person of the managing director, were allowed
to take decisions.[98]

Again management was concerned with emphasizing that within
the framework of the new works community, profit in common and
not individual interests were 'to be the basis of everyone's dealings',
after 'the class struggle contradictions, which have for too long
prevented bosses and workers from working profitably together,
have been removed'.[99] Once again efforts were made with unproven
and partly false statements to influence the workforce – now called the

95. Ibid.
96. Ibid., vol. 5 (1934), no. 10, p. 150.
97. Ibid.
98. Stocker, 'Bergbau', p. 257.
99. Ibid., vol. 9 (1938), no. 8, p. 114.

works 'following' – in favour of the bosses and the new rulers. A second major enemy was portrayed alongside marxism, the Jews. Both were connected with one another in the Nazi litany of hostility, for example in works newspaper references to 'Jewish marxism', the aim of which was the extermination of the entrepreneurs, 'and not the upward mobility or progress of the workers'.[100] Only the plant manager, not the socialist worker representatives, was entrusted with the responsibility for the existential security of the 'people working under him'. In contrast with the works communities introduced by the Ständestaat in which the differences between employee and employer were not only recognized, but, because of the opposing professional organizations, also turned out to be unbridgeable, the plant communities to be introduced under the Nazis were to show no such contradictions.[101]

The plant managers were, therefore, required to pay more attention than before to their social responsibilities, and this went together with an idealistic exaggeration of the value of everybody's work 'to the common advantage of people and state'.[102] 'This means that working is regarded from the elevated position of the interests of people and state, thereby removing working from the realm of mere material services rendered and giving it a higher meaning'.[103] The fact that the trustee councillors, who were not elected from among the staff but appointed, were only permitted to advise the plant manager and not to practise any form of co-determination, showed who decided what the interests of people and state were.[104] The higher value put on work as such, in itself something to be welcomed as a counter-balance to the overestimation of capital's importance, was connected with a downgrading of the workers themselves and their subjugation to the interests of the people or the firm. The trustee council, unlike the earlier works councils, was 'not supposed to speak up one-sidedly for the interests of the works following, but to represent the plant's interests as a whole'. Its duty was 'to intensify the mutual trust within the plant community' and -- as a logical consequence – 'to work towards a banning of all disputes within the plant community'.[105]

100. Ibid.
101. Ibid.
102. Ibid.
103. Ibid., vol. 9 (1938), no. 12, p. 162.
104. Ibid., vol. 9, no. 8, p. 114.
105. Ibid., vol. 9, no. 12, p. 163.

The practical execution of these theoretical programme aims had
the plant management really trying harder to expand the social
benefits passed on by the plant in earlier years: plant-based specialist training was promoted especially, so that in 1939 a further
works school was set up in Seegraben and, a little later on, a new
seminar boarding home of the WTK in Ampflwang. Housing
projects and different forms of leisure activities, whether for physical and sporting or for artistic and literary purposes, were also
expanded.[106]

Extensive housing programmes were planned for the various
works branches of the Alpine Montangesellschaft, which together
with its junior company, GKB, was merged with the newly
founded state-owned Hermann-Goering-Werke.[107] These plans
were put rapidly into action. In the Köflach-Voitsberg district alone
there were plans for 330 dwellings. In Fohnsdorf, first of all,
building began on a village-type works estate, where the various
families had either a one-storey house with an attic extension or a
one-bedroom flat with kitchen, bathroom, lavatory and hall. Here
the inside bathroom of the flats counted especially as a major
novelty compared with the earlier kitchen-bedsitters.[108] An estate
with a similar design or layout was built in Ampflwang by the
WTK, which from the 1920s on belonged to the Upper Austrian
state government.[109] The first houses here were ready for moving
into at the start of 1940.[110] By mid-April 1941, 255 of the 640
dwellings planned in Seegraben, Köflach, Voitsberg and Fohnsdorf, in the Styrian coal mining districts belonging to the Alpine
and GKB, were ready.[111]

Although the economic aspect of the house-building activity was
still always emphasized – 'healthier living quarters were seen as one
of the possible routes to a productivity increase in German mining'[112]
– the workers' families enjoying thereby the benefits of new flats and
houses had their living standards greatly improved. The *Siedlerstim-
men*, or 'voices of the estate inhabitants', that one sees in the works

106. *Werkszeitung der Betriebsgemeinschaft Alpine Montan Aktiengesellschaft 'Hermann Göring' Linz*, vol. 13 (1939), no. 9, p. 138; *Glückauf. Werkszeitschrift der
Wolfsegg-Traunthaler Kohlenwerks AG*, vol. 1 (1940), nos. 10/11, p. 6.
107. Mathis, *Big Business*, p. 27.
108. Hinner, *Arbeit und Leben*, pp. 146ff.
109. Mathis, *Big Business*, p. 368.
110. *Glückauf*, vol. 1 (1939), no. 1, p. 13, and nos. 3/4, p. 15.
111. *Werkszeitung der Alpine*, vol. 15 (1941), no. 7, p. 108.
112. *Glückauf*, vol. 1 (1939/40), nos. 3/4, p. 15.

newspapers have to be read with a certain amount of scepticism, but that they had some foundation in fact is understandable. The estates were, just to name some of their advantages, near to the place of work, they offered more living space and, on the land belonging to the houses, there were facilities for market gardening, which was especially valued by the 'settlers'.[113] Sport and fitness were held in higher regard now than before and the organization *Kraft durch Freude* was responsible for it all nationwide.[114] The sports fields already in existence were improved or extended, new ones were laid out, and in the works newspapers, too, works sports activities were reported on more regularly and frequently than before. The existing works libraries or those newly opened formed part of the policy of spreading facilities as broadly as possible, and this policy extended to the organization of diverse cultural events with the occasional film-showing. At the WTK after 1940, there was a specially-built cinema for the purpose.[115] In general it is noticeable that, by contrast with the Styrian districts, which, influenced by the DINTA and their mostly German owners, had already begun with such measures in the late 1920s, the Upper Austrian WTK only began now to try and catch up, putting a lot of effort into doing so.

Finally the plant management paid special attention to hygiene and health questions and to family welfare. They saw to it that – in cases where no such institutions existed – baths, laundries, works infirmaries and treatment rooms were set up, that works employed their own doctors and supported works nurseries. All this was supposed to help stabilize the plant community, so dear to the firm management and to the party. It is, therefore, understandable, that efforts were made to inform the employees via the works newspapers – which since 1939 had also been published at the WTK – about the various social benefits, and the diverse financial aid measures with which, in special cases, they were complemented. All this was done in the hope that 'every honest, unprejudiced person who judges it all for himself must admit, after all, that a new wind is blowing at the WT [WTK]'.[116] The staff was also expected to be prepared for 'every sort of co-operation', an expectation that the chief plant representative, the local NSDAP trustee, also shared:

113. Cf., for example, *Glückauf*, vol. 1 (1940), nos. 10/11, p. 16.
114. Stocker, 'Bergbau', p. 255.
115. *Glückauf*, vol. 2 (1941), no. 2, p. 9.
116. *Glückauf*, vol. 1 (1940), nos. 10/11, p. 6.

'We've racked our brains, moaned and rebelled in the past, mostly with justification, sometimes without, as well – but enough of that. Now we spit lively in our hands and pull all together on the same line: for a true, genuine, national socialist plant production community'.[117]

In wartime the workers in coal mining were required to work more than before and were to be motivated not just through references to the achievements of the plant management, but also just as much through massive propaganda. Constant blaming of the war enemies, who 'threatened the life rights of the German people' served to conceal the state's real war aims and references to the 'work front line' and to the central importance of coal production were meant to achieve an improvement in productivity to match the efforts of the soldiers on the military front line.[118] 'The soldiers at the front line are keeping the borders safe, while the soldiers at work deep in the earth produce, shift after shift the huge coal tonnages our boom economy needs'.[119] As a consequence, as early as October 1939, the miner had to work one hour more per day at the WTK, for example, and probably in other districts as well, in order to compensate for the dip in productivity after conscription had started. The extra-time work was a clear success.[120] Daily overtime was extended as a result and extra Sunday working followed, so that a further major production increase was possible.[121] The best daily production figures claimed in the Hausruck district were achieved during a so-called 'Panzer' shift in February 1943, when, on the last Sunday of the month, even plant managers and plant and department heads joined the miners at the face to work a nine-hour shift 'voluntarily' and for nothing.[122] The increasing length of the war led to a constant increase in pressure to raise production, so that miners were called on to raise their productivity even more. Top achievements by individual miners were praised and mentioned as examples fit for imitation – as with the two WTK miners, who, starting with a normal production rate at a normal hewing site, concentrated their physical and will power within a half-shift so that their productivity per man and shift was almost

117. Ibid.
118. Ibid., vol. 1 (1939), no. 1, pp. 10f.
119. Ibid., vol. 1 (1939/40), nos. 3/4, p. 4.
120. Ibid., vol. 1 (1939/40), no. 2, p. 3.
121. Ibid., vol. 3 (1942), nos. 1/3, p. 3.
122. Ibid., vol. 4 (1943), nos. 1/3, p. 2.

doubled, a result that was even bettered during the next half-shift.[123] Regardless of how voluntary such efforts were, the social programme pursued by the National Socialists was to a great extent offset by the forced increases in productivity and the political disenfranchisement of the miners. To take a more differentiated point of view of the latter, however, it must be taken into account that the disenfranchisement of the miners was a direct result of the National Socialist political programme, whereas the extension of working hours was most of all a result of the war and, therefore, only indirectly involved with Nazi rule. It is more relevant for our theme that the partnership idea, which had been dictated from above since the setting up of the Ständestaat, with its simultaneous suppression of the class struggle, was continued after the Annexation and efforts were made to force it through with positive incentives and repressive measures.

The Completion of Partnership Integration After 1945: Results and Costs

The end of the war brought about a new situation in Austrian coal mining in various respects. The form of industrial relations was decisively determined by nationalization – which was finalized by law as early as 1946 and involved almost all primary industries, including mining – and by the continuous participation in government of the Socialists, which was only interrupted for brief periods in the years between 1966 and 1970. The Socialists were the ones who had demanded, at least in their political programmes and according to marxist theory, the nationalization of the means of production, and it seemed logical to entrust them with the ministry responsible for nationalized industry.

As far as industrial relations in mining were concerned, this arrangement meant that the party of the employees, the one which had tried, since its early days, to promote their interests in the ongoing dispute with the bosses, now moved up in one fell swoop into the position of a boss and a government in one. The result had two aspects to it: firstly, the Socialists felt obliged to run the enterprises entrusted to them within a still-capitalist economic system according to the principles of a market economy, meaning

123. Ibid., vol. 5 (1944/45), no. 4.

in a profit-oriented way; secondly, they had to be interested – as representatives of the owners or as managers and as government, just like their predecessors – in preventing their plants from suffering under class struggle problems, so that they worked in a trouble-free and peaceful way, in other words, in a partnership-oriented atmosphere. Such a situation was clearly laden with potential for internal party conflict. In order to cope with the dilemma of being simultaneously entrepreneur, government and representation for the workers, they had to find a compromise policy to suit the interests of as many of the disputing parties as possible and to reduce the number of people who were dissatisfied and neglected to a minority. The following is intended to show how the Socialists attempted to achieve this demanding aim.

The system of freely elected works councillors which had been accepted by the employers without enthusiasm and had been mostly removed during the Ständestaat period was reintroduced by the Works Council Act of 1947, but the remaining conditions for a partnership-oriented policy in the first postwar years were not unfavourable. In the face of public need and the necessity for reconstruction work, the workers, especially in coal mining, had to be persuaded to give up for the time being a rapid improvement of their material situation – but, on the other hand, they were readier to accept arguments put to them in this connection if they were presented by their own party, and not, as before, by private entrepreneurs.

The party itself was to find it a great advantage that, after the foundation of the Austrian Union Federation (ÖGB) – which had its predecessor in the Ständestaat period – a central negotiating partner with final powers of decision existed, and the necessity of having to deal with each single union department or craft union was obviated.[124] From the start, however, the heads of the Austrian Union Federation pursued a policy that regarded a flourishing economy as a basic precondition for an improvement of the employees' lot and at the same time worked towards a sort of strike monopoly on top of the monopoly on collective bargaining that already existed.[125] As a result, there was very much a parallel

---

124. For the importance of Ständestaat institutions for the period after 1945, cf. Alfred Klose, 'Geistige Grundlagen der Sozialpartnerschaft im katholischen Sozial-denken', in *Wurzeln der Sozialpartnerschaft*, pp. 60f., and Talos, 'Sozialpartnerschaft', p. 55.

125. Prader, *Probleme*, p. 324; Joachim Bergmann, 'Gewerkschaften – Organisa-

between the policies of the Socialist Party leadership, which, because of its government coalition with the bourgeois People's Party, was forced to accept compromises, and those of the Socialist-dominated Union Federation, which regarded social partnership as the best way to care for the interests of the working class within the realm of feasibility. The deviation from the class struggle line did not need to be dictated from above any more, as in the Genossenschaft Act of 1896 and the relevant ordinances of the Ständestaat and the Nazi regime; it came about as a conscious voluntary decision taken by the union and the party leadership. In order to pursue such a policy over a length of time, the union had to have a majority of the employees behind it. It exerted the necessary influence on the latter in order to win over the majority, employing the works councillors as links to the workforce for the purpose. This proved especially effective where the workforce really stood ideologically close to the Socialist Party. The Union Federation and the Socialist works councillors, who were interested in their own re-election, saw the benefits of reducing the influence of any Communist groups oriented towards the class struggle. This led to a situation in many ways reminiscent of the interwar period, so that it was not just the bosses alone who tried to keep down the excessive or perhaps even justified demands of individual employee groups, but also the employee representatives close to the Socialist Party.

The same phenomenon turned up in other branches, but Austrian coal mining was a specially apt example of such persuasion, which was in the end successful. The results of the works council elections and the contents of the various works newspapers are evidence.[126] In advocating social partnership, it proved most difficult to influence the miners of Fohnsdorf, of the Pölfing districts of the GKB in South Styria, at the 'Marienschaft' belonging to the Steirische Kohlen AG, which had also been nationalized, and in the Grünbach coal mine. In Fohnsdorf, the Communists were the victors in five of the six works council elections between November 1945 and November 1951. Later, the Socialists got the upper hand, just about in 1953 and 1959, and quite clearly in all the other

tionsstruktur und Mitgliederinteressen', in G. Endruweit *et al.* (eds), *Handbuch der Arbeitsbeziehungen Deutschland Österreich Schweiz*, Berlin/New York 1985, p. 92; Pelinka, *Modellfall*, pp. 50f.

126. The works council election results mentioned in the following were put at my disposal by the Metal–Mining–Energy Union in Vienna and Graz.

election years till the closure of the Fohnsdorf pits in 1978.[127] During this period, their share of the vote was sometimes four, five or six times the share that the Communist ticket got.[128] In the Pölfing district the works council was, until the beginning of the 1950s, mostly Communist. Thereafter the Socialist representatives took over, and they then dominated the works council by a similar margin to that in Fohnsdorf. The smaller Marienschaft in Voitsberg saw both groups evenly balanced with one another up to about 1950; in the following years the Communists lost ground and in 1955 they vanished from the works council. The Communist works councils were, on the other hand, quite strong in Grünbach, where the coal mine stayed under Soviet administration till the signing of the State Treaty. It would, however, be wrong to assume that undue pressure on the part of the Soviets was responsible for this state of affairs, as has been done for other enterprises administered by the Soviets.[129] Even in 1963 – two years before the plant closed down and some time after the departure of the Soviets – almost half the work force voted for the Communist ticket, 'Union Unity'.

In all the other larger and in most of the smaller mining districts, the works councillors friendly to the Socialist Party had clear majorities at their disposal early on, leaving only a few votes and seats to the Communist tickets. This was especially true of the GKB's Köflach–Voitsberg district, for which Franz Zwanzger, a long-serving convenor of the Austrian miners, who worked there, was responsible. In Seegraben one notices again that the Communist works councillors also stayed in the minority, but that their voting results hardly changed between 1949 and 1961 (mining stopped in Seegraben in 1964, because of pit exhaustion)[130] while Socialist voting numbers dropped by more than a third. Similar results are noted, albeit for a later period, for the WTK, where the number of Communist ticket voters in Ampflwang stayed at a constant, even if somewhat low, level between 1966 and 1978, while the Socialists, on the other hand, lost more than half their votes. The results were reversed in the following years.

In short, the social partnership policy of the Union Federation did not meet with by any means unanimous approval among the

127. Stocker, 'Bergbau', p. 269.
128. Ibid., p. 265.
129. William B. Bader, *Austria between East and West 1945–1955*, Stanford, Calif. 1966, pp. 147, 149, 151 and 153.
130. Richter and Kirnbauer, *Seegraben*, p. 22.

Austrian coal mining workers in the first postwar years, and the adherents of the class struggle line were even able to count on a majority in some plants and pits, and were altogether a minority not to be ignored. Only in the 1950s did the Socialists manage to take over the works councils dominated by the Communists. Several factors were responsible for this state of affairs.

As far as the development of real wages was concerned, no increase worth mentioning occurred between roughly 1950 and 1954, since inflation more or less kept up with the average nominal wages.[131] The buying power of coal mining wages only managed to rise in later years: whereas the cost of living index tripled between 1960 and 1984, the average hourly wage rates of the coal mining workers rose more than sevenfold in the same period.[132] Whether the wage development of the early 1950s can be regarded as a reason for the weakening of the class struggle stance in Austrian coal mining, then, seems to be at the least questionable.

The reductions in working hours for the same pay, negotiated in collective bargaining agreements between the ÖGB's specialist (craft) unions and the corresponding association of the Federal Economic Chamber (BWK), and introduced progressively, are comparable with wage increases – but such reductions also started relatively late. After a time when the old forty-eight-hour week was still the rule, the forty-five-hour week was introduced in 1958 in coal mining.[133] Twelve years later, a new phase of paced-out reductions in working hours started, leading to the introduction of the forty-hour week in 1975.[134] The week of 38.5 hours practised now – counted, since 1972, without the breaks – was fixed in the collective agreement of 1988.[135]

Further improvements were made as regards the Christmas bonus, the holiday bonus and the periods of notice to be kept to by the employer – to mention just some of the more important collective agreement terms.[136] The Christmas pay bonus, still fixed

131. *Statistisches Handbuch der Republik Österreich*, new series 6 (1955), pp. 156f.; *Rot-Weiß-Rote Kohle*, p. 100.
132. From 11.12 to 82.60 schillings. *Österreichisches Montan-Handbuch 1961–1985*, Vienna 1961–85; *Jahrbuch der österreichischen Wirtschaft 1984*, Vienna 1985, p. 293.
133. *Die Inlandkohle. Nachrichtenblatt des österreichischen Kohlenbergbaues*, vol. 6 (1958), nos. 10/11, p. 1.
134. Cf. Klenner, *90 Jahre Kampf*, p. 137.
135. *Arbeiterrkollektivvertrag für die eisen- und metallerzeugende und -verarbeitende Industrie vom 1. November 1988*, Vienna 1988.
136. For the following cf. *Kollektivvertrag für den österreichischen Kohlen- und Eisenerzbergbau vom Oktober 1948*, *Kollektivvertrag für den österreichischen Kohlen- und*

in 1948 at the equivalent of a week's wage, was later on regulated at
various levels based on years of service, until it was consolidated
from the 1970s on at the equivalent of a month's wages (except for
those employed at the plant for less than a whole year).

Holidays were extended several times under the Workers' Holiday Act, and finally reached the level of thirty working days
(Monday to Saturday) for employees with less than twenty-five
years of service, and of thirty-six working days for those with
more. Since the 1950s the collieries have paid out holiday grants on
their own initiative, amounting at first to three and later to four
weeks, or, in the end, one month's wages. The periods for giving
notice of dismissal also stayed the same for a relatively long time
and were changed only in the 1960s from five to six weeks, first for
those employees with at least ten years' service and later for all the
others. The latest regulation consists of ten weeks' notice for those
with at least twenty-five years of service, eight for those with at
least fifteen years and six for those with at least five years, while the
remaining workers have been left with starters' terms of two weeks
or with one week for those with less than one year's service.

In contrast to the somewhat belated rises in real wages and
salaries and the other improvements, where the beneficial difference was quite marked compared with previous decades, the remaining factors promoting the partnership idea were already part
and parcel of the traditional state of affairs. Neither the influence of
the newspapers nor the expansion of social benefits internal to the
plant nor the occasional exertions of pressure were new: only the
actors had changed. Instead of the works newspapers published by
the management of the firm, as known in the interwar years, there
were publications of the Socialist works council majority or of the
Metal and Mineworkers' Union. They were mostly uncritical,
keeping a balance between information and entertainment and
serving to mask any class conflicts. The role of the more anti-enterprise newspapers published by the unionized workers was
taken over by the more aggressive, critical writings of the Communist 'Union Unity', which did not show any satisfaction with
the achievements either of management or of the works council,
which for the most part co-operated with the former.

*Eisenerzbergbau vom Oktober 1948 unter Einbeziehung aller Vereinbarungen und
Änderungen bis 1. Oktober 1958, Kollektivvertrag für den österreichischen Bergbau vom
November 1974, Arbeiterkollektivvertrag für die eisen- und metallerzeugende und -verarbeitende Industrie vom 1. November 1988.*

The striking similarity between the works newspapers of the 1920s and the 1930s and the roneoed flysheets and leaflets of the works councils of the late 1950s and early 1960s is not to be missed, in, for example, *Kumpel, Werkszeitschrift des Kohlenbergbaues Fohnsdorf*, published by the cultural department of the Fohnsdorf workers' works council.[137] Apart from reports on its own activities, on the works council's demands and its meetings, there are notes on the Workers' Chamber, on works sports and works dwellings, all jumbled up with stories, humour and comic strips, obituaries, wedding news, etc., while criticism of firm or branch management is mostly excluded. Without any kind of class struggle stance, the newspaper or magazine had to have a stabilizing effect, just what management used to wish for, strive for and still does. Anyone criticizing the moderate policy of the union was attacked more massively than was management. Without wondering whether a more aggressive strategy would bring about more improvements, one of the leading Fohnsdorf works councillors writes disparagingly in 1960 about the demands of the Communists for a more radical redistribution of income: 'What use is it, to play the wild man constantly, just because the managing director earns more than the worker, the Lord Mayor of Vienna more than the managing director and the Federal Chancellor more than the Mayor of Vienna – relations such as these are the same in East and West. I do not believe the rumour that the Mayor of Moskow walks around the Kremlin with Nikita, the Soviet boss, every evening, collecting alms in order to have something to eat the next day. And were the three luxury Cadillacs bought in the USA meant for the workers?'[138]

Just like the Socialist workers of earlier decades, who were attacked by the employers for, among other things, their 'class hatred and marxist agitation', the Communists now became the victims of Socialist attempts – quite successful indeed – to picture them as a dangerous alternative, and at the same time to associate them by hints with the 'real' Marxism in the Soviet Union, a real frightener. The fact that the usually pro-capitalist *bon mot* – 'In capitalism man is exploited by man, in communism it's the other way around' – was published in the *Kumpel*, shows how much the Socialist works council majority in Fohnsdorf had reneged on the

137. *Der Kumpel. Werkszeitschrift des Kohlenbergbaues Fohnsdorf*, published by the culture committee of the workers' works council, April 1956 to November 1960.
138. *Der Kumpel*, November 1960, p. 3.

class struggle or any form of criticism of capitalism.[139]
Polemical articles against rival works council factions and a lack
of criticism of management were also very much characteristic of
other magazines; there was hardly any difference to be seen between the publications of the Austrian Workers' and Office
Workers' Federation, the employees' representative wing in the
Austrian People's Party, and those of the Socialists.[140] Both shared
in common the characteristic of reporting much more on affairs
outside their factories or mines than inside, and thus – consciously
or sub-consciously – of distracting people from thinking about
problems in their own enterprises, or not allowing the workers to
become aware of such problems.

The works newspapers of Union Unity however, were con-
ceived of as 'fighting papers of the workers', not as 'gossip columns'.[141]
Opposing works council groups were attacked in them as well, but
mostly together with management, and in both cases – works
council and management – the complaint was that both did too
little for the workers. Unlike other magazines the internal affairs of
the respective workplaces were quite clearly the focal point of
attention. There are repeated charges of desolate living quarters and
not enough building of homes at the WTK, for example, complaints about inadequate supplies of pit-timber and works clothing,
about the use of force against workers and arbitrary dismissals.
Add to that the criticism of the same company's reduction of coal
allowances in kind, of the men being urged to work faster, leading
to greater danger of accidents, of insufficient accident prevention
measures in general, and of the unsatisfactory share in the fruits of
their increased productivity allotted to the workers in the form of
higher wages, and one has a good picture of the reporting done in
the Communist works press. The Grünbach miners were already
demanding the reduction of weekly working hours to forty at a

---

139. *Der Kumpel*, October 1958.
140. Cf., for example, the *Mitteilungen des ÖAAB für die Bergleute des Lavanttaler
Kohlenbergbaues 1961*, the *Mitteilungen der Sozialistischen Betriebsräte der Lavanttaler
Kohlenbergbau GmbH St. Stefan 1960*, the *WTK-Reporter 1965*, published by the
ÖAAB, or the *Metall- und Bergarbeiter* as the press organ of the Austrian Metal- and
Mineworkers' Union for 1946 and later.
141. *Der Zangthaler Knappe. Nachrichtenblatt der Fraktion der Gewerkschaftlichen
Einheit*, vol. 1 (1955), October. Cf., too, the *Zentralsortierer. Organ der Gewerkschaft-
lichen Einheit bei der GKB 1957–1961*, *Der Seegrabner Bergarbeiter. Glück-Auf 1962–1964*
and *Der Bergarbeiter. Organ für die Einheit der Arbeiter und Angestellten des Wolfsegg-
Traunthaler Kohlenreviers 1951*.

time when the forty-five-hour week had not even been negotiated, together with the lowering of the pension age to sixty and wage rises reaching up to the Styrian level. They criticized the excessively softly-softly approach of the ÖGB and the Socialist Party and informed the workers about their rights.[142] Finally, in Pölfing–Bergla, South Styria, the feastday of St Barbara was used to set the achievements of the workers against those of the managers.[143] Regardless of whether or not the criticism, complaints and demands were justified, they show in any event that the Communists – for whatever reasons – were trying hard for improvements in the workers' position, ones which went beyond what the ÖGB had for the most part achieved in agreement with management.

Nevertheless, the Communists did not manage to increase their following in the plants or, in many cases, even to keep it at the previous level. This was the result not just of the opposition's propaganda but also of the social benefits maintained and expanded by management.[144] The miners were paid, over and above the wages fixed in the collective agreements, various extras and allowances, such as family allowances, merit grants, achievement bonuses and long-service bonuses. When necessary, works old age grants complemented the pensions paid out by the miners' pension fund. The preventive health care also provided by the miners' insurance fund was, as it had been before, complemented by the activity of doctors and health care institutions paid for by the works.[145] The policy line on council housing started by the Nazis, which consisted of building loosely connected estates instead of flats squashed together on small building plots, was continued. The offer or the opening up of suitable building land and the payment of building grants and loans all helped to promote the foundation of charity-status estate co-operatives and to support private self-help home-building initiatives.

By 1954, the old and the new coal districts – temporarily extended after the war as a result of increased demand and, apart from

142. *Der Wegweiser. Zeitung der Gewerkschaftlichen Einheit Grünbach 1956.*
143. *Das Grubenlicht. Organ der Gewerkschaftlichen Einheit in Pölfing-Bergla*, vol. 1 (1954), no. 1.
144. On the following cf. Fachverband der Bergwerke und Eisenerzeugenden Industrie (ed.), *Kohle. Unseres Landes Hauptquelle für Wärme und Kraft*, Vienna 1963, pp. 107–10; *Rot-Weiß-Rote Kohle*, pp. 100ff.; *Der österreichische Bergbau 1945–1955*, Vienna 1955, pp. 152f.
145. Cf. Gernot Weber, *Die Entwicklung des Alpine-Konzerns seit 1945*, doctoral thesis in economics, Graz 1969, pp. 186ff.

the Salzburg district in Trimmelkam, closed down bit by bit from the 1960s on – saw the construction of not fewer than about two thousand flats and a number of one- and two-unit dwelling houses.[146]

Efforts in the field of family welfare and the extension of the leisure and education programmes were continued. The enterprises provided kindergartens, old people's homes, playgrounds, libraries, social counselling services, sports grounds and swimming baths for such purposes. Even the traditional social democratic May Day celebrations were – and this was symptomatic of the farewell to class struggle – given up in 1945 in favour of sports events; at the same time, the feast of St Barbara, which was also a sign of the partnership idea, was celebrated not, as before, just by the office workers but by the workers as well.[147] Special care was lavished on apprenticeships in mining, carried out initially in the apprentice schools of all districts and in training courses, and afterwards in the Mining and Ironworks School in Leoben, where one could go on after the mining apprenticeship to acquire a foreman's qualification at the firm's expense.[148]

Though all these measures were based not on social motives, but on the principle of preserving a high achievement capability,[149] they contributed none the less not only to an improvement in living standards among the miners, but also undoubtedly to a reduction in their readiness to pick a quarrel. So long as the Communists did not succeed in pointing out a concrete example of where communist programmes had been realized and had actually led to a genuine improvement of the workers' lot, they found it impossible to get the majority of the workforce on their side.

The undoubted improvement in the lot of the miners, by comparison with the past, was something they enjoyed as a result of the partnership policies of the enterprises and the union itself, and it contributed, together with the occasional pressure exerted on the Communists, to a destabilizing and weakening of the latter's position in the pits, a position that had not been strong to start with. The most famous example of repressive treatment by the coalition government and the ÖGB was their reaction to the biggest strike of

146. *Rot-Weiß-Rote Kohle*, p. 102.
147. Hinner, Lackner and Stocker, 'Bergarbeiterkultur', p. 313.
148. *Rot-Weiß-Rote Kohle*, p. 108f.
149. Ibid., p. 102.

the postwar period, in autumn 1950.[150] The smaller workplace disputes straight after the war had been disowned by the ÖGB as being 'wild-cat strikes', so that it was no wonder that the ÖGB and the government tried to put down the spontaneous general strike that broke out after the announcement of the fourth wage–price agreement, a strike mobilizing at the start about 120,000 strikers.

Instead of seeking out the real causes of dissatisfaction among the workers, or admitting any such dissatisfaction at all, they blamed disinformation and communist agitation for it all. They even went one step further and denied that the strike had any basis in the situation of the strikers themselves; they characterized it instead as a political putsch attempt of the Communists, although even then individual socialist unionists saw through the 'putsch lie' for the fiction that it was, and the strike was primarily regarded as a result of the workers' dissatisfaction with their union leadership. Once more the Communists, because they questioned what they considered to be the excessively soft policies of the ÖGB, together with its representation monopoly, were discredited with warnings about an East European-type people's democracy.

The attacks did not stop at the merely verbal, but often tried to destroy the Communist plant organization and to purge the nationalized enterprises of the opposition Communists. Soon after, at the Socialist Party conference in the same year, Karl Waldbrunner, the minister responsible for the nationalized industries, justified the purges by saying, 'The answer to this sabotage can only be the sacking of the rabble-rousers from these communist terror gangs';[151] one feels reminded of times when the Social Democratic unionists themselves were combated in the same way. One of the reasons for the drastic reduction in the influence of the Union Unity Organization after the start of the 1950s must have been, besides the social and material benefits accruing to the miners, this policy of attrition.

On the Essence of Social Partnership

The reduction of the Communist opposition meant that – as in other industries – the way was open for social partnership in Austrian coal mining, too. The monopoly position of the ÖGB in

150. On the following cf. Prader, 'Probleme', pp. 266f., 324, 334 and 338f.
151. Ibid., p.339.

the representation of employees' interests, a rare thing in Western democracies, led to the relegation of any special interests associated with the individual or politically oriented unions, whose officials were even nominated by the ÖGB, below those of the top central union leadership.[152] The ÖGB was one of the four pillars of what was called the Joint Parity Commission. It was set up as a parity-based decision-taking body formed by the employer and employee sides to deal with wage and price questions. The Joint Parity Commission did not have a foundation in any sort of constitutional clause or special law, but in a declaration of intent formulated by the then heads of the Union Federation and the Federal Economy Chamber (BWK). The ÖGB and the Workers' Chambers represented the employees' side in the commission, the Federal Economy Chamber and the Agricultural Chambers, as the most important organizations, the employers' side. Any conflicts or disputes, for example during collective agreement negotiations, which were only allowed after permission was given by the Joint Parity Commission, were subject to peace-making attempts of an informal sort carried out only by the chairmen of the four organizations.[153]

Their efforts at consensus, as a matter of principle, and at compromise, as a matter of necessity, for the purpose of keeping the peace were complemented and supported at the party political level by the two major parties, the Austrian People's Party (ÖVP) and the Socialist Party of Austria (SPÖ). In general they were always ready for some form of consensus over and beyond the period of the Grand Coalition. Personal contacts and connections among the top levels of the organizations and the parties also played a major part in this nexus.[154] Such a climate oriented to peace and no strikes in the long term[155] certainly favoured the SPÖ's efforts to join together employee and employer interests in the nationalized coal mining industry.

How is the social partnership policy to be judged in relation to its development in coal mining? Was the social partnership policy – as one would expect from the meaning of the term – a relationship that bestowed anything like the same amount of advantages and

152. Bergmann, *Gewerkschaften*, p. 92; Pelinka, *Modellfall*, pp. 50f.
153. Ibid., pp. 7f.
154. Ibid., pp. 16ff.
155. Gerhard Botz, 'Streik in Österreich 1918 bis 1975. Probleme und Ergebnisse einer quantitativen Analyse', in *Bewegung und Klasse*, pp. 807–32.

disadvantages on both sides or was it rather a one-sided affair, whereby one side gave up more, the other less? The answer to these questions depends primarily on what aims the opposing classes, who were at the start enemies, were trying to achieve, and which ones were actually achieved. The aims of the bosses were, to put them somewhat simply, essentially the exclusive right to decision-taking, the cheapest possible production methods and undisturbed production runs; those of the employees were a say in the running of the plant, higher wages and salaries, shorter hours and, at the beginning, the nationalization of the means of production. What was the net result after about a century? The question of who takes the decisions in the non-state firms of the Austrian economy was, despite all the co-determination clauses laid out in the Works Council Act of 1974, still to be answered with 'the owner'.[156] In nationalized coal mining, the answer is not so simple. As a result of the SPÖ's participation in the government, both the employees' and the owner's representatives were close to the party leadership; and the leadership was able to restrict the powers of decision of the one or the other side, if necessary, in the interests of its own aims. Whether the interests of the managers or those of the workers were more often adopted by the party leadership will be discussed later on.

As far as the cheapest possible production methods are concerned, the higher wages and shorter working hours would seem at first to mean higher production costs, which could have had a negative effect on profits and competitiveness. The simultaneous rise in productivity, however, was quite able to compensate for production cost inflation (see table 8). Competitiveness and profit rates depended to a much greater extent on the relation between work productivity and wage costs than on the wage costs as such. Irrespectively of whether the relative wage costs had risen or not – a question to be examined on its own – it was clear that the coal mine managers did not have to suffer any income losses because of the wage rises that seriously affected their lifestyle. It would be just as difficult to defend the opinion that those Austrian coalpits that closed down had to be closed down because excessively high wage costs had reduced their competitiveness. Higher wages and shorter working hours were feasible thanks to increased work productivity, and came about without the managers having to give up any

156. Cf. Wolfgang-Ulrich Prigge, 'Unternehmer und Arbeitgeber', in *Handbuch der Arbeitsbeziehungen*, p. 48.

income increases worth talking about, and without any basic
reduction in competitiveness.

Social partnership in Austrian mining was in the end then a
one-sided, unequal affair. The employers did not have to give up
their higher income, nor did they have to tolerate a weakening of
their competitiveness which would threaten the existence of their
enterprises in order to achieve the target they had been aiming at
for decades, undisturbed production runs rid of class struggle
politics. On the contrary, the employees were, as before, denied
any say in internal firm matters in the form of co-determination;
and their higher wages and lower working hours were commensurate with their increasing productivity and no more, since their
union representatives wanted mostly to have as little as possible to
do with a redistribution policy aimed at cutting down existing
income differences.[157]

The only achievement which could have forced the employers to
tighten their belts was, therefore, the government's external say in
the firm's internal affairs, which was its privilege as owner's representative and the result of nationalization. This say, however,
could only have an effect favourable for employees and restrictive
for employers if – as was almost always the case in postwar Austria
– the SPÖ took part in government as the employees' party.
Whether in fact the employees' interests were more and those of the
employers less represented, will have to be the subject of the
following paragraphs. How far did the statement that 'SP officials
taking over the function of bosses in the nationalized industries led
to the rejection of a works representation body that put the
workers' interests before those of the firm' apply to coal mining?[158]

Since it would break the mould set for this contribution to
analyse every decision and negotiating position to do with mining
on the part of the SPÖ as government party, the question is to be
discussed by taking an early example as a model, to wit the origins
of the first collective agreement after the war, in 1948. If one
compares the drafts submitted by both sides – that of the Austrian
Mining and Ironworks Industrial Branch Association for the Mines
Federation, and that of the ÖGB's Metal- and Mineworkers

---

157. The sacrifice of a genuine redistribution policy is also described as a basic
component of the Austrian social partnership in Pelinka, *Modellfall*, pp. 56f. and
95ff., and Wolfgang C. Müller, 'Die Rolle der Parteien bei Entstehung und Entwicklung der Sozialpartnerschaft', in *Sozialpartnerschaft in der Krise*, p. 173.
158. Prader, 'Probleme', p. 338.

Union – with one another and then with the collective agreement, as finally signed and sealed, two things stand out: on the one hand, the employers' draft contained a whole list of conditions which were less favourable than those of the union's draft; and, on the other hand, the final agreement was not a compromise between the two drafts at all, but almost completely identical with the employers' draft. That shows that, at least to start with, both parties to the agreement had very different ideas, although they belonged to or were close to the same party, and that in the end it was the employers who got their way and not the employees. It would be difficult to decide whether the Socialist Party leadership was not able to act otherwise because of its bourgeois coalition partner, or whether it did not want to act otherwise.[159]

In detail one can discern differing points of view in such matters as working hours, overtime rules, the Christmas bonus, sick pay, the legal point of time for entering into a contractual relationship and the periods for notice of dismissal. While the employers held out for the forty-eight-hour week, the union went in for the forty-two-hour week, which was only introduced twenty-four years later. The union wanted overtime work to be agreed with the works council in advance, the Mines Branch Association only wanted to have the works councils notified. One side suggested two weeks' wages as the Christmas bonus, the other one week's. The union demanded sick pay after fourteen days of employment, the employers offered it after a month's. Similar differences obtained in the question of the legal point of time for entering into the contractual relationship. One side demanded that this occur after a probation period of two weeks, the other only after a period of four weeks. Finally, the employees demanded a period of notice of six weeks for employees with ten years' uninterrupted service, while the employers offered only five weeks for the same category.

These last conditions seem to be relatively unimportant for an enterprise, so that the rejection of the union demands in this respect seems all the more surprising. In any case, this example indicates that, at the start, employers' interests were dominant in Austrian coal mining as well – and that this was obviously accepted by the union. The chance that nationalization of the mines offered for a

---

159. The drafts of the union and the branch association and the final version of the collective agreement of 1948 are available for study at the offices of the branch association in the BWK.

fundamental improvement of the miners' position was probably only partially exploited. As far as management was concerned, the pro-employee voice of the government did not reach very far, relatively speaking, and this made the social partnership seem even more one-sided and unequal.

Although as a result the balance of social partnership is more favourable for the employers' side than for the employees', there is still the question to be answered as to whether a class struggle policy would have had fundamentally different results, and ones positive for the Austrian miners. Anton Pelinka's statements would seem to suggest that no significantly higher rate of redistribution towards achieving a reduction in social inequality was possible in countries with more strife-laden industrial relations.[160]

This lack of convincing and realized alternatives is primarily responsible for the final acceptance by most Austrian coal miners of the social partnership, despite its weak points and its inadequacies.

---

160. Pelinka, *Modellfall*, p. 98.

8
Industrial Relations in the Coal Industry in Twentieth-Century Europe and the USA: A Historical and Comparative Perspective[1]

Gerald D. Feldman

As one enters the last decade of the twentieth century, it is difficult to recapture the passion invested in the control of coal during the first half of this era. Where in today's advanced industrial societies would a social democrat produce an election pamphlet with the title 'The Miner's Distress and Coal Profiteering' with any expectation of gaining a significant and responsive audience? When the Austrian marxist, Otto Bauer, addressed this subject in 1911 and compared the profits of the great mining capitalists – Petschek, Weinmann, Rothschild, Gutmann – with the miseries of the 'army of poor miners' who served them, he could confidently view the coal industry, with its mounting profits and its exploitation of both workers and consumers, as the cutting edge of the inevitable crisis of capitalism and call for the expropriation of the coal barons. For the reader of today, to be sure, there is some ambiguity in this rather quaint pamphlet. On the one hand, Bauer pointed out that the workers were not alone in their resentment of the coal capitalists and that the other branches of the economy dependent on coal were heartily disgusted with the prices they were paying and quite ready to accept state control over the industry if only they could avoid setting a precedent for the entire private economy. On the other hand, in more conventional marxist fashion, Bauer concluded with a declaration that the 'hour will come when the working class will break the chains of the capitalist class state and will conquer the power to expropriate the capitalists. In this world historical hour, the capitalistic coal monopoly will collapse along

---

1. Since this essay seeks to derive some general conclusions from the preceding discussions of the various national coal industries, it presupposes that the reader is familiar with them. I will refrain from providing citations to materials in this book.

with monopoly capital in general. The expropriation of the great capitalist expropriators is the final goal of Social Democracy'.[2]

What collapsed seven years later, of course, was not the capitalist class state, but rather the multinational state, and the multinational coal industry against which Bauer's attack was directed was supplanted by the private national coal industries of the successor states. When these were finally nationalized over a quarter of a century later, it was in a 'world historical hour' that had precious little to do with what democratic socialists like Bauer had in mind. This was as true in Bauer's native Austria as in the post-1945 socialist states. Where nationalization in the latter was more or less the product of communist seizures of power, nationalization in the former, as in other capitalist societies, constituted a public solution to a national economic problem, a 'socialization of losses' rather than an 'expropriation of the expropriators'.

The environment of such nationalizations was thus a very different one from that implied in Bauer's argument that even the propertied classes would be willing to accept socialization to end their exploitation as consumers. The idea that the coal industry was 'ripe' for socialization, which played such an important role in the nationalization movements at the end of the First World War in Germany and England, rested on the assumptions that the coal industry was a commanding height of the economy and one that was at once vital and profitable. The industry's history during the following decades was to call these assumptions into serious question, and the efforts to organize and reorganize the coal industries of the leading industrial nations, whether by private means or by nationalization, have in most cases arisen not because the industry was economically 'ripe', but rather because it was economically backward or even obsolete, and uniquely troubled in its industrial relations to boot.

Bauer, and those who shared his beliefs, had imagined a history very different from the one contained in this volume and in contemporary studies dealing with the coal industry. They envisaged socialized coal industries that would retain their positions as the key suppliers of energy and that would fulfil this function at cheap prices while providing their labour forces with the safe mines, healthy working conditions, secure employment and the decent

---

2. Otto Bauer, *Bergmannsnot und Kohlenwucher*, Vienna 1911, reprinted in Otto Bauer, *Werkausgabe*, Vienna 1975, vol. 1, pp. 799–813, quote on p. 813.

wages of which profit-hungry capitalists had allegedly deprived them. They could not imagine a situation in which French miners in the Aubin Basin would feel compelled to occupy their mines and go out on strike in 1961–2 to protest against a decision to close down mines in a nationalized mining industry and that the minister in charge would legitimize his decision by arguing that '[I]t would be neither reasonable nor humane to make men continue to carry out, at several hundred meters underground, the rough and often dangerous work of the mine to produce a fuel which is practically useless and can, in any case, be easily replaced by others which cost French miners less effort'.[3] As the historian Donald Reid has suggested, there was more than a little irony in this mode of argumentation, especially when similar language was used to justify the closing of fully modernized mines in the region: 'In its efforts to develop public support for the mine closings, the government lost no opportunity to evoke now anachronistic *Germinal*-like conditions that had once won the miners the special solicitude of the state'.[4]

These initial reflections suggest not simply the difficulty but the genuine impossibility of attempting to treat the history of twentieth-century coal mining and its industrial relations, whether nationally or comparatively, as a continuous whole. While one may speak of a general decline in its relative economic importance since 1913, its central significance to national economies has been sporadically resurrected by two world wars and periodic peacetime energy crises. Although it is possible to detect a common trend in the direction of public control and even ownership in most of the countries discussed here, the forms that these have taken have been quite diverse, and the contemporary trend toward privatization, especially in Great Britain, may introduce a new element of discontinuity. Similarly, the power and influence of labour and the character of industrial relations have varied enormously over time and from country to country, and it is difficult to make meaningful generalizations about the patterns of management–labour relations in the coal industry in nations with such varying traditions as the six countries under consideration here.

3. Quotation and discussion from Donald Reid, *The Miners of Decazeville. A Genealogy of Deindustrialization*, Cambridge, Mass./London 1985, p. 200.
4. Ibid., pp. 199f., 314. Emil Zola's novel *Germinal* appeared in 1885. It depicted the condition of the miners and the events surrounding the Anzin strike of the previous year and strongly influenced public opinion.

Nevertheless, despite the diversity of the various coal mining industries and their patterns of industrial relations, there are certain features of the industry which have made it 'special' everywhere. At the simplest level, there is the work of coal mining itself – its unique location underground, the singular physical stress of mining which was at best alleviated but by no means eliminated by mechanization, the distinctive dangers and diseases to which miners are subjected, and the uncommon dilemmas of sanitation and cleanliness miners confront. Labour in the mines has largely been characterized by a peculiar mixture of individualism, small group interdependence, and collective solidarity which has helped to reinforce the tendencies toward a special miner subculture within the broader working-class culture of modern industrial nations. Here, too, the progress of mechanization and modernization have relativized these special characteristics, but the 'bizarre mixture of autonomy and discipline, individualism and group cohesiveness, rare skills and brute strength, hierarchy and equality'[5] have indelibly stamped industrial relations in mining with an exceptional quality. Difficulties of supervision at the workplace have promoted special tensions between miners and management; quality control and modes of compensation have provided the stuff of constant controversy; payment for time spent in travel in and out of the mine, in preparatory work, and in washing up have made the calculation of work time an especially thorny issue in many instances. Thus, only a utopian would fail to recognize that the fact that industrial conflict was endemic to mining was a natural and necessary consequence of the work itself.

A second 'special' characteristic of mining was the degree and extent of state intervention in the affairs of the industry. Naturally, such intervention had deep historical roots because of the strong concern of rulers and the state with the ownership and exploitation of mineral rights. State supervision and control, particularly well developed in Central Europe in the so-called *Direktionsprinzip*, is to be found in various forms throughout Europe, including the United Kingdom. Although the scope and frequency of intervention in the twentieth century varied from nation to nation, at least three types of considerations have triggered state action. The first of these, already evident in the nineteenth century, related to the problems already discussed of the nature of mining work and the tense industrial

---

5. Barry Supple, *The History of the British Coal Industry*. Vol. 4. *1913–1945. The Political Economy of Decline*, Oxford 1987, p. 432.

relations in the industry. Thus, mining was one of the earliest industries in which governmental authorities, whether at the national or state level, sought to impose safety and health standards, regulate the age of workers, limit the employment of women, control the hours of work and intervene in major labour disputes.

The other two impulses to state intervention, the contrary and often alternating conditions of shortage and glut at levels dangerous to national security and economic stability, are unique to this century. Wartime and postwar shortages have required special measures to maintain the labour force and promote production, and also have frequently necessitated controls over allocation and distribution. Of far greater long-term significance, however, has been the dethronement of coal as the chief source of energy for industrial societies, a situation already evident in the interwar period and, as noted earlier, a major impulse to nationalization and mixed economic enterprise after the Second World War. The glut of coal and the problems of producing it profitably have turned the industry into an ideal-typical illustration of the dilemmas of effective market control, the problems of rationalization and technological unemployment, the advantages and disadvantages of subsidization, and the conundrum of how to deal with older industries while implementing a modernizing industrial policy. One of the distinguishing characteristics of industrial relations in the twentieth-century coal industry is that they are not simply about wages, working conditions, work rules and the other 'traditional' issues concerning which labour and management regularly negotiate. Industrial relations in this industry have also had to focus on the fundamental problems of the economic fate and organization of the industry itself, and these problems are issues of public as well as private policy. In so far as the public policy issue cannot be regulated within the private sector, be it through the extraordinary influence of a trade union leader like John L. Lewis, or the adjudication of a banker like Hermann Abs and the bankers and industrialists with whom he was allied in creating the Ruhrkohle AG in 1967/8, the task will inevitably fall to the state. In any case, some variety of 'tripartism' has characterized the history of twentieth-century coal mining in every major coal-producing country.

All this does not explain very much, however, and the historian's task is to make the varieties of developments and strategies employed to deal with coal intelligible. One of the more frustrating aspects of such a task is that, with a few notable exceptions, so little work has been done on the history of mining firms and on owners and managers. The truth is that we know a great deal more about the social history of miners and their communities, trade unions, strikes and government policies than we do about owners and managers and their decision-making processes. Oddly enough, this seems especially to be the case with the greatest and most individualistic of all the capitalist countries, the United States. There, wide dispersion of ownership, and the advantages of 'competitive equality', produced especially chaotic circumstances which were overcome, in so far as they could be overcome, by the towering figure of John L. Lewis, a trade-unionist who taught the industry the virtues of organization and mechanization. His great challengers, in so far as they were not the Presidents of the United States, were the great leaders of US Steel, Ben Fairless, George Love and Harry Moses, who certainly were less anonymous than the mass of coal mine operators for whom he apparently had so much contempt. Is one then to compare the role of Lewis with that of the union-hating reactionary founder of the Rhenish-Westphalian Coal Syndicate, Emil Kirdorf? Functionally, the fact that they played vaguely analogous roles in promoting the stabilization of the industry is not uninteresting, but to raise the question of why the strongest impetus to market organization came from a trade union leader in America whereas it came from major coal industrialists in Germany would be to oversimplify much more complicated histories. German trade-unionists were champions of rationalization and concentration in the Weimar and Bonn Republics and, in both Germany and the United States, businessmen outside the coal industry have played no small role in promoting these processes. How does one find a coherent way of explaining different patterns of discontinuities in what is the same industry over space and time? Indeed, it is possible to argue that the most interesting outcome of a comparative exercise in this case should be the explanation of differences rather than the explication of superficial similarities. As Jonathan Zeitlin has argued, 'international differences in industrial relations can best be explained not by variations in social and economic structure, but rather by historical divergences in institutional development resulting from the organization and strategies of trade unions, employers and the state'.[6]

6. Jonathan Zeitlin, 'From Labour History to the History of Industrial Relations', *Economic History Review*, 2nd series, 40, no. 2 (May 1987), pp. 159–84, quote on p. 178.

Having thus begun with the identification of certain common denominators that define the terms on which a comparative perspective is possible, the object of the discussion which follows will be to identify patterns of difference among the national cases discussed here and to present them as a spectrum of historically determined solutions to a set of common but not entirely similar problems and issues.

Stages in the Evolution of Industrial Relations: Germany and England

The two leading candidates for any effort to compare industrial relations in the mining industry must inevitably be Great Britain and Germany. While all nations are 'exceptional', there are stronger arguments for American 'exceptionalism' than for that of most other nations, as should be evident from David Brody's contribution to this volume. It is most practical to begin with a comparison of the two leading European producers. Furthermore, there is, of course, a long tradition of comparing every significant aspect of the divergent developments of Germany and England for purposes of illuminating the German 'special path' (*Sonderweg*), although the validity of this approach has been called into question during the past decade. In the case of industrial relations in coal mining, the comparison was made very early in the 1960s by Gaston Rimlinger in terms that served to support the notion of deficient German development. Rimlinger did not, to be sure, romanticize the British mining experience, which certainly would have been a very dubious enterprise. Rather, he stressed the purported advantages of British development, especially the capacity of the workers to appropriate the techniques employed by the middle class in its struggle against the aristocracy. Only in the 1880s did the miners break with their middle-class models, identify with a working-class ideology, and develop goals at once influenced by Methodist traditions and ethics and an undoctrinaire British socialism:

> The fact that workingmen could develop a protest movement in an opportunistic fashion, and largely with the ideological arsenal of their opponents, is the outstanding feature of the British case. It should be noted also that the ideological break with the employers grew not while the workers were weak and disorganized but while they were strong and

368 Gerald D. Feldman

368 Gerald D. Feldman

368 Gerald D. Feldman

united. It was a sign of strength rather than of revolutionary alienation. British society was able to allow industrial protest to develop without sowing the seeds of social revolution. The workers were able to organize a protest movement to which the employers and the state could adjust without causing irreconcilable rifts in the body politic.[7]

For Rimlinger, the very fact that the German miners had enjoyed a much more honorific and privileged position that was protected by the state and enjoyed special benefits and status proved to be a disadvantage once the industry was largely privatized in the 1860s. Habituated to dependence on the state and its paternalism, the German miners lacked the incentive to develop habits of self-help enjoyed by their more 'savage' brethren across the Channel:

> In the German social and political climate the quoting of the Bible or opportunistic arguments about the rights of the individual, ideas which the British miners used so effectively, would not have been much of an ideological weapon, especially not in the service of a worker who may have been hard put to it to convince even his fellow workers. Workers with a legacy of dependence on their social superiors tend to be reluctant to shift their loyalty to a member of their own group. This shift can occur only once they have accepted the fact that they must look out for themselves. As a result, the start of the German miner's protest movement depended on the assistance of individuals of higher social status who could fashion a compelling protest ideology.[8]

For Rimlinger, this explained both the influence of the Socialist and Christian trade unions and the fact that labour protest in Germany, in contrast to England, proved threatening to the political system itself. The decisive element was the quality of the opposition the workers met in the course of their self-assertion. Where, as in England, protest against authority had been legitimized by the middle class itself, labour could make its claims by appealing to the very standards earlier used by the middle-class employers in their struggle with the state. Where, as in Germany, protest and challenges to authority had not been legitimized through historical struggle and experience, labour's challenge faced more formidable

---

7. Gaston V. Rimlinger, 'The Legitimation of Protest: A Comparative Study in Labor History', *Comparative Studies in Society and History*, 2nd Series, April 1960, pp. 329–43, quote on p. 337.
8. Ibid., p. 340.

obstacles and was necessarily more radical in character. Reconciliation with labour, Rimlinger concluded,

> is much more difficult in a rigidly stratified society where the employers are allied with a centralized ruling bureaucracy, as was the case in Germany, than in a more fluid society where the employers at times seek the support of the workers to challenge the power of an entrenched aristocracy, as was the case in Britain. Therefore, in the crucial stages of early industrialization it may not be possible to direct protest into those channels which would be most desirable in the long run.[9]

Since these lines were penned, the question of what is 'desirable', be it in the short run, be it in the long run, have become much less clear and obvious. Writing in the heyday of what might today be called classic modernization theory, Rimlinger's intrinsically unobjectionable contention that the 'successful adaptation' of a collective labour protest movement 'to the industrial society depends on radical changes in outlook and attitude among workers as well as among employers and government officials'[10] presumed transformations at once more unilinear and definitive than had ever actually taken place. The outcome, in any case, was a British model or an Anglo-American model of industrial relations involving 'some balance between industrial democracy and industrial conflict',[11] and a reasonably well-functioning collective bargaining system operating in the context of a stable system of representative political democracy. This was still a time when 'corporatism' was associated with fascism and when no respectable political scientist or historian would have ventured to speak of a 'German model' of anything but organized terror.

Scholarly views have manifestly changed and, of course, continue to change. 'Corporatism', broadly defined as the institutionalized collaboration and compromise among the organized forces of industry and/or management, labour and government to deal with economic and structural as well as social issues in an industry and often in the economy as a whole, has come to be viewed by many analysts as the most satisfactory means of organizing interests and managing the problems of modern industrial societies in a democratic rather than in an authoritarian context. Not surpris-

9. Ibid., p. 343.
10. Ibid., p. 342.
11. Ibid., note 67.

ingly, the more organized an economy and society, the more easily
corporatist solutions can be found, and Germany, despite her
unenviable political history, has become *the* model of 'organized
capitalism' for those interested in the comparative history of industrial systems thanks to the high degree of her industrial organization in trade and employer associations, cartels and large
enterprises and, if one conveniently overlooks the Third Reich, her
strong, relatively centralized trade union movement. In this context, the shaky 'tripartism' described by Roy Church would seem
to confirm that Britain is what two recent analysts have called
'capitalism's weakest link'.[12]

The journey from the British to the German 'models' or even
ideal types of industrial relations behaviour has, of course, much to
do with contemporary perceptions of the central issues in industrial
relations. Where democratization and problems of co-determination and participation assumed centre-stage in the 1960s and
early 1970s, increasing concerns about economic growth and industrial policy have provided industrial relations discussions with a
different context since that time. In so far as historians have not
simply abandoned industrial relations issues altogether in favour of
Alltagsgeschichte and 'rank and file' romanticism, an especially strong
tendency in the environment of the 'new cultural history' and the
powerful influence of E. P. Thompson,[13] they have had to pay
more attention to the relative success of various societies in managing economic and structural problems, and this has relativized the
once so stark contrasts between Britain and Germany.

It would most assuredly be a great mistake to see the German
development in an excessively rosy glow, as Bernd Weisbrod's
essay should make amply clear, and it is also important to recognize
that the condition of British coal since the end of the First World
War is not simply a product of the 'British disease'. Germans
deserve no special credit for having the chief coalfields concentrated
in the Ruhr and for being able to concentrate their production very
heavily in a few limited sorts of coal, the most important and useful
of which was the coal employed in coking. The locational and

12. Scott Lash and John Urry, *The End of Organized Capitalism*, Cambridge 1987,
especially pp. 17ff., 269ff.
13. For Thompson's views, see E.P. Thompson, *The Poverty of Theory and Other
Essays*, London 1978; on the 'rank and filists', see Zeitlin, 'From Labour History',
pp. 165–7; on the 'new cultural history', see Lynn Hunt, *The New Cultural History*,
Berkeley 1989.

transport advantages of the Ruhr valley are matters of good fortune, not achievement. Conversely, the scattered nature of the British fields and the greater variety of British coal types were also determined by geology rather than human choice, and there were thus natural barriers to the kind of concentration of mining and heavy industry that took place in the Ruhr, Saar and Upper Silesia. Voluntary cartelization and vertical concentration made more sense in Germany than in British, American and French heavy industry. Thus, there was a 'natural' foundation for Germany's 'organized capitalism' that has to be taken into account before one begins exploring the extent to which the historical traditions of the country have genuine explanatory power. To credit German coal industrialists with making the most of the opportunities offered them by nature and circumstance is surely important since there is nothing written in the stars that says that men must act wisely. At the same time, it is necessary to recognize that German methods did not offer the same advantages to British industrialists.

Probably the only sensible way out of the trap of declaring now one, now another country to be a 'model' as defined by current interests and concerns is to recognize that the processes of modernization have involved very different and varied kinds and rates of 'learning' on the part of the social actors – management, labour and the 'state' – and that the 'lessons' learned are always partial in character and more useful in certain circumstances than in others. From this perspective, Rimlinger's early comparative analysis continues to have much to offer, especially when, as has been the case in other important studies, the focus shifts somewhat from labour protest to owner and manager behaviour and to the interaction between entrepreneurial and worker responses to advancing industrialization.

The problems of socio-politically conditioned entrepreneurial learning behaviour are central themes of the pioneering studies of prewar industrial relations in Upper Silesia and the Ruhr by, respectively, Lawrence Schofer and Elaine Glovka Spencer. The former, while cautioning against 'expecting too much from even the most advanced entrepreneurs', has insisted that their adaptation needs to be measured 'against their own goals' and that, when it came to accepting and adjusting to bureaucratization of management and working circumstances, the acceptance of work rules, clear lines of authority and more impersonal relations between workers and management, the workers 'were certainly more "modern" than their employers'[14] in defining their aims in terms of bread and butter issues rather than protesting changes imposed by advancing technology and concentration. Schofer, like Spencer, suggests that the resistance to the granting of citizenship rights and social rights to the German workers drove labour in a more revolutionary and hostile direction than in England and the United States. Indeed, in comparing German heavy industrialists with their counterparts in England and the United States, Spencer is quite explicit in her rejection of the notion that the German coal mine owners and managers were more than very selective modernizers or had much to offer by way of solution to their industry's industrial relations problems. While they certainly were most self-congratulatory about the superiority of their paternalistic policies and welfare programmes, these were clearly reaching the limits of their usefulness without providing either the labour peace or stability for which they were intended. Their intransigent opposition to collective bargaining was justified either by arguments that trade unions in England were sufficiently responsible so that one could negotiate with them while their German counterparts were allegedly such menaces to state and society that one could not deal with them, or by claims that the labour unrest in Britain, especially the strike of 1912, demonstrated the futility of collective bargaining. Spencer thus argues that the Ruhr employers 'were most willing to learn from foreign experience as it related to technology and to the organization and direction of production. They found foreign experience least relevant in addressing societal problems that transcended the corporation'.[15]

One of the most important advances of recent research, however, is to make clear that although the coal mining industry was especially notable for its repressive policies toward organized labour, its use of organized employer labour exchanges and black lists and its adamant refusal to negotiate, there were, on the one hand, cracks in the employer position and, on the other, increasing pressure from the state and from public opinion that had put the employers on the defensive. Coal mine employers had 'danced out of line' and made concessions to the workers in the strike of 1889,

14. Lawrence Schofer, *The Formation of a Modern Labor Force. Upper Silesia, 1865–1914*, Berkeley 1975, p. 163.
15. Elaine Glovka Spencer, *Management and Labor in Imperial Germany. Ruhr Industrialists as Employers, 1896–1914*, New Brunswick 1974, pp. 141–6, quote on p. 146.

and, probably more significantly in the long run, the increasing depersonalization of the mining enterprises and the creation of mixed concerns shifted the locus of conflict to supervisory personnel and spawned a new breed of managers who were less fanatical in their opposition to negotiation with trade unions. Some found the worker committees established after the 1905 strike useful, and there were sporadic efforts to woo the Christian Unions. The notable establishment of 'yellow' company unions after 1912 could be read not only as hostility towards the striking trade unions but also as recognition that some sort of worker organization was unavoidable. Prior to 1914, however, the solidarity of the mineowners was firm and they easily earned their reputation as the chief exponents of the *Herr-im-Haus* attitude. This stiff-necked attitude did expose them to increasing criticism and made them feel as if they were on the defensive, both politically and socially.[16]

This, in turn, relates to another potentially mitigating feature of the otherwise very bleak picture that must be painted of industrial relations in the pre-1914 mining industry. The 'alliance' between employers and government was not as clear-cut in Germany as Rimlinger's old comparison has suggested, and the situation cannot simply be judged from the perspective of government collaboration with industry in putting down the 1912 strike, a strike whose effectiveness certainly was as much placed in question by the divisions between the Socialist and Christian trade unions, or by the conservative socio-political turn of 1913–14, which was too short-lived to form the basis of any long-term prediction. The strike of 1905, which had led to government intervention in favour of the miners, an intervention compounded by the government's nearly contemporaneous effort to influence coal syndicate prices by trying to acquire a controlling interest in the Hibernia mining company, had placed the mine-owners on the defensive, especially since important segments of bourgeois opinion were also highly critical of the mine-owners. If the latter could continue to count on the governmental authorities prior to 1914, the experience of 1905 and subsequent differences of opinion as well as strong public criticism left the mine-owners feeling as if they were in a constant battle to prevent state interference and involvement in their affairs

16. This more differentiated approach to Ruhr coal mine owner attitudes is well summarized in Elaine Glovka Spencer's 'Employer Response to Unionism: Ruhr Coal Industrialists before 1914', *Journal of Modern History* 48, September 1976, pp. 397–412.

except when it suited them.

When all is said and done, however, the historical balance in terms of the legacy of industrial relations in German coal mining prior to the great transformations wrought by the war is negative. The coal mine owners had demonstrated their greatest imagination in the industrial relations field, in so far as this was not exercised in repressive practices, by seeking to evade recognition of trade unions and collective bargaining through more advanced paternalistic practices in housing and social policy and by creating company unions. That is, they became well practised in trying to find alternatives to collective bargaining before having any experience with that system, and this opportunism combined with a historically conditioned predisposition towards undemocratic alternatives arising from the Second Empire's political milieu, did not die with their very belated acceptance of collective bargaining in the autumn of 1918. Not accidentally, they blamed both the losing of the war and the breakdown of their efforts to stop the spread of collective bargaining during the war on state concessions to labour and weakness, and these attitudes were carried into the new conditions of the Weimar Republic and the various crises of the German coal industry of the interwar period.

The British point of departure clearly was different. Recognition of trade unions and collective bargaining were not contested issues in British coal mining before the First World War. Increased organization of both sides was marked by the tendency to establish boards of conciliation and arbitration, that is to rationalize industrial relations. This is in sharp contrast to Germany, where the organization of industry and labour certainly involved some measure of rationalization of industrial relations and was a necessary condition for collective bargaining in the future, but where this organizing process nevertheless took place in the context of fundamental and undisguised antagonism. Similarly, the fact that state mediation in the great labour disputes of 1893 and 1912 in Britain functioned in the context of an established system of collective bargaining also made a world of difference in terms of the kinds of issues which the state addressed in its intervention. Where the Prussian government in 1905 had to impose a system of worker representation in the mining industry that gingerly side-stepped the issue of the role of trade unions, the British government could move forward with the creation of a National Board of Conciliation in the wake of the 1893 conflict. Similarly, where the German coal strike of 1912 ended in a total defeat for the striking unions thanks in no small measure to government support for the employers, the British strike of that year terminated in minimum wage legislation imposed by the government.

Thus, while both governments felt compelled to adopt a strategy of intervention in major coal disputes prior to 1914 and recognize that a complete *laissez-faire* posture toward coal mining was not consonant with public order and security, the political systems and environments of the two countries produced very different outcomes. The political and social systems in Germany were linked in such a way that the government could not act in favour of the unions without endangering the political system itself. The grotesque possibilities of this situation were amply demonstrated during the war when the military and civilian authorities felt compelled to deal with the mine unions and constantly mediated price–wage trade-offs with strong inflationary consequences. Yet, the authorities were forced to act as if these conditions had no implications for the future structure of industrial relations in the industry. The coal mine owners, despite some exploratory secret negotiations with trade union leaders, refused to engage in formal collective bargaining. The British government, in contrast, did not have to suffer from inhibitions about intervening to influence both the mechanisms and the substance of collective bargaining in the coal industry, and could continue to take decisive action to maintain labour peace in coal mining during the war by forcing the coal mine owners to make concessions and even taking control of the industry.

These divergent experiences reflected the fact that coal was a special case of more general patterns of wartime development in the two countries. Paradoxically, the constitutional regime of the Kaiserreich, trapped between parliamentary and worker pressure, on the one hand, and the authoritarian and militaristic programmes of the Supreme Command and the political Right, on the other, destroyed its own credibility with both labour and industry and finally collapsed under the weight of its effort to achieve total victory against impossible odds while maintaining a social and political order which the war effort undermined. Since the army and the government could be blamed for the mismanagement of the food supply and the coal supply systems and for the wearing out of the miners and the wear and tear on the mines, both the trade unions and the employers could join together in excoriating the government while seeking to manage the crisis created by military collapse and revolution. While it certainly would be a gross exaggeration to say that the German coal mine owners and managers emerged from the war unscathed by their past records as the most notorious hard-liners (*Scharfmacher*) in industry and as profiteers in peace and war, their reputation for technical and managerial expertise and business acumen emerged not only untarnished but probably heightened by their wartime performance while their prewar political opportunism served them in particularly good stead under the new circumstances.[17] These factors, as much as the deficient political skills and fearful insecurity in economic matters of Germany's new leaders in 1918, help to explain the enormous influence exercised by Hugo Stinnes, whose imagination and dynamism went unmatched on either the Left or the Right, and his colleagues during and after the revolution.

The British coal mine owners were not blessed with an incompetent government and failed regime to veil their deficiencies and relativize the legitimacy crisis of the privatized coal industry. Whatever the virtues of collective bargaining in British coal, one of them was most decidedly not good relations between industry and labour. The stiff-necked attitude of the producers and the long record of exploitation and deprivation suffered by the miners, above all in South Wales, combined with the miners' intention to maximize their advantages at all costs, seriously called into question the viability of a private coal industry. Furthermore, the British coal mine owners did not enjoy the prestige of the German *Bergassessoren* (state-trained mining official) and were plagued by localistic habits which contrasted sharply with the organized approach and corporatist inclinations of the Germans.

A comparison between the performance of the two groups of employers in the face of the nationalization threats after the First World War is instructive. R. H. Tawney's evaluation of the British mine-owners' testimony before the Sankey Commission in 1919 as 'extraordinarily incompetent, not to say stupid',[18] neatly sums up both contemporary and subsequent judgement. The employers

---

17. For a discussion of wartime and postwar developments in Germany, see Gerald D. Feldman, *Army, Industry, and Labor in Germany, 1914–1918*, Princeton 1966, and 'Das deutsche Unternehmertum zwischen Krieg und Revolution: Die Entstehung des Stinnes-Legien-Abkommens', in *Vom Weltkrieg zur Weltwirtschaftskrise*, Göttingen 1984, pp. 100–27.

18. Quoted in Supple, *History*, p. 127.

were unable to defend the existing organization of the industry, legitimize their high wartime profits, justify the existing wages, working and social conditions of the miners, or attain the slightest moral or intellectual advantage from the hearings. In Germany, where socialization was a widespread demand and appeared virtually inevitable at an early stage of the revolution, the coal mine owners and managers along with other industrialists ended up dominating the hearings of the two socialization commissions which met between 1919 and 1921. Most remarkable was Hugo Stinnes's audacious attempt to use the socialization discussion to promote his pet schemes of vertical concentration on the basis of mixed economic enterprises dominated by private interests, a pre-emptive strike which made some of his more timid and worried colleagues very nervous. The manner in which the leading coal industrialists assumed centre-stage in discussing the organizational and technical problems of the industry, their attention to the financial structure of the industry, concern with creating worker housing, willingness to toy with schemes for worker share-holding and general capacity to overwhelm the entire discussion with a shrewd combination of technically-grounded reservations and new ideas gave them an intellectual edge over their opponents throughout the discussion. Finally, everyone settled for the creation of a Reich Coal Council and a system of collective bargaining dominated by the existing organizations of labour and industry. While government interference in the early Weimar Republic certainly was more substantial than it had been in the Kaiserreich, especially in matters of prices, the corporatist management of the industry, such as it was, reflected the domination of the owners and managers.

In the end, of course, neither the British nor the German industries were nationalized after 1919, a parallel development which should not be taken to demonstrate that the incompetence of British owners was just as well rewarded as the cleverness of the Germans. The dangers confronting the German owners were far more serious than those facing the British on two counts. First, Germany had undergone a revolution under socialist auspices, and the unrest in the Ruhr and Upper Silesia posed a constant threat to the industry and pressure for socialization. Second, coal was the centrepiece of the reparations problem, on the one hand because of the great importance of German coal to Franco–Belgian reconstruction and, on the other hand, because coal production was essential to German reconstruction and to the industrial activity required if Germany were to produce the export goods needed to generate the hard currency to pay for imports and reparations in cash. Theoretically, the key function of coal in reconstruction and reparations could serve as a compelling argument for nationalization. In practice, it had precisely the reverse effect because the industrialists were able to instrumentalize the reparations issue as an excuse to maintain the private character of the industry. Not only could they argue that the Allies could seize the German coal mines as a security for German reparations payments if they were nationalized, but they could also insist that the external and internal requirements for increased production made all 'experimentation' dangerous. These claims had resonance in socialist as well as bourgeois circles. In the last analysis, the fundamental argument for coal socialization in Germany was not economic but rather political, that is the improved chances for German democracy that would arise by breaking the power of the more authoritarian wing of German industry. But the German employers could capitalize on the failures of wartime government interference in the economy and the lacklustre performance of the new regime to argue the case for leaving the control of coal to those who knew the business and had demonstrated their capacity to manage it before 1914.

The twin claims of national emergency and entrepreneurial competence were valuable advantages for the German coal mining employers in setting some of the terms of the new system of collective bargaining as well. Efforts to introduce the six-hour shift in 1919–20 were defeated by the combined forces of the employers, the state and the leading trade unions on productivist grounds, and the entire effort by the more radical workers to achieve this much reduced shift was used to drive a wedge between the coal miners and large portions of the public, including important labour groups, who felt that the miners were taking unfair advantage of their stranglehold over the economy. Similar feelings were generated over the claim of the miners to the highest wage scale and to special food supplements during this period. Although certainly not planned, the involvement of the miners in the inflationary wage–price spiral between 1918 and the end of the hyperinflation did much to undercut their special claims for consideration. When stabilization came at the end of 1923, the questioning of the special prerogatives of the German miners with respect to hours of work and wages during the inflation could be effectively carried on by demanding longer hours at lower pay in order to make the industry competitive and pay reparations. Even if the government refused to dismantle the national system of collective bargaining in coal created after the revolution, it did use its powers of compulsory arbitration in rolling back wages and increasing hours. Basically, the stabilization settlement replaced a rather shaky and increasingly industrialist-dominated corporatism in mining with a tripartite arrangement under which the government encouraged and subsidized ruthless rationalization while permitting a measured recovery of labour in the realm of wages and working conditions. All this, of course, did not prevent the employers from likening the continued existence of the Reich Coal Council and the role of the state in pricing and wages to the system employed in the Soviet Union.[19]

Compared to Germany, England experienced much more of a 'restoration' and did so much more rapidly. Political revolution and the peace treaty made it impossible for Germans to pretend that one had returned to prewar conditions let alone do so under existing conditions, even if one could fatuously fantasize about the possibility. Folly was the luxury of the victorious. Once the initial postwar upset in England was past, it was much more feasible to entertain illusions and even to act on them, as was most dramatically demonstrated by the successful drive to restore the gold standard at prewar parities between 1918 and 1925. The policy of decontrol, deflation and the restoration of market forces swept all in its wake and was particularly fateful in coal. If the Sankey Report did not move the employers to offer a programme of structural reform for their industry, it did bestir them to organize more effectively in defence of the status quo and to succeed in using their political influence to reverse the sentiment in favour of nationalization. The polarization between trade-unionists and their supporters insisting on nationalization and those favouring restoration of the old order had the untoward effect of submerging proposals for

19. On the problems of socialization and industrial relations in the coal mining industry in the Weimar Republic, see Peter Wulf, 'Die Auseinandersetzungen um die Sozialisierung der Kohle in Deutschland 1920/1921', *Vierteljahrshefte für Zeitgeschichte* 25, 1977, pp. 46–98; Gerald D. Feldman, 'Arbeitskonflikte im Ruhrbergbau 1919–1922. Zur Politik von Zechenverband und Gewerkschaften in der Überschichtenfrage', *Vierteljahrshefte für Zeitgeschichte* 28, no. 2, April 1980, pp. 168–223. Also see Gerald D. Feldman and Irmgard Steinisch, 'Die Weimarer Republik zwischen Sozial- und Wirtschaftsstaat: Die Entscheidung gegen den Achtstundentag', *Archiv für Sozialgeschichte* 18, 1978, pp. 353–439, and 'Notwendigkeit und Grenzen sozialstaatlicher Intervention: Eine vergleichende Fallstudie des Ruhreisenstreits in Deutschland und des Generalstreiks in England', *Archiv für Sozialgeschichte* 20, 1980, pp. 57–118.

reform and leaving long- and short-term legacies with unhappy consequences. On the one hand, labour felt 'cheated' by the collapse of the nationalization programme, although its own intransigence had largely been responsible for throwing out the baby of reform with the bath water of nationalization. On the other hand, labour systematically exploited its market advantages during the war and in 1918–19 and concentrated on immediate wage increases, first, to the neglect of structural reforms and, subsequently, as a kind of compensation for the abortion of reform. These policies revenged themselves with the onset of the depression of 1920–1, which also marked the unanticipated permanent structural crisis on the coal market of which Britain was to be the greatest victim. The miners' cause suffered a substantial loss of public support and the employers were able to pursue a militant policy and deal heavy blows to the miners in the strike and lockout of April 1921 and then the great conflict of 1926. If anything, the employer victory was too complete, for wages were so low and unemployment in the industry so high after 1926 that employers dared not press their advantage while public opinion turned more sympathetic and the government was careful to permit no significant actions to the disadvantage of the miners. Unhappily, this also included a failure to press for rationalization and structural reform which, without government subsidization and assistance, would have involved further unemployment in the industry. The social and economic stalemates in British mining thus reinforced one another because the social and economic costs were too high for the government to bear under existing political and ideological conditions, because of the powerlessness of labour and because owners and managers were stuck in a rut which they seemed to find altogether too comfortable.

Thus, by the mid-1920s, industrial relations could not be called satisfactory in either British or German coal but, if one had to choose, the Germans had not only caught up in collective bargaining but had also forged ahead in establishing something resembling a 'system' in the form of the Reich Coal Association and the Reich Coal Council involving the participation of industry, labour and the state. Even if one had few illusions about the role of labour or the state in actually regulating the industry, there did seem to be genuine self-regulation by industry, and the reform-minded British industrialist, Sir Alfred Mond, could not help contrasting Britain's lack of system with that of Germany, where 'coal is not sold by a disorderly mob of people throwing their coal into the market without any relation to consumption or the demands of the market'.[20] Even more importantly, comparisons of productivity, rationalization and mechanization all came down in favour of the Germans, who had reconstructed their industry along the most modern lines. Productivity in the Ruhr increased 23 per cent between 1913 and 1939, while coal production increased 9 per cent and the number of employed decreased 11 per cent.[21] In fact, British productivity compared unfavourably with every European producer. Where output per manshift in Britain increased only 1 per cent between 1913 and 1929, it increased 34 per cent in the Ruhr, 12 per cent in Polish Upper Silesia, 10 per cent in the Pas-de-Calais, 9 per cent in Belgium and 52 per cent in Holland. All this had a devastating effect on exports in an already declining market, British coal exports dropping from 73.4 million tons in 1913 to 57.8 million in 1929.[22]

Thus, British coal mine owners seemed not only anarchic, but also backward. This argument is one that has recently received strong support from the historian Michael Dintenfass, who has challenged both the labour-cost explanation of British failure to modernize, which was of course a great favourite of the coal mine owners, and the explanations in terms of the structure of the industry and industrial relations already discussed. Dintenfass points out that there were important cases of British coal mine owners who managed to engage in concentration, mechanization and quality control despite all the alleged British disadvantages while other producers simply failed to do so. Certainly such examples of British enterprise do raise serious questions about the quality of British entrepreneurship, and it is also possible to point to sharp differences between overly conservative and cautious British coal mine owners and more technically-minded engineers who served as managers. Nevertheless, Barry Supple seems more convincing in his arguments that invidious comparisons between Britain and Germany overlook the differences in geographical

20. Quoted in Supple, *History*, p. 258.
21. The numbers for 1913–37 are even more impressive, production per worker rising by 68 per cent, coal production by 11 per cent, while the number of employed in the industry declined by 33.4 per cent. See Irmgard Steinisch and Klaus Tenfelde, 'Technischer Wandel und soziale Anpassung in der Schwerindustrie während des 19. und 20. Jahrhunderts', *Archiv für Sozialgeschichte* 28, 1988, pp. 27–74 and 40f.
22. Supple, *History*, pp. 189, 192.

concentration and variety of coal between the two countries and understate the manner in which poor market conditions and prospects, in the context of Britain's peculiar institutional constraints, served as a rational reason for British coal men to be cautious: 'The most potent obstacle to enterprising reorganization, industrial concentration, and large-scale investments was the market situation of the coalmining *business*.'[23]

Indeed, by concentrating on the British failures, it is easy to forget how troubled the industry was everywhere and the extent to which reduced demand for coal domestically in every country and competition from new producers and new sources of energy depressed the industry everywhere and encouraged overcapacity in the absence of satisfactory national policies to regulate production and markets. The international record of the industry after the First World War shows each nation in turn profiteering from another's miseries. The British suffered from German reparations obligations but benefited enormously from the French occupation of the Ruhr, while everyone, but especially the Germans, derived profit from the 1926 British coal strike. Germany's famed rationalization did not stand up terribly well under close inspection. Robert Brady, in his classic study of the early 1930s, concluded that

> it seems impossible to avoid the conclusions that a great deal of what has been accomplished, including much that has been called 'rationalization,' is really not rationalization at all. Excess plant capacity prevails in coal mining, coke production, gas production, and in the output of many of the more important by-products. The syndicates have been unable either to equate supply to current needs or to bring any degree of order out of the existing marketing chaos. Duplicate wholesale and retail selling agencies, duplicate railroad switches, yards, bunkers, offices, trucks, and other marketing equipment, are the rule, not the exception, in every section of the country. Taking the industry as a whole, it is probable that this duplication would sum to a total from 50 to 100 per cent in excess of the requirements which a planned and efficient organization would have stipulated. While no worse than the case of the same industries in England and the United States, one may doubt whether the coal situation in Germany has been helped at all because of the existence of the *Reichskohlenverband*.[24]

23. Ibid., p. 408. For Dintenfass's important arguments, see 'The Interwar British Coal Industry and the Case for Entrepreneurial Failure', *Business History Review* 62, spring 1988, pp. 1–34.

24. Robert Brady, *The Rationalization Movement in German Industry. A Study in the Evolution of Economic Planning*, Berkeley 1933, pp. 100f.

Written in the depression, Brady's picture is particularly bleak, and
a more balanced appraisal, especially before 1930 or in the later
1930s when high coal production was at a premium again for
military reasons, probably would make for a more favourable
comparison between Germany and the other two chief coal producers with respect to rationalization and concentration. As the
recent very convincing analysis of Rudolf Tschirbs has shown, the
rationalization process in Germany permitted the industry to
reduce its labour costs very effectively while at the same time
enabling the miners to reach their prewar standard of living.[25] The
real problem lay in the mentality of the *Bergassessoren* who, as John
Gillingham has shrewdly noted, 'had been schooled in the virtues
of technical excellence and an authoritarian approach to management. Believing coal mining to be a quasi-sacred task, they viewed
themselves as custodians of a precious and depletable asset. Their
overriding concern was not return on investment but yield per ton
of coal output'.[26]

In any case, the more relevant question here pertains to the
relationship between the progress of rationalization in the industry
and industrial relations in Germany. Here the record is, to say the
least, 'mixed'. The Communists made much of Socialist trade
union weakness in the face of employer policies, the technological
unemployment of the period, the loss of status of the miner
through mechanization and rationalization and the purportedly
mistaken policy of dependence on the compulsory arbitration
powers of the government. None of this appears very impressive
historically. The more sober leaders of the Socialist trade unions
saw at least some of the handwriting on the wall even during the
height of the German inflationary 'boom' in 1921 and warned that
'just as our English comrades have guiltlessly and against their will
come to experience the consequences of the Treaty of Versailles,
the collapse of the German currency and reparations deliveries, so
will we in the course of the regular interactions among the individual national economies come sooner or later here to experience the
same phenomena which now make themselves felt in Belgium,
France and England'.[27] Their consolation, provided by Rudolf

25. Rudolf Tschirbs, *Tarifpolitik im Ruhrbergbau 1918–1933*, Berlin/New York
1986, p. 368.
26. John R. Gillingham, *Ruhr Coal, Hitler and Europe. Industry and Politics in the
Third Reich*, New York 1985, p. 12.
27. *Protokoll vom 1. Reichsbetriebsrätekongreß für den Bergbau, abgehalten am 6. und 7.*

Hilferding and his followers, was that, in Germany at least, these phenomena were taking place in the context of the development of an 'organized capitalism' in which the workers had an institutional voice and which might pave the way toward an 'economic democracy' that would, in its turn, show the path toward a democratic and socialist organization of the economy.

This vision helped the German trade unions to accept the processes of rationalization and accompanying deskilling and unemployment of the mid-1920s and depend on government arbitration in their conflicts with management. It did not help them to persuade the coal mine industrialists to try to attain some kind of international arrangement in coal before the crisis of overproduction became enmeshed in the world economic crisis.[28]

This faith in an 'organized capitalism' that was about to collapse and in the possibilities of economic democracy under interwar conditions was a risky business, above all because the coal mine owners and managers had by no means accepted as permanent the social constitution created at the beginning of the Weimar Republic or abandoned the opportunism that had characterized their approach to labour relations even before the war. Allegedly stalwart defenders of the 'freedom of the economy', they none the less cheerfully accepted government subsidization to keep the industry afloat and strongly resisted any and all assaults on their pricing powers and policies, even from so friendly a source as the first depression chancellor Brüning. Throughout the interwar period, they had convinced themselves that the answer to Germany's competitive problems lay almost solely in the reduction of labour and social costs, ignoring the successes of their rationalization efforts and never pondering the implications of continuous expansion of productive capacity in an international market that was palpably in crisis. At the same time, they were at the forefront of the anti-union *Werksgemeinschaft* programme and the efforts to 'educate' a new breed of worker that would conform to the will of management through the German Institute for Technical Labour Instruction (*Deutsches Institut für Technische Arbeitsschulung* – DINTA). 'Social

    28. Gerald D. Feldman, 'Wirtschaftspolitischen Vorstellungen des "Alten Verbandes" in der Weimarer Republik', in Hans Mommsen and Ulrich Borsdorf (eds), *Glück auf Kameraden! Die Bergarbeiter und ihre Organisationen in Deutschland*, Cologne 1979, pp. 301–24.

rationalization' would thus accompany economic rationalization.[29] In the German context, therefore, the issue of concentration and rationalization in mining cannot be separated from political and ideological struggles that were fundamentally systemic in character. As beaten and repressed as the English coal miners were, the limits of their repression were defined by parliamentary and democratic politics and the dilemma of what to do about mining was perceived in the context of the existing political system. In Germany, the same issues were fraught with ideological and political content, and the alternative solutions had profound implications for the political system. The struggle over 'economic democracy' could be fought in the context of the existing parliamentary system, but the triumph of the *Werksgemeinschaft* depended, even if this was not always consciously understood by the employers, on changes in the political system itself. In Great Britain, a well-entrenched system of political democracy and collective bargaining may in no way have protected labour from grim defeats and high unemployment in the interwar period, but they did define 'rules of the game' for all concerned. In Germany, democracy and collective bargaining had become linked and the one could not survive without the other.

These differences were made abundantly clear during the Great Depression and the period leading up to the Second World War. The German coal industrialists used the crisis to roll back wages. As in 1923–5, the state's binding arbitration powers were instrumentalized for this purpose, but the industrialists at the same time resisted price reductions while remaining on a confrontation course with the trade unions. Needless to say, neither this anti-labour policy nor a new wave of rationalization measures saved them from further reductions in their competitive ability, especially after the devaluation of the pound sterling in 1931, and it was not long before they were blaming the trade unions and 'socialist' governments of Weimar for having compelled industry to undertake an excessive and over-hasty rationalization by a high wage and high social cost policy. While the employers moved steadily to dismantle the collective bargaining system well even before Hitler created 'ideal' political conditions by smashing the trade unions, the trade unions began to fall back on their old demand for the socialization of the mining industry, not so much because it was 'ripe', but

29. Tschirbs, *Tarifpolitik*, pp. 339ff.

rather because the public interest demanded some protection against the consequences of false economic policies being placed solely on the backs of the miners.[30]

The advent of Hitler, of course, made all such discussions academic and, if one excludes some marginal difficulties with the NSBO (National Sozialistische Betriebszellen Organisation) and the Labour Front, 'industrial relations' after 1933 were characterized by employer domination, assisted where necessary by the Gestapo and Nazified company welfare policies. It is both significant and indicative, however, that the German coal mine owners appear to have perceived the regime as an opportunity to return to tried and true methods rather than to undertake significant structural reforms. Fundamentally the *Bergassessoren* wished to be left as Herr-im-Haus, and while they accepted the Nazi trade policy of bilateral trade agreements which saved German exports in the first years of the regime, they were neither especially significant as leaders in the Nazi economic mobilization prior to the war, nor overly interested in exploiting Nazi conquests except to relieve the pressures for higher production and more efficiency emanating from Berlin through the administration of mines in the conquered territories and the employment and exploitation of foreign and slave labour. Through co-operation with the regime and manipulation of its competing elements, they were able to hold out against threats of genuine interference and restructuring.

The private British coal industry, in contrast, did not emerge from the depression and Second World War in comparably autonomous shape. Survival in the 1920s had depended on a goodly measure of state inattention to the inefficiency and structural problems of the industry, but the depression created a new atmosphere of state involvement. Furthermore, an argument can probably be made that the relative unambiguity of the coal mine owners success against labour prior to the depression helped them to realize that the old formula of reducing wages had reached both the limits of its usefulness and credibility. Thus, even before the Coal Mines Act of 1930, employer resistance to marketing and production controls was breaking down. The development of such controls during the 1930s, often under government pressure, and the acceptance of the government policy of trade agreements and efforts to establish international quotas certainly marked a break with the past, albeit

30. Ibid., pp. 429ff.

not enough of one to prevent the employers from putting up strong resistance to pressures for amalgamation. In the end, however, it was not the glut of coal but its shortage that led to the nationalization of coal in 1947. The war not only confirmed the inability of the privately controlled mining industry to meet the needs of the nation, but also clearly demonstrated that the co-operation of the miners in dealing with the problems of the industry could only be attained through nationalization. The bitter industrial relations in the industry meant that structural reform, if it were to be attained at all, could only be achieved under state auspices. Nationalization, therefore, had little to do with ideology or even with serious concepts of worker control. One of the most striking aspects of the 1947 nationalization, in fact, was how little attention the trade unions paid to questions of worker control and how much they concentrated on bread and butter issues throughout the process.[31]

In truth, there is a good deal of continuity in the patterns of industrial relations in Great Britain and West Germany after 1945, and the manner in which historical tendencies reasserted themselves after so great an upheaval does bring the historian some undeniable gratification by attesting to the usefulness of historical analysis even if the consequences of this continuity are sometimes somewhat depressing. There have, of course, been very important changes, especially with respect to the final dethronement of coal as the major energy industry in both countries, a much greater and more systematic mechanization and rationalization of the industry, radical reductions in the number of coal miners and a displacement of the old owner and managerial elites who had dominated the industry. These processes of change have, of course, been accelerated or slowed down in response to market conditions, and they have probably been faster and more thoroughgoing in Great Britain, where the industry's crisis was deeper and more clearly recognized and where continuity of the political system permitted speedy action after the war, first to deal with the crisis of production and then to confront the continuation of decline that began in the late 1950s.

Nevertheless, the extent to which industrial relations in coal mining in Britain has remained all too familiar under nationalization is quite remarkable. The industry has continued to be notorious as a locus of labour–management confrontation and great

31. See Supple, *History*, pp. 410ff. and chapters 11ff.

strikes such as those which occurred in 1972 and 1984–5. As in
1926, these strikes have been viewed by the governments in power
as challenges to the authority of the government and constitutional
practice and have served as an excuse for the recent promotion of
restrictive legislation seeking to regulate labour unrest. Nationalization, to be sure, did make possible a rationalization and mechanization of the industry under humane conditions, primarily because
the government willingly paid the financial price of smoothing out
potential frictions between labour and management through concessions to labour, but it has increasingly abandoned this role,
permitted regional differences to intensify once again the disorder
in the industry, and seems inclined not only to let management and
labour fight it out in the old unsocial manner but to look to
privatization and American managers and models of union-
busting.

Naturally, one can adopt a variety of things from the United
States, not all of them as bad as the ones the British seem to be
embracing, and it is most significant that the process of what
Volker Berghahn has called 'Americanization' in German industry
has probably promoted a socially and economically happier read-
justment of the German coal mining industry to structural changes
than elsewhere.[32] The 'new managerial culture' that suppressed the
Bergassessoren and reorganized the industry in 1966–7 was, as Weisbrod notes, not simply a break with managerial and economic
predispositions that had ceased to be viable, but also with an
authoritarian tradition of a very special sort. At the same time, the
importance of the role played by organized labour in the present
organization of the coal industry is not to be overlooked. As in
1918–19, so in the immediate post-1945 period, the goal of socialization was never attained. One important reason was severe coal
shortages and a good deal of labour–management collaboration for
the purpose of the industry's recovery. Another reason was because
the trade-unionists and employers had a sufficient common interest
in protecting the integrity of the industry from Allied decartelization plans to accept co-determination. Lastly, there were external
pressures, this time American occupation rather than the reparations problem, and Cold War politics.

Any comparison between Germany and England, however,

---

32. Volker Berghahn, *Unternehmer und Politik in der Bundesrepublik*, Frankfurt am
Main 1985, pp. 12ff., 283ff.

must also bear in mind the very positive contribution of the trade unions and 'corporatist' traditions to the management of structural problems in the German economy. Unlike the British trade unions, the German trade union leadership paid far more attention to organizational and structural issues than did their British counterparts and was indeed way ahead of the employers in realizing the need for a consolidated mining company in the Ruhr. They never focused narrowly on bread and butter issues and, indeed, sacrificed a good deal for broader reconstruction goals. At the same time, they demanded and received a much greater voice in industrial affairs through co-determination. Their predisposition toward concerted action on the basis of co-determination to solve the problems of the industry antedated that of either the heavy industrialists or the pre-Great Coalition governments and remained far more stable. The tendency to backslide on co-determination came from industry, just as the inclination to roll back collective bargaining came from industry during Weimar. Nevertheless, the German coal mine unions have been far more successful in holding their own than they were in Weimar and than their British counterparts have been since 1945. In truth, there has been an 'astonishing continuity'[33] in German coal mine union policy since the 1940s characterized by the call for a unified corporation with co-determination that would extend to the plant level, and these goals were achieved at the end of the 1960s. Whatever the industry's difficulties, the management of the postwar problems of German coal mining has been far more collaborative and successful in the German than in the British case.

It certainly is true that one can overdo the corporatist model, just as one can overdo the tendencies toward the 'rationalization of industrial conflict' in modern industrial societies and forget the political foundations of industrial relations. Nevertheless, it is reasonable to argue that the absence of the kind of free market ideology in the German tradition that was and is so important in England and the United States, the tendency to view 'economic freedom' in terms of 'industrial self-regulation', a term which presupposes some form of regulation, the long tradition of interest group development and activity, and even the unsuccessful experiments with corporatism

33. See Martin Martiny, 'Die Durchsetzung der Mitbestimmung im deutschen Bergbau', in Mommsen and Borsdorf (eds), *Glück auf Kameraden!*, pp. 389–414, quote on p. 408.

in the Weimar period, do give German businessmen, trade-unionists and government bureaucrats a predisposition to tripartite relationships that are more corporatist than those found in Great Britain.

A Spectrum of Solutions

What makes the mining industry 'special' is that the character of work and its structural problems at one and the same time produce especially conflictual relationships between labour and industrial management and special requirements of organization and collaboration. Our comparison of the British and German cases has shown how different the handling of the industry's problems have been in the two leading European producers, in part because of different geological and economic conditions, but largely because of very different political traditions and patterns of industrial relations. It remains to be asked where the other coal producing nations discussed here fit into what has been earlier described as a 'spectrum' of industrial relations in twentieth-century coal mining.

The greatest and the smallest of the coal-producing nations discussed here, the United States and Austria, certainly belong at the far ends of the spectrum in their styles of industrial relations as well as in their productive capacities. The Austrian experience seems to be almost an extreme version of the German one in many respects, except that the ideology of 'social partnership' has been taken more literally, whether imposed from above, as it was from Dollfuss through Hitler, or more or less accepted from below, as has been the case in the nationalized coal industry. As Franz Mathis shows, 'social partnership' in the nationalized industry has been only relatively more favourable to labour than in the private controlled industry, and this suggests that what has made the difference has been the character of the political regime in the face of an astonishing continuity in the character of industrial relations themselves.

At the other end of the spectrum stands the United States. Far more even than Great Britain, the American coal mining industry is a case of individualist, disorganized capitalism in which the chief rationalizing pressures came either from allied industries, first iron and steel and then more recently the utilities industries, or from organized labour. Typically enough, even the role of organized labour was highly individualistic, as demonstrated by the key personage of John L. Lewis in David Brody's account. Nevertheless, it is easy to overdo the 'wild west' quality of the American industry. Industrial relations in American coal would seem in fact to have reflected the American 'business civilization' in which hard bargaining on the basis of mutual interest, cut-throat competition and a substantial amount of brutality all have their place.

In the United States, there appears to have been a good deal less of the kind of conscious class hostility and stubborn refusal to find workable solutions that have often characterized the British industry. The pragmatic mixture of 'adversarial and mutualistic' relations in the interstate joint conference before the First World War and the system of 'competitive equality' were a uniquely American answer to problems dealt with by collective bargaining in Britain and cartelization in Germany, but is hard to imagine either the adversaries in British coal mining or the *Bergassessoren* engaging in arrangements with so much 'give and take'. At the same time, the very pragmatism described also seems to have given American industrial relations and its peculiar brand of tripartism an extremely partial and transient character. This was reflected in the non-participation and anti-unionism of the West Virginia coal operators before the First World War, the government's alternations between almost complete *laissez-faire* and the kind of massive intervention that occurred during the First World War, the breakdown of the interstate system between the wars and the extraordinary role played by the unions, in alliance with the government, in forcing through market controls and rationalization of the industry during the New Deal. Typically, however, the Second World War and the immediate postwar period transformed the alliance between government and unions into hostility and then to a restoration of industry–labour collaboration to maintain wages and promote modernization in the postwar period. As Brody shows, however, the very progress of this modernization along with the competition from other forms of fuel and the further diversification of the employer interests involved in mining have ultimately served to undermine collective bargaining and even unionization in the industry. Where American 'business principles' have helped the Germans to overcome the power of the *Bergassessoren* in the context of a corporatist tradition involving a goodly amount of co-determination and governmental attention to problems of structural and regional change, those same principles in the United

States tend to produce union decline in a climate of public indifference and may be influencing Great Britain in the same direction.

Finally, where do the French and Belgian industrial relations
experiences in coal mining fit in? It is possible, but not very useful
or enlightening, to place them in the 'middle' of the spectrum
simply because they obviously display a variety of features that
resemble sometimes more and sometimes less vaguely those of the
countries already discussed. The French case is especially difficult
because of the profound differences among the various coal mining
regions and the fact that, in the most important of these, the
Pas-de-Calais, industrial relations before the First World War did
not even conform to what one writer has called the 'archaeo-
liberalism' of the French employers.[34] These employers did negotiate with unions, in contrast to their colleagues in other industries,
and were especially inspired to do so as a means of resisting state
intervention on behalf of the miners. While these efforts were only
partially successful and while collective bargaining in no way
eliminated severe conflicts with the workers, the coal mine owners
not only withstood the unrest immediately after the First World
War but also took advantage of the devastation wrought by the war
to rebuild along extremely rational and modern lines. In this they
certainly differed from the British and were closer to the Germans,
and it is significant that the crisis that developed in French industrial
relations during the Popular Front, where there was actually much
catching up with industrial relations long existing in the mines, was
especially bitter because it involved a crisis in authority relations
between engineers and supervisory personnel, on the one hand, and
the workers, on the other. Technocracy had become the enemy.
The problem was not solved by nationalization, which was in
important respects a response to owner and manager collaboration
with the Nazi occupation authorities; as elsewhere, it was not long
before it became clear that, as in other nationalized industries, the
miners had only changed bosses. What is most interesting in the
French case, however, is that the very technocracy embodied in the
authority and work rules of the engineers and supervisors in the
interwar period began to appear as an enemy to both workers and
their supervisors as the state moved, after the late 1950s, to shut

---

34. Patrick Fridenson, 'Herrschaft im Wirtschaftsunternehmen. Deutschland und
Frankreich 1880–1924', in Jürgen Kocka (ed.), *Bürgertum im 19. Jahrhundert. Deutschland im europäischen Vergleich*, vol. 2, Munich 1988, pp. 65–91.

down unprofitable mines and to operate on the basis of considerations no different from those of private entrepreneurs. The central issue of the industry had thus become the voice of those whose livelihood was dependent on mining in the fate of an industry in the process of decline. Industrial relations, after all, cannot be carried on in shut-down enterprises.

This is as true for the nationalized industry of France as it is for the still largely private industry of Belgium, which is in many respects very similar in its history to the German case. As in Germany, banks and iron and steel have played a very important role in the development of the mining industry and its policies; and collective bargaining was very slow to develop and, before the war, was tied up with the question of universal suffrage. That is, the issue of social rights was tied to that of full citizenship rights. If the mining interests created their employer organization later than in Germany, the model used was the German one. As in Germany, the trade union movement was divided, with Christian unions playing a very important role, and as in Germany, collective bargaining and a substantial public intervention through mediation and arbitration in a highly politicized environment were outcomes of the First World War and its political upheavals.

Nevertheless, Belgian circumstances differed in important respects from those of Germany and well illustrate the particularly important role which the European coal market and its evolution played for particular national industries. As in France, regional differences made the achievement of a national marketing policy very difficult and, as in England, government pressure was required to establish more orderly market conditions during the 1930s. The international market and the price of coal were central issues in collective bargaining arrangements during the 1920s and 1930s. Significant parallels with Germany continued after the war in the defeat of nationalization in the context of a 'battle for coal' and in an increased 'Americanization' of management, as well as in a 'concerted action' to deal with the crisis of the industry after 1958. The role of the state has unquestionably been greater in Belgium than in Germany and has become more parallel to that of France because of the Common Market arrangements and the continued decline of the industry.

It would be a mistake to conclude this rather loose comparative effort with the suggestion that the great passion which has been invested in the industrial relations of coal mining is gone forever.

Recent events in England show that this is not the case, while developments in the United States may demonstrate that even collective bargaining is not to be taken for granted. Clearly, however, coal has ceased to be a central motor of industrialization and the central focus of industrial relations it often was. Furthermore, the increasing approximation of coal mining work to factory work has also reduced some of the aura surrounding mining. Obviously, mining retains a 'special' character, but much of what has made it 'special' is now of greater historical than immediate interest. The historical significance of the mining industry, its industrial relations and its fate in the twentieth century, however, goes well beyond the peculiarities of mining and the specific patterns of industrial relations to be found in the coal-producing countries. As Barry Supple has pointed out, 'the depression of coalmining in the context of the interwar economy provided a critical bridge between an era of private enterprise and the emergence of a mixed economy. And the larger symbolic role of the industry in the twentieth century therefore lies in its stimulus to the transfer and centralization of decision-making in business and industrial policy'.[35] Historically, the tripartite relations of industry, labour and the state in coal mining have served as a bridge between the traditional issues of industrial relations, having to do with hours, wages, working conditions and safety, and the burning question of industrial restructuring which now face advanced industrial societies.

35. Supple, *History*, p. 693.

Appendix 1

Production (mi. t.) and employment (1,000s) in Western coal mining during the twentieth century

Year	Great Britain Prod.	Great Britain Employment	USA Prod.	USA Employment	France Prod.	France Employment	Belgium Prod.	Belgium Employment	Germany Prod.	Germany Employment	Austria Prod.	Austria Employment
1900	751.2	229	212.3	304.4	33.4	162	23.5	132.7	109.3	414		
1901	778.7	223	225.8	340.2	32.3	163	22.2	134.1	108.5	448		
1902	796.3	231	260.2	370.1	29.9	164	22.9	134.9	107.5	451		
1903	813.8	234	282.7	415.8	34.9	167	23.8	139.6	116.6	470		
1904	818.8	236	278.7	437.8	34.1	171	22.8	138.6	120.8	491		
1905	829.2	240	315.1	460.6	35.9	175	21.8	134.7	121.3	493		
1906	852.2	255	342.9	478.4	34.1	178	23.6	139.4	137.1	511		
1907	909.2	272	394.8	513.3	36.7	183	23.7	142.7	143.2	545		
1908	957.7	266	332.6	516.3	37.3	195	23.6	145.3	146.1	591		
1909	983.1	268	379.7	543.1	37.8	190	23.5	143.0	147.0	613		
1910	1.018.2	269	417.1	555.5	38.3	196	23.9	143.7	151.1	621		
1911	1.035.5	276	405.9	549.8	39.2	200	23.1	144.1	158.6	628		
1912	1.057.8	265	450.1	548.6	41.1	202	23.0	145.7	174.9	640		
1913	1.095.2	292	478.4	571.9	40.8	203	22.8	145.3	190.1	654	0.1	0.6
1914	1.133.0	270	422.7	583.5	27.5		16.7	92.2	161.4			
1915	953.6	257	442.6	557.5	19.5		14.2	86.1	146.9			
1916	998.1	260	502.5	561.1	21.3		16.9	88.1	159.2			
1917	1.021.3	253	551.8	603.1	28.9		14.9	75.6	167.7			
1918	1.008.9	231	579.4	615.3	26.2		13.8	75.5	160.8		0.1	1.9

Year												
1920	233	1.248.2	568.7	639.5	25.2	207	22.4	159.9	114.8		0.1	1.7
1921	166	1.144.3	415.9	663.8	28.9	219	21.8	161.1	145.8		0.1	2.1
1922	154	1.162.8	422.3	688.0	31.9	229	21.2	152.8	141.2		0.2	2.5
1923	281	1.220.4	564.6	704.8	38.5	264	22.9	160.0	71.5		0.2	2.3
1924	296	1.230.2	483.7	619.6	44.9	285	23.4	172.3	132.8	559	0.2	1.9
1925	247	1.117.8	520.1	588.5	48.0	298	23.1	160.4	145.6	557	0.1	1.8
1926	128	1.128.2	573.4	593.6	52.4	307	25.2	160.2	159.0	515	0.2	1.6
1927	255	1.005.0	517.7	593.9	52.8	313	27.6	174.5	167.2	562	0.2	1.2
1928	241	921.3	500.7	522.2	52.4	292	27.6	163.2	164.0	518	0.2	1.0
1929	262	939.4	535.0	503.0	54.9	287	26.9	151.3	177.0	517	0.2	1.1
1930	248	914.3	467.5	493.2	55.0	292	27.4	155.1	155.9	469	0.2	1.1
1931	223	851.6	382.1	450.2	51.0	253	27.0	152.1	130.0	372	0.2	1.1
1932	212	803.6	309.7	406.4	47.2	242	21.4	130.1	115.2	309	0.2	1.2
1933	211	773.6	333.6	418.7	47.9	230	25.3	134.5	120.5	310	0.2	1.3
1934	224	772.8	359.4	458.0	48.6	230	26.4	125.1	136.2	331	0.2	1.3
1935	226	754.3	372.4	462.4	47.1	219	25.5	120.2	144.7	387	0.3	1.5
1936	232	751.7	439.1	477.2	46.1	215	27.9	120.5	158.3	394	0.3	1.5
1937	244	776.1	445.5	491.9	45.3	220	29.9	124.9	184.5	450	0.2	1.4
1938	231	781.7	348.5	441.3	47.5	232	29.6	130.3	151.3	418.9	0.2	1.3
1939	235	766.3	394.9	421.8	50.2		29.8	128.7	152.9	414.4		
1940	228	749.2	460.8	439.1	40.9		25.5	117.2	149.7	418.9		
1941	210	697.6	514.1	457.0	43.8		26.7	125.4	153.8	432.1		
1942	208	709.3	582.7	462.0	43.8		25.1	121.7	153.2	454.8		
1943	202	707.8	590.2	416.0	42.4		23.7	122.3	153.4	508.1		
1944	196	710.2	619.6	393.3	26.5		13.5	97.5	130.3	499.1		
1945	186	708.9	577.6	383.1	35.0		15.8	89.0	38.9	317.4		
1946	193	696.7	533.9	396.4	49.2	330	22.9	96.4	61.8	365.2	0.1	1.4

Production (mi. t.) and employment (1,000s) in Western coal mining during the twentieth century
continued

Production (mi. t.) and employment (1,000s) in Western coal mining during the twentieth century

	Great Britain		USA		France		Belgium		Germany		Austria	
Year	Prod.	Employment	Prod.	Employment	Prod.	Employment	Prod.	Employment	Prod.	Employment	Prod.	Employment
1947	200	703.9	630.6	419.2	47.3	330	24.4	118.4	81.6	446.0	0.2	1.9
1948	211.5	716.5	599.5	441.6	45.1	291	26.7	144.2	99.5	499.7	0.2	1.6
1949	218.2	712.5	437.9	433.7	53.4	284	27.9	146.3	117.4	522.8	0.2	1.5
1950	219.6	690.8	516.3	415.6	52.5	258	27.3	135.8	125.7	536.8	0.2	1.4
1951	225.6	692.6	533.7	372.9	54.9	255	30.0	134.3	135.0	548.8	0.2	1.4
1952	228.4	709.7	466.8	335.2	57.3	248	30.4	136.1	139.4	569.6	0.2	1.5
1953	227.1	711.5	457.3	293.1	54.3	235	30.1	132.1	140.7	588.9	0.2	1.4
1954	227.2	701.8	391.7	227.4	56.3	227	29.2	120.1	144.7	588.1	0.2	1.3
1955	225.2	698.7	464.6	225.1	57.3	218	29.9	116.6	147.9	586.9	0.2	1.3
1956	225.6	697.4	500.9	228.2	57.3	215	29.5	112.9	151.4	592.9	0.2	1.4
1957	227.2	703.8	492.7	228.6	59.0	216	29.0	112.3	149.4	604.0	0.2	1.5
1958	219.3	692.7	410.4	197.4	60.0	213	27.1	104.7	148.0	599.1	0.1	1.5
1959	209.4	658.2	412.0	179.6	59.7	205	22.8	81.7	141.7	557.5	0.1	1.2
1960	196.7	602.1	415.5	169.4	58.2	191	22.5	71.5	142.3	505.0	0.1	1.0
1961	193.6	570.5	403.0	150.5	55.2	182	21.5	64.0	142.7	476.6	0.1	0.9
1962	200.6	550.9	422.1	143.8	55.2	178	21.2	59.0	141.1	446.4	0.1	0.9
1963	251.5	517.0	458.9	141.6	50.2	172	21.4	58.6	142.1	421.3	0.1	0.9
1964	195.6	491.0	487.0	128.7	56.2	167	21.3	58.5	142.2	405.7	0.1	0.8
1965	185.7	455.7	512.1	133.7	54.0	161	19.8	52.0	135.1	387.7	0.1	0.7
1966	175.8	419.4	533.9	131.8	52.9	163	17.5	43.8	126.0	353.9	0.02	0.2
1967	173.8	391.9	552.6	131.5	50.5	142	16.4	38.4	112.0	304.8	0.01	0.1
1968	163.2	336.3	545.2	127.9	45.1	126	14.8	33.1	112.0	272.2		
1969	149.8	305.1	560.5	124.5	43.5	114	13.2	35.0	111.6	257.7		
1970	144.7	287.2	602.9	140.1	40.1	104	11.4	30.0	111.3	249.7		

Year										
1971	122.3	281.5	552.2	145.7	35.7	98	11.0	27.8	110.8	247.8
1972	140.5	268.0	595.4	149.3	32.7	89	10.5	26.3	102.5	229.7
1973	108.8	252.0	591.0	157.8	28.4	80	8.8	23.0	97.3	210.3
1974	127.2	246.0	603.4	166.7	25.6	76	8.1	20.5	94.9	202.8
1975	125.8	247.1	648.4	189.8	25.6	74	7.5	20.0	92.4	204.0
1976	120.8	242.0	678.7	202.3	25.0	70	7.2	18.3	89.3	198.1
1977	120.9	240.5	691.3	221.4	24.3	65	7.1	17.2	84.5	193.1
1978	119.9	234.9	665.1	242.3	22.4	61	6.6	16.1	83.5	187.1
1979	123.3	232.5	776.3	224.2	21.0	56	6.1	15.5	85.8	182.8
1980	128.2	281.6	719.1	248.7	18.1	59.0	6.3	21.4	86.6	184.1
1981	125.3	266.5	697.6	229.3	18.7	56.5	6.4	21.3	87.9	187.3
1982	121.4	254.9	707.2	217.1	16.9	56.3	6.5	21.0	88.4	185.7
1983	116.5	233.3	662.6	175.6	17.0	53.7	6.1	20.5	81.7	181.0
1984	49.5	217.4	750.1	177.8	16.6	48.9	6.3	19.0	78.9	171.6
1985	90.8	185.6	741.1	169.1	15.1	44.3	6.2	18.6	81.8	167.0
1986	104.6	151.4	740.2		14.4	38.9	5.6	17.4	80.3	164.4
1987	101.6	130.4	763.2		13.7	33.5	4.4	10.5	75.8	159.5
1988	101.4	108.9	795.0		12.1	27.9	2.5	7.1	72.9	151.4

For the years up to 1979, the table is based on statistics that have been gathered by the authors of this volume and compiled by the editors. These figures include the national production yields and the number of employed miners. Brown coal is not included. As far as possible, the variations in the statistical database are indicated below.

The editors would like to thank the Statistik der Kohlenwirtschaft eV, which with the help of the Ruhrkohle AG, Essen, compiled the statistics for all the countries from 1980 on. All the employment figures are year-end figures (USA: the average daily manpower level of the bituminous coal mines plus the number of workers in the anthracite coal mines). The USA production figure for 1988 is an estimate. The British hard coal production figure is the total of the output of all the pits, including strip mines; however, the employment figures do not include those working in strip mines. The statistical database differs for some countries particularly for the years preceding 1980, making a comparison between the figures up to 1979 and those from 1980 on impossible. As far as possible, the sources for the years up to 1979 are indicated here: *Britain*: Roy A. Church, 'Production, Employment and Productivity in the British Coalfields, 1830–1913. Some Reinterpretations', *Business*

continued

History Review, forthcoming; Barry Supple, *The History of the British Coal Industry*, Volume IV: *The Political Economy of Decline*, Oxford 1987, table 1; W. A. Ashworth and Mark Pegg, *The History of the British Coal Industry*. Volume V: *The Nationalized Industry*, Oxford 1986, table 1. For 1946–79, output figures include strip mining production which ranges from 9 to 15 million tons. The employment statistics do not include coke workers and those not directly employed at the pits. *USA*: Morton S. Baratz, *The Union and the Coal Industry*, New Haven 1955, pp. 40f.; Charles R. Perry, *Collective Bargaining and the Decline of the United Mine Workers*, Philadelphia 1984, p. 9. *France*: *Annuaire statistique de la France* (INSEE), different issues. Employment figures indicate only the number of workers, including those of related industries (coke production etc.). *Belgium*: 'Statistique des industries extractives et métallurgiques et des appareils à vapeur en Belgique', as published yearly in *Annuaire statistique de la Belgique*: *Statistiques économiques belges 1919–1928, 1929–1940, 1941–1950, 1950–1960, 1960–1970*; special issues of the *Bulletin d'Information et de Documentation de la Banque Nationale de Belgique*. Output figures are based on information provided by the mining industries and are not homogeneous. Employment figures are yearly averages. *Germany*: All data, even before 1945, refer to the territory of the Federal Republic of Germany, including the Saar region even for those years in which it did not belong to the German Reich or to the FRG. Unfortunately, employment figures are only available starting with the year 1936. The data was provided by the Statistik der Kohlenwirtschaft eV. *Austria*: The territory under consideration is always present-day Austria. *Österreichisches Montan-Handbuch*; *Mitteilungen über den österreichischen Bergbau* (various issues); *Kohle. Unseres Landes Hauptquelle für Wärme und Kraft*, ed. Fachverband der Bergwerke und Eisen erzeugenden Industrie, Vienna 1963, pp. 64f.; *Rot-Weiß-Rote Kohle*, ed. Kohlenholding GmbH, Wien, Vienna 1956, pp. 97f.; Karl Bachinger and Herbert Matis, 'Strukturwandel und Entwicklungstendenzen der Montanwirtschaft 1918 bis 1938. Kohlenproduktion und Eisenindustrie in der ersten Republik', in Michael Mitterauer (ed.), *Österreichisches Montanwesen*, Vienna 1974, pp. 115ff., 128.

Appendix 2

The development of industrial relations in Western coal mining since the end of the nineteenth century: a synopsis

Year	Great Britain	USA	France	Belgium	Germany (FRG)	Austria
Before 1900	• 1854 Mining Association of Great Britain (MAGB) • 1872 Regulation of Coal Mines Act • 1889 Miners' Federation of Great Britain (MFGB) • 1893 'National' strike by the MFGB	• 1890 United Mineworkers of America (UMWA) • 1897 National bituminous strike • 1898 Chicago Joint Interstate Conference	• 1883 Fédération Nationale des Mineurs • 1887 Comité Central des Houillères de France (CCHF) • 1889 Strike in Pas-de-Calais • 1891 First wage agreement, Union des Houillères du Nord • 1892 Fédération Nationale des Mineurs	• 1885 Chevaliers du Travail (Charleroi) • 1889 Fédération Internationale des Mineurs (Centre) • 1890 Fédération Nationale des Mineurs (POB) • 1895 Fédération des Francs-Mineurs (chrétien) dans le Centre;	• 1865 General Mining Law • 1868 on Regional attempts at trade union organization • 1889 Major miners' strike; Verband deutscher Bergarbeiter • 1890 Anti-strike league of owners in Ruhr coal mining	• 1854 General Mining Act (to be amended 1884) • 1889 Law on sickness funds • 1890 First congress of Austrian miners • 1893 Zentralverband der Berg- und Hüttenarbeiter • 1895 Allgemeiner Bergarbeiter-Verein

continued

The development of industrial relations in Western coal mining since the end of the nineteenth century: a synopsis

Year	Great Britain	USA	France	Belgium	Germany (FRG)	Austria
	• 1896 Conciliation (Trades Disputes) Act: National Conciliation Board		• 1894 Law on sickness and retirement funds	• Comité Central Industriel (owners in mining, iron and glass) • 1898 Commission syndicale in the POB • 1899 First national wage strike	• 1892 Amendment to the General Mining Act: safety regulations etc. • 1893 Ruhr Coal Syndicate • 1895 Gewerkverein christlicher Bergarbeiter	• 1897 Zentral-verein der Bergwerks-besitzer
1901					• ZZP; Union of the Polish Miners	
1902		• National strike in anthracite coal mining	• General strike in support of the eight-hour day	• General suffrage strike		
1903			• Law on eight-hour shift			• Union der Bergarbeiter
1905				• National wage strike; • Law on safety regulation	• National strike; • General Mining Act amended;	

Year			
1906	• Workmen's Compensation Act; • Labour Party adopts nationalization of the mines as party policy	• Catastrophe of Courrières (1100 casualties), general strike in the North; • Anti-strike agreement of the CCHF	• Compulsory work councils
1907	• MFGB becomes truly national, incorporating unions in Scotland, South Wales and the North-East		
1908	• Coal Mines Regulation Act (eight-hour shift for underground workers)		• Reforms of sickness and retirement laws; • Zechenverband of the Ruhr coal owners; • Catastrophe at the Radbod

continued

The development of industrial relations in Western coal mining since the end of the nineteenth century: a synopsis

Year	Great Britain	USA	France	Belgium	Germany (FRG)	Austria
1909				• Fédération des Associations Charbonnières; • Nine-hour shift introduced by law	mine (344 casualties); • Amendment to the General Mining Act: safety inspectors	
1910		• US Bureau of Mines				
1911	• Coal Mines Act					
1912	• National stoppage; • Coal Mines (Minimum Wages) Act		• Extension of the eight-hour day	• Retirement funds; • Strike in the Borinage	• Strike in the Ruhr	
1913				• General suffrage strike; • Union of Christian trade unions		

Year					
1915	• Coal Mining Organization Committee; • South Wales miners strike				
1916	• Government control of the South Wales coal industry under the Defence of The Realm Act extended nationally 1917		• Auxiliary Service Act		
1917	• National Coal Association; • Washington Interstate Agreement; • Fuel Administration	• 1917 on Numerous wage agreements		• 1917 on Strikes	• 1917 on Strikes
1918	• Coal Mines Control Agreement (continuation) Act	• Groupement charbonnier (owners)	• Negotiations for the first time between unions and entrepreneurs; • Revolution; • Stinnes–Legien	• Revolution	

continued

The development of industrial relations in Western coal mining since the end of the nineteenth century: a synopsis

Year	Great Britain	USA	France	Belgium	Germany (FRG)	Austria
1919	• Sankey Commission	• 1 Nov., National strike prohibited; • John Lewis becomes acting president of the UMWA		• Centrale Nationale des Mineurs (POB); • wage strikes; • Commission Nationale Mixte pour l'étude de la réduction du travail dans les mines	Agreement, formal working groups on industrial relations; • Numerous strikes, trade union opposition, work councils movement; • Commission on socialization of the mining industry	• Work council and collective bargaining law
1920	• Coal Mines (Emergency) Act; • National strike; • Mining	• Standard day rate ($7.50) established by arbitration	• General strike	• First collective agreement	• Kapp Putsch; • Civil war in the Ruhr area	

1921	1922	1923	1924	1925
• Industry Act (welfare fund); Reorganization of the MAGB • State of emergency; National Wages Agreement; Coal Mines (Decontrol) Act	• National strike	• Occupation of the Ruhr; • End of working community policy; • Lockout in Ruhr coal mining	• National Wages Agreement • Jacksonville Interstate Agreement • Lewis: 'The Miners Fight for American Standards'	• Mining Industry (Welfare Fund) Act; • Government sets up Court • Strike in the Borinage
• Resumption of collective bargaining • Fédération des Charbonnages de Belgique; • Eight-hour day; • Abrogation of Article 310 of Code Pénal limiting the right to strike				

The development of industrial relations in Western coal mining since the end of the nineteenth century: a synopsis

Year	Great Britain	USA	France	Belgium	Germany (FRG)	Austria
	of Inquiry to adjudicate on wages					
1926	• General Strike and national coal strikes; • State of emergency; • Coal Mines and Mining Industry Acts; • Nottinghamshire miners form breakaway union			• Fédération des Associations Charbonnières de Belgique	• Reform of sickness and pension funds (Reichsknappschaft)	
1927		• Collapse of the in-state bargaining system	• Rationalization begins			
1928	• Industrial Transference Board; • Lord Mayors' Fund	• Watson Coal Stabilization Bill			• 'Ruhreisenstreit': lockout of steel workers	

Year				
1929				• Metall- und Bergarbeiterverband
1930	• Coal Mines Act (market regulations); • Coal Mines Reorganization Commission		• Comptoir Belge des Charbons Industrielles; • Syndicate Belge des Cokes et des Charbons à Coke	
1932			• Strikes in the Borinage, later also in the Centre and Charleroi	
1933		• New Deal: National Recovery Acts; • UMWA is reformed; • National Coal Control Association, Coal Code;		• NS seizure of power, all unions prohibited

continued

The development of industrial relations in Western coal mining since the end of the nineteenth century: a synopsis

Year	Great Britain	USA	France	Belgium	Germany (FRG)	Austria
		• Appalachian Collective Bargaining Agreement				
1934		• Seven-hour day for bituminous coal miners			• Work Order Act, Deutsche Arbeitsfront; • Reforms of the 'Knappschaft'	• Reorganization of the industry and union system
1935	• Royal Commission on the Safety of Mines, to 1938	• National Labor Relation Act; • Committee for Industrial Organization (CIO)		• Office Nationale des Charbons (cartelization)		
1936			• Popular Front; • Forty-hour week	• Strikes		
1937		• Guffey–Vinson Coal Stabilization Act		• Office Belge des Charbons		
1938	• Coal Act creates Coal Commission to		• 30 Nov., General strike			• Austria incorporated in the Reich;

Year				
1939	• nationalize coal royalties			• NS reorganization of the industry
1940	• Coal Production Council			• Goering decree to promote efficiency
1941	• Essential Work (Coal Mining Industry) Order	• North–South wage uniformity	• Union des Travailleurs Manuels et Intellectuels (UTMI)	• Reforms of the 'Knappschaft'
1942	• Greene Inquiry and pay award; • Mines Medical Service; • Nationalization of mining royalties	• National War Labor Board, Little Steel Formula;	• Comptoir Belge des Charbons (COBECHAR)	
1943	• Scheme for Joint National Negotiating Committee and a Reference Tribunal	• Confiscation of the mines; • Lewis–Ickes Agreement		

continued

The development of industrial relations in Western coal mining since the end of the nineteenth century: a synopsis

Year	Great Britain	USA	France	Belgium	Germany (FRG)	Austria
1944	• Strikes; • Porter inquiry and pay award		• Provisional nationalization	• Pacte de Solidarité Sociale		
1945	• National Union of Mine Workers (NUM); • Technical Advisory Committee report; • National wages agreement	• Pittsburgh Coal Consolidation Company		• Fédération Générale du Travail de Belgique (FGTB)	• North German Coal Control; • Unions refounded on local and regional level, Gewerkschaften	• Österreichischer Gewerkschaftsbund (ÖGB)
1946	• Coal Industry Nationalization Act; • 'Miners' Charter' presented	• Government seizure of the mines and measures against the UMWA	• Charbonnages de France; • Statute of the Miners; • 'Coal battle'	• Fédération des Industries Belges	• 1946–47 Debate on socialization; • Dec., German Industrial Trade Union for Mining	• Nationalization of coal mining; • Work councils reintroduced
1947	• National Coal Board (NCB) takes control; • National pit conciliation	• Resumption of collective bargaining; • Taft-Hartley Act	• Strikes in defence of nationalization	• Wage strikes; • Conseil National des Charbonnages	• Deutsche Kohlenbergbauleitung (DKBL)	

Year	Britain	United States	France	Belgium	Germany
	• British Association of Colliery Managers (BACM); • National Association of Colliery Overmen, Deputies and Shotfirers (NACODS)				
1948			• Wildcat strikes; • CGT banned from work councils		• Workers' co-determination introduced
1949					• IG Bergbau (German Industrial Trade Union for Mining) reorganized; • Deutscher Gewerkschaftsbund (DGB)
1950	• First *Plan for Coal* published	• National Bituminous Wage Agreement; • Coal Operators' Association		• Strikes against the return of the King;	• General strike

continued

The development of industrial relations in Western coal mining since the end of the nineteenth century: a synopsis

Year	Great Britain	USA	France	Belgium	Germany (FRG)	Austria
1951				• First elections of union delegates in the enterprises	• Coal and steel co-determination	
1952	• NCB extended to opencast mining		• Strikes against pit closures			
1953				• Fédération Charbonnière de Belgique (FEDECHAR)		
1955	• New NUM 'Miners' Charter'; • National Day Wage Structure Agreement					
1957	• Reconstruction of the NCB					• Parity Commission
1958	• Sick pay scheme					
1959	• Revised *Plan for Coal* adopted		• 'Jeanneney' plan to decrease output	• Strikes in the Borinage against pit closures	• Notgemeinschaft deutscher Steinkohlenbergbau,	

Year					
1960–1962	• National Conciliation Scheme amended; • Inter-divisional Transfer Scheme	• Lewis retires	• Strike in Decazeville against pit closure	• FCTB strikes; • Directoire de l'Industrie Charbonnière	Five-hour week; • Demonstrations against pit closures, 'March on Bonn'
1963			• Last collective wage strike		
1965	• Coal Industry Acts; reconstruction of the NCB				
1966	• National Power Loading Agreement			• Strikes	
1967	• Collins Report implemented			• Kempense Steenkoolmijnen	• 'Concerted action' for coal; • Three-phase programme of reorganization, 'Kohlegesetz'

continued

The development of industrial relations in Western coal mining since the end of the nineteenth century: a synopsis

Year	Great Britain	USA	France	Belgium	Germany (FRG)	Austria
1968			• General strike			
1969	• National strike of surface workers; • Unofficial strikes among Yorkshire miners	• Coal Mine Health and Safety Act			• Ruhrkohle AG; • After the formation of this overall company various 'plans for adjustment', including pit closures	
1970	• Redundant Mineworkers' Payment Scheme			• Strike in Campine		
1971	• National Day Wage Structure Agreement to apply to all mineworkers; • Coal Industry Act; • Many unofficial strikes		• Strike in Faulquemont against pit closure		• Pit closures also in the Saar area	

1972	• National strike; • State of emergency		
1973	• Pay settlement following Lord Wilberforce Inquiry; • National overtime ban		
1974	• State of emergency; • Conservative general election defeat; • *Plan for Coal*; • Coal Industry Tripartite Group	• Strike in Faulquemont against pit closure	
1977			• Gewerkschaft Metall-Bergbau-Energie
1979	• New Conservative government		
1980	• Coal Industry Act		
1982	• Employment Act makes secondary picketing illegal		

continued

The development of industrial relations in Western coal mining since the end of the nineteenth century: a synopsis

Year	Great Britain	USA	France	Belgium	Germany (FRG)	Austria
1984–1985	• National strike; • NUM action declared unlawful by court judgment; • Union of Democratic Mineworkers formed in Nottinghamshire					
1987	• NCB becomes British Coal					
1988	• Conservative government declares intention to privatize the coal industry					

The chronology is based on information gathered by the authors of this volume.

Notes on Contributors

David Brody, born in 1930, is Professor of History in the Department for Industrial Relations at the University of California at Berkeley. He completed his education at Harvard University in 1958 with a dissertation on the history of the American steel workers, taught at Columbia University from 1961 to 1964, then at Ohio State University from 1965 before taking over his present position at the University of California at Davis in 1967. His book *Steel Workers in America* appeared in 1960, followed by *The Butcher Workmen* in 1964, a study on *Labor in Crisis* in 1965 and most recently *Workers in Industrial America* in 1988. David Brody has published numerous articles in journals and edited volumes on labour and working-class history, on general political history and on economic and social history in the twentieth century.

Roy Church, born in 1935, now teaches as Professor of Economic History at the University of East Anglia in Norwich. He was educated at the University of Nottingham and gained his Ph.D. there in 1960 with a dissertation on Victorian Nottingham. Before going to Norwich, he was assistant professor at Purdue University and at the University of Seattle in the United States, as well as at the University of British Columbia in Canada. He then taught as a lecturer at the University of Birmingham in England. His dissertation, entitled *Economic and Social Change in a Midland Town, 1815–1900: Victorian Nottingham*, was published in 1966; this was followed by *Kenricks in Hardware, 1791–1966: A Family Business* (1969), two books on British economic history in the nineteenth century, *The Great Victorian Boom, 1850–1873* (1975) and *The Dynamics of Victorian Business* (1979), and then *Herbert Austin: The British Car Industry, 1896–1940* (1980) and, most recently, *Victorian Pre-eminence* (1986) on the period between 1830 and 1913 as the second volume of the large-scale *History of the British Coal Industry*. Numerous articles have also appeared on these topics of research. Roy Church is currently working on a research project on the history of strikes by British miners since 1893.

Gerald D. Feldman, born in 1937, is Professor of History in the Department of History at the University of California at Berkeley. He studied at Columbia University in New York (B.A., 1958) and at Harvard University (M.A., 1959; Ph.D., 1964). He gained his doctoral degree with the study *Army, Industry and Labor in Germany, 1914–1918*, which was published in 1966, and the book was later revised and published in German under the title *Armee, Industrie und Arbeiterschaft in Deutschland 1914 bis 1918* in 1985. G. D. Feldman has written numerous articles and books on German and European economic and social history. In 1977, the study *Iron and Steel in the German Inflation, 1916–1923* appeared, as well as *Industrie und Inflation. Studien und Dokumente zur Politik der deutschen Unternehmer 1916 bis 1923* (Industry and Inflation. Studies and Documents on the Politics of German Entrepreneurs, 1916–1923) (together with Heidrun Homburg). In 1984, a volume of his own essays entitled *Vom Weltkrieg zur Weltwirtschaftskrise. Studien zur deutschen Wirtschafts- und Sozialgeschichte 1914–1932* (From World War to World Depression. Studies on German Economic and Social History, 1914–1932) was published, followed a year later by a collection of documents, *Industrie und Gewerkschaften 1918–1924. Die überforderte Zentralarbeitsgemeinschaft* (Industry and the Unions, 1918–1924. The Overtaxed Central Labor Community) (together with Irmgard Steinisch). G. D. Feldman has edited, among other works, *German Imperialism, 1914–1918* (1972), two volumes of the collection *A Documentary History of Modern Europe* (1972 on) and several volumes from a large research project on the history of the German inflation. Currently he is working on a general comprehensive history entitled *The Great Disorder: A Political and Social History of the German Inflation, 1914–1924*, as well as on a biography of the business leader Hugo Stinnes.

Ginette Kurgan-van Hentenryk, born in 1938, is Professor of History at the Université Libre de Bruxelles. She completed her studies in Brussels in 1970 with a dissertation on Belgian financial interests in China that was published in 1972 under the title *Léopold II et les groupes financiers belges en Chine. La politique royale et ses prolongements (1895–1914)* (Leopold II and the Belgian Financiers in China. Royal Policy and its Continuation, 1895–1914). In additional studies, primarily on banking and financial history, she focused on *Rail, finance et politique: les entreprises Philippart 1865–1890* (Rail, Finance and Politics: The Enterprises of Philippart, 1865–1890)

(1982), and *Un siècle d'investissements belges au Canada* (A Century of Belgian Investment in Canada), written together with J. Laureyssens and published in 1986. Ginette Kurgan has often contributed to journals and volumes on business history in Belgium and on international financial history, as well as on the social history of Belgian emigration and on the Belgian petty bourgeoisie.

Franz Mathis, born in 1946, is Professor of Economic and Social History at the Institute for History of the University of Innsbruck. He completed his studies at Innsbruck by gaining his master's degree in 1971 and his doctoral degree in 1973 with a study on English–Austrian relations during the War of Spanish Succession. He has since turned to economic and social history and published a study on the *Auswirkungen des bayerisch-französischen Einfalls von 1703 auf Bevölkerung und Wirtschaft Nordtirols* (The Consequences of the Bavarian–French Invasion of 1703 on the Population and Economy of Northern Tirol). In 1977 *Zur Bevölkerungsstruktur österreichischer Städte im 17. Jahrhundert* (The Population Structure of Austrian Cities in the Seventeenth Century) was published, a work with which Franz Mathis earned his habilitation. He has to his credit numerous articles on diplomatic history, latterly an increasing number on urban history, and on trade and business history; in 1987 these were followed up by a book entitled *Big Business in Österreich. Österreichische Großunternehmen in Kurzdarstellungen* (Big Business in Austria. A Brief History of Major Austrian Firms). A second volume of this work is currently in the press. Franz Mathis was a visiting professor at the University of Salzburg in 1981 and at the University of New Orleans, USA, in 1983.

Joël Michel, born in 1951, has been Secrétaire des débats de l'Assemblée in Paris since 1987. At the University of Lyon II he earned his agrégé d'histoire and then his docteur d'Etat; his habilitation thesis (thèse d'Etat) was devoted to a comparison of the history of miners and the history of the mineworkers' movements in England, France, Belgium and Germany. This work was completed in 1987 but has not yet been published. Joël Michel has written numerous articles on comparative studies of the history of mining and mineworkers, on Polish workers in France, and on the history of the co-operative movement.

Jean Puissant, born in 1942, is chargé de cours at the Free University in Brussels. Following his education in Brussels, he worked in various research projects, including time in Africa, and earned his doctoral degree in 1974 with a study of the history of the socialist workers' movement in Borinage. In 1982, *L'évolution de mouvement ouvrier socialiste dans le Borinage (1880–1940)* (The Evolution of the Socialist Worker's Movement in the Borinage) was published, followed by a co-authored history of the Belgian metal workers from 1887 (1987). Jean Puissant has distinguished himself by his numerous articles on the history of the Belgian workers' movement, as well as on the history of working-class culture, on general political history and on industrial archaeology. He is currently working on a biographical encyclopaedia of the history of the Belgian workers' movement.

Klaus Tenfelde, born in 1944, is Professor of Economic and Social History at the Institute for History of the University of Innsbruck. Following a period of employment, he completed the alternative course of university-preparatory education (Zweiter Bildungsweg) and studied at the University of Münster, as well as in the USA; he finished his dissertation on the *Sozialgeschichte der Bergarbeiterschaft an der Ruhr im 19. Jahrhundert* (Social History of Miners in the Ruhr Valley in the Nineteen Century) in 1975 and it was subsequently published in 1977, with a second edition in 1981. His habilitation was earned in 1981 at the University of Munich with a study entitled *Proletarische Provinz. Radikalisierung und Widerstand in Penzberg/Obb. 1900 bis 1945* (Proletarian Province. Radicalization and Resistance in Penzberg/Obb. 1900–1945) which was published the following year. Together with Gerhard A. Ritter, he edited a *Bibliographie zur Geschichte der deutschen Arbeiterschaft und Arbeiterbewegung 1863 bis 1914* (Bibliography of the History of the German Labour Force and the Labour Movement, 1863–1914) which appeared in 1981; in the same year he edited a volume entitled *Streik. Zur Geschichte des Arbeitskampfes in Deutschland während der Industrialisierung* (Strike. The History of the Labour Disputes in Germany during Industrialization) together with Heinrich Volkmann. In 1986 a small volume on *Arbeit und Arbeitserfahrung in der Geschichte* (Work and Work Experience in History) and a collected edition of research reports on *Arbeiter and Arbeiterbewegung im Vergleich. Beiträge zur internationalen historischen Forschung* (Workers and the Labour Movement Compared. Reports on international historical research) were published. Klaus Tenfelde is currently working on a comprehensive handbook on the history of the German worker and the labour movement. He has published many articles on the general economic and social history of the nineteenth and twentieth centuries, on mining history in the early modern era, on the history of working-class culture, on subjects of historical demography and on the history of resistance to National Socialism.

Bernd Weisbrod, born in 1946, is Professor of Modern History in the History Department of the University of Bochum. Following his studies in Heidelberg and Berlin, he gained his doctorate with a study on *Die Schwerindustrie in der Weimarer Republik* (Heavy Industry in the Weimar Republic) which was published in 1978. He contributed to the creation of the collected edition entitled *Industrielles System und politische Entwicklung in der Weimarer Republik* (The Industrial System and Political Development in the Weimar Republic) which was published in 1976, was academic associate (Wissenschaftlicher Mitarbeiter) at the German Historical Institute in London from 1980 to 1982, and then earned his habilitation in Bochum with a study on poverty, social reform and social control in Victorian England. This work was published in 1989 under the title *'Victorian values'. Arm und Reich im Viktorianischen England* ('Victorian values'. The Poor and the Rich in Victorian England). In addition, Bernd Weisbrod has written many articles on the history of heavy industry, of industrial relations and social policy in the twentieth century in Germany, and on the history of poverty, childhood and social policy in England in the nineteenth century.

Select Bibliography

Abelshauser, Werner, 'The First Post-Liberal Nation: Stages in the Development of Modern Corporatism in Germany', *European History Quarterly* 14 (1984), pp. 285–318
——, *Der Ruhrkohlenbergbau seit 1945. Wiederaufbau, Krise, Anpassung*, Munich 1984
Ashworth, W. A. and Pegg, Mark, *The Nationalized Industry*, vol. 5 of *The History of the British Coal Industry*, Oxford 1986
Baratz, Morton S., *The Union and the Coal Industry*, New Haven 1955
Benson, John, 'The Coal Industry', in Chris Wrigley (ed.), *A History of British Industrial Relations, 1875–1914*, London 1982
Berg, Werner, *Wirtschaft und Gesellschaft in Deutschland und Großbritannien im Übergang zum 'organisierten Kapitalismus'. Unternehmer, Angestellte, Arbeiter und Staat im Steinkohlenbergbau des Ruhrgebietes und von Südwales, 1850–1914*, Berlin 1984
Berger, Suzanne (ed.), *Organizing Interests in Western Europe: Pluralism, Corporatism, and the Transformation of Politics*, Cambridge 1981
Bergmann, Joachim and Shigeyoshi, Tohunaga (eds), *Economic and Social Aspects of Industrial Relations. A Comparison of the German and the Japanese Systems*, Frankfurt/New York 1987
Berkovitch, Israel, *Coal on the Switchback: The Coal Industry since Nationalization*, London 1977
Brunier, M. (ed.), 'Techniques, pouvoirs, main d'oeuvre et reconversion dans les mines de charbon d'Europe occidentale', *Revue belge d'histoire contemporaine* XIX (1988), special issue
Burgess, Keith, *The Origins of British Industrial Relations. The Nineteenth Century Experience*, London 1975
Charles, Roger, *The Development of Industrial Relations in Britain, 1911–1939: Studies in the Evolution of Collective Bargaining at National and Industry Level*, London 1973
Church, Roy, Hall, Alan and Kanefsky, John, *Victorian Preeminence*, vol. 3 of *The History of the British Coal Industry*, Oxford 1986
Clegg, H., *The Changing System of Industrial Relations in Britain*, Oxford 1979

Cole, G. D. H., *The National Coal Board*, Fabian Research Series 10 (1948/49), pp. 218–65.

Coppe, A., *Problèmes d'économie charbonnière*, Bruges 1940

Davidson, Roger, 'Social Conflict and Social Administration: The Conciliation Act in British Industrial Relations', in T. M. Smout (ed.), *The Search for Wealth and Stability. Essays in Economic and Social History presented to M. W. Flinn*, London 1979, pp. 175–97

Dejonghe, Etienne, 'Les houillères à l'épreuve 1944–1947', *Revue du Nord*, octobre–décembre 1975, pp. 643–66

Delattre, A., *Souvenirs*, Cuesmes 1957

de Leener, G., *Étude sur le marché charbonnier belge*, Brussels/Leipzig 1908

Dethier, N., *Centrale syndicale des travailleurs des mines de Belgique. 60 années d'action 1890–1950*, Brussels 1950

Diefenbacher, Hans and Nutzinger, Hans G. (eds), *Mitbestimmung. Probleme und Perspektiven der empirischen Forschung*, Frankfurt/New York 1981

——, *Mitbestimmung: Theorie, Geschichte, Praxis. Konzepte und Formen der Arbeitnehmerpartizipation*, Heidelberg 1986

Dix, Keith, *Work Relations in the Coal Industry: The Hand-Loading Era, 1890–1930*, Morgantown (West Virginia) 1977

Dubofsky, Melvin and Van Tine, Warren, *John L. Lewis: A Biography*, New York 1977

Erdmann, Karl Dietrich, 'Eigentum, Partnerschaft, Mitbestimmung. Zur Theorie des Sozialstaats in Österreich und der Bundesrepublik Deutschland', *Geschichte in Wissenschaft und Unterricht* 39 (1988), pp. 393–412

Feldman, Gerald D., *Army, Industry and Labor in Germany 1914–1918*, Princeton 1966

——, 'Arbeitskonflikte im Ruhrbergbau 1919–1922. Zur Politik von Zechenverband und Gewerkschaften in der Überschichtenfrage', *Vierteljahrshefte für Zeitgeschichte* 28 (1980), pp. 168–223

Feldman, Gerald D. and Steinisch, Irmgard, 'Die Weimarer Republik zwischen Sozial- und Wirtschaftsstaat: Die Entscheidung gegen den Achtstundentag', *Archiv für Sozialgeschichte* 18 (1978), pp. 353–439

——, 'Notwendigkeit und Grenzen sozialstaatlicher Intervention. Eine vergleichende Fallstudie des Ruhreisenstreits in Deutschland und des Generalstreiks in England', *Archiv für Sozialge-*

schichte 20 (1980), pp. 57–118

——, *Industrie und Gewerkschaften. Die überforderte Zentralarbeitsgemeinschaft*, Stuttgart 1985

Fine, Ben, O'Donnell, Kathy and Prevezer, Martha, 'Coal after Nationalization', in Ben Fine and Laurence Harris (eds), *The Peculiarities of the British Economy*, London 1985, pp. 167–202

Flanders, Allan, *Management and Unions. The Theory and Reform of Industrial Relations*, London 1970

Forsyth, David J. C. and Kelly, David M., (eds), *Studies in the British Coal Industry*, London 1969

Fosh, Patricia and Littler, Craig R. (eds), *Industrial Relations and the Law in the 1980s: Issues and Future Trends*, Aldershot 1985

Garside, W. R., 'Management and Men: Aspects of British Industrial Relations in the Inter-War Period', in Barry Supple (ed.), *Essays in British Business History*, Oxford 1977, pp. 244–67

Gebhardt, Gerhard, *Ruhrbergbau. Geschichte, Aufbau und Verflechtung seiner Gesellschaften und Organisationen*, Essen 1957

Gillet, Marcel, *Les charbonnages du nord de la France au 19ème siècle*, Paris 1973

Hainsworth, R., 'The Trade Union Movements and Labour–Management Relations in the Coal Mines of the Nord and Pas-de-Calais Departments of France during the Depression 1930–1936', Ph.D. Thesis, London School of Economics 1979

Handy, L. J., *Wages Policy in the British Coalmining Industry: A Study of National Wage Bargaining*, Cambridge 1981

Hardy, Odette, *Industrie, patronat et ouvriers du Valenciennois pendant le premier vingtième siècle*, Lille 1985

Hardy-Hemery, Odette, 'Rationalisation technique et rationalisation du travail à la Compagnie des Mines d'Anzin (1927–1928)', *Le mouvement social*, juillet–septembre 1970, pp. 3–48

Hyman, Richard, *Industrial Relations. A Marxist Introduction*, London 1975

Jain, Ham C. and Giles, Anthony, 'Workers' Participation in Western Europe. Implications for North America', *Relations industrielles* 40 (1985), pp. 747–74

Johnson, James P., *The Politics of Soft Coal: The Bituminous Industry from World War I through the New Deal*, Urbana 1979

Kirby, M. W., 'The Politics of State Coercion in Inter-war Britain: the Mines Department of the Board of Trade, 1920–1942', *Historical Journal* 22 (1979), pp. 373–96

Koch, Max Jürgen, *Die Bergarbeiterbewegung im Ruhrgebiet zur Zeit*

Wilhelms II (1889–1914), Düsseldorf 1954

Krieger, Joel, *Undermining Capitalism*, Princeton 1983

McCormick, B. J., *Industrial Relations in the Coal Industry*, London 1979

Michel, Joël, 'Un maillon plus fertile du syndicalisme minier. La Fédération Nationale des mineurs belges avant 1914', *Revue belge de philologie et d'histoire* 55 (1977), pp. 425–73

——, 'Le mouvement ouvrier chez les mineurs d'Europe occidentale (Allemagne, Belgique, France, Grand-Bretagne). Étude comparative des années 1880 à 1914', thèse d'Etat, Lyon II 1987

Middlemas, Keith, *Politics in Industrial Society. The Experience of the British System since 1911*, London 1979

Mitchell, B. R., *The Economic Development of the British Coal Industry, 1800–1914*, Cambridge 1984

Mommsen, Hans and Borsdorf, Ulrich (eds), *Glück auf, Kameraden! Die Bergarbeiter und ihre Organisationen in Deutschland*, Cologne 1979

Moons, J., *De economische structuur van de Kempische steenkolenmijnnijverheid*, Hasselt/Leuven 1957

Moutet, Aimée, 'La rationalisation dans les mines du Nord à l'épreuve du Front Populaire. Etude d'après les sources imprimées', *Le mouvement social*, avril–juin 1986, pp. 63–99

Müller-List, Gabriele (ed.), *Montanmitbestimmung. Das Gesetz über die Mitbestimmung der Arbeitnehmer in den Aufsichtsräten und Vorständen der Unternehmen des Bergbaus und der Eisen und Stahl erzeugenden Industrie vom 21. Mai 1951*, Düsseldorf 1984

Nocken, Ulrich, 'Corporatism and Pluralism in Modern German History', in Dirk Stegmann *et al.* (eds), *Industrielle Gesellschaft und politisches System. Beiträge zur politischen Sozialgeschichte*, Bonn 1978, pp. 37–56

Pelinka, Anton, *Modellfall Österreich? Möglichkeiten und Grenzen der Sozialpartnerschaft*, Vienna 1981

Perry, Charles R., *Collective Bargaining and the Decline of the United Mine Workers*. Industrial Research Unit, University of Pennsylvania, Philadelphia 1984

Puissant, Jean, *L'évolution du mouvement ouvrier socialiste dans le Borinage*, Brussels 1982

Reid, Donald, 'Industrial Paternalism: Discourse and Practice in Nineteenth-Century French Mining and Metallurgy', *Comparative Studies in Society and History* 27 (1985), pp. 579–607

——, *Miners of Decazeville. Genealogy of Deindustrialization*, Cambridge (Mass.) 1985

Rimlinger, Gaston V., 'Die Legitimierung des Prostests. Eine vergleichende Untersuchung der Bergarbeiterbewegung in England und Deutschland', in Wolfram Fischer and Georg Bajohr (eds), *Die soziale Frage*, Stuttgart 1967, pp. 284–304

Ritter, Gerhard A., *Der Sozialstaat. Entstehung und Entwicklung im internationalen Vergleich*, Munich 1989

Saville, John, 'Staat, Unternehmerschaft und Gewerkschaften in Großbritannien 1870 bis 1914', in Wolfgang J. Mommsen and Hans-Gerhard Husung (eds), *Auf dem Wege zur Massengewerkschaft. Die Entwicklung der Gewerkschaften in Deutschland und Großbritannien*, Stuttgart 1984, pp. 389–99

Sells, Dorothy, *British Wages Boards: A Study in Industrial Democracy*, Washington 1939

Spencer, Elaine Glovka, 'Between Capital and Labour: Supervisory Personnel in the Ruhr Heavy Industry before 1914', *Journal of Social History* 9 (1975), pp. 178–92

——, 'Employer Response to Unionism: Ruhr Coal Industrialists before 1914', *Journal of Modern History* 43 (1976), pp. 397–412

Stourzh, Gerald and Grandner, Margarete (eds), *Historische Wurzeln der Sozialpartnerschaft*, Vienna 1986

Streeck, Wolfgang, 'Staatliche Ordnungspolitik und industrielle Beziehungen. Zum Verhältnis von Integration und Institutionalisierung gewerkschaftlicher Interessenverbände am Beispiel des britischen Industrial Relations Act 1971', in Udo Bermbach (ed.), *Politische Wissenschaft und politische Praxis. Tagung der Deutschen Vereinigung für Politische Wissenschaft in Bonn, Herbst 1977*, Opladen 1978, pp. 106–39

——, *Industrial Relations in West Germany. A Case Study of the Car Industry*, London 1984

Supple, Barry, *The Political Economy of Decline*, vol. 4 of *The History of the British Coal Industry*, Oxford 1987

Tarling, R. and Wilkinson, F., 'The Movement of Real Wages and the Development of Collective Bargaining in the U.K.: 1855–1920', *Contributions to Political Economy* 1 (1982), pp. 1–23

Tenfelde, Klaus, 'Probleme der Organisation von Arbeitern und Unternehmern im Ruhrbergbau 1890–1918', in Hans Mommsen (ed.), *Arbeiterbewegung und industrieller Wandel. Studien zu gewerkschaftlichen Organisationsproblemen im Reich und an der Ruhr*, Wuppertal 1980, pp. 38–61

Teuteberg, Hans-Jürgen, *Geschichte der industriellen Mitbestimmung in*

Deutschland. Ursprung und Entwicklung ihrer Vorläufer im Denken und in der Wirklichkeit des 19. Jahrhunderts, Tübingen 1961

Thum, Horst, *Mitbestimmung in der Montanindustrie. Der Mythos vom Sieg der Gewerkschaften*, Stuttgart 1982

Tolliday, Steven and Zeitlin, Jonathan (eds), *Shop Floor Bargaining and the State: Historical and Comparative Perspectives*, Cambridge 1985

Tomlins, Christopher L., *The State and the Unions: Labor Relations, Law, and the Organized Labor Movement in America, 1880–1960*, Cambridge 1985

Trempé, Rolande, *Les mineurs de Carmaux 1848–1914*, Paris 1971

Tschirbs, Rudolf, *Tarifpolitik im Ruhrbergbau 1918–1933*, Berlin/ New York 1986

Van Den Driessche, E., *La Centrale des Francs-Mineurs. Son histoire, sa vie*, Brussels 1984

von Alemann, Ulrich, 'Auf dem Weg zum industriellen Korporatismus? Entwicklungslinien der Arbeitsbeziehungen in der Bundesrepublik Deutschland und Großbritannien', *Gewerkschaftliche Monatshefte* 30 (1979), pp. 552–65

—— (ed.), *Neokorporatismus*, Frankfurt/New York 1981

von Bandemer, Jens Dither and Ilgen, August Peter, *Probleme des Steinkohlenbergbaus. Die Arbeiter- und Förderverlagerung in den Revieren der Borinage und Ruhr*, Basle/Tübingen 1963

von Beyme, Klaus, *Gewerkschaften und Arbeitsbeziehungen in kapitalistischen Ländern*, Munich 1977

Weisbrod, Bernd, *Schwerindustrie in der Weimarer Republik. Interessenpolitik zwischen Stabilisierung und Krise*, Wuppertal 1978

Wendt, Bernd-Jürgen, 'Industrial Democracy. Zur Struktur der englischen Sozialbeziehungen', *Aus Politik und Zeitgeschichte* 46 (1975)

Windmuller, John P. and Gladstone, Alan (eds), *Employers Associations and Industrial Relations: A Comparative Study*, Oxford 1984

Wisotzky, Klaus, *Der Ruhrbergbau im Dritten Reich. Studien zur Sozialpolitik im Ruhrbergbau und zum sozialen Verhalten der Bergleute in den Jahren 1933 bis 1939*, Düsseldorf 1983

Wrigley, Chris (ed.), *A History of British Industrial Relations, 1875–1914*, Brighton 1982

——, *A History of British Industrial Relations, 1914–1939*, Brighton 1986

Zeitlin, Jonathan, 'Sindacati e controllo del posto di lavoro: una critica dell'interpretazione spontaneistica', *Quaderni storici* 62

(1986), pp. 595–612

——, 'From Labour History to the History of Industrial Relations', *Economic History Review* 40 (1987), pp. 159–84

Index

431

Beeringen mine, 216
Belgian Coal Agency, 256, 270
Belgian Coke and Coking Coal
 Union, 240, 241
Belgian–German commercial
 treaty, 241
Belgian Industrial Coal Agency, 239,
 240, 242
Belgian Workers Party (BWP), 226,
 227, 228, 230, 232, 234, 244, 247,
 248, 252, 258, 261
Belgium, 81, 211, 212, 215, 216, 218,
 222, 223, 227, 231, 233, 237, 244,
 249, 254, 257, 263, 266, 267, 268,
 269, 303, 383, 393, 396, 398,
 401–18
Belgium–Luxembourg Economic
 Union, 225
Bergag, 153
Bergassessoren, 376, 383, 386, 388,
 391
Bergbau Verein, 128, 129, 130, 132,
 133, 134, 135, 136, 138, 151, 170,
 171, 184, 193
Berg-Capelle, 334
Berggeist, 334
Berghahn, Volker, 388
Bergregal (mining royalties), 125
Berlepsch, Hans Frhr. von, 135
Berlin, 11, 133, 134, 135, 158
Bessèges (firm), 285, 303
Béthune, 277, 281, 298
Betriebsgemeinschaft (company
 community), 179
Bevin, Ernest, 191
Bismarck, Prince Otto von, 132, 151
Bituminous Coal Operators
 Association (BCOA), 110, 111,
 112, 113, 114, 115, 116, 117
Bizone Treaty, 192
black lists, 143
Blanzy, 273, 274, 289, 290, 292
Blum, Leon, 299, 300
Board of Trade (Great Britain), 37,
 41
Bochum, 132
Böckler, Hans, 200
Boël, P., 235
Bohemia, 323, 326
Bois-de-Luc, 209
Bolshevists, 161
Bonn, 366
Borinage (coalfield), 203, 204, 205,
 206, 207, 209, 210, 212, 214, 217,

228, 231, 235, 236, 243, 250, 251,
 261, 264, 268, 409
Borsdorf, Ulrich, 193
Bouveri, 283
Brandi, Ernst, 170
Brauns, Heinrich, 165, 166
Brentano, Lujo, 146
British Association of Colliery
 Managers (BACM), 413
British Coal, 71, 73
Brady, Robert, 382
Brody, David, 367, 391
Bruderladen (Friendly Societies), 320,
 321, 327
Brüggemeier, Franz-Josef, 141
Brüning, Heinrich, 174, 175, 176,
 180, 384
Brufina (holding co.), 216, 217, 219,
 220, 262
Brufina-Coppée, 219
Brussels, 208, 212, 265
Bruay, 280
Bülow, Bernhard von, 13
Bundesbahn, 301
Bureau of Internal Revenue
 (USA), 98, 99
Burgenland, 318
Burgfrieden, 152
Burns, A., 63
Busson, Felix, 334
BWP, see Belgian Workers Party

Cabinet (Great Britain), 31, 32, 36,
 71
Cambrian Collieries, 19
Cambrian Combine, 19
Campine (coalfield), 203, 216, 217,
 222, 223, 225, 241, 244, 248, 249,
 250, 251, 256, 258, 261, 262, 263,
 266, 270
Canada, 8
Capiau, Herman, 242
Carmaux, 273, 292
Catholic congress (Belgium), 209
Catholic Party (Belgium), 212, 226,
 235
CCHF, see Comité Central des
 Houillères de France
CD, see Christian Democrats
CDF, see Charbonnages de France
CDU, see Christian Democratic Union
Central Association of German
 Industrialists, 136
Central Association of the Mineowners

446

Index